ORAL BACTERIAL ECOLOGY:
The Molecular Basis

Edited by

Howard K. Kuramitsu
State University of New York
Departments of Oral Biology/Microbiology
Buffalo, NY 14214 USA

and

Richard P. Ellen
University of Toronto
Faculty of Dentistry
Department of Graduate and Postgraduate Studies
Toronto, ON Canada M5G 1G6

Copyright © 2000
Horizon Scientific Press
P.O. Box 1
Wymondham
Norfolk NR18 0EH
England

www.horizonpress.com

British Library Cataloguing-in-Publication Data

A catalogue record for this book is available from the British
Library

ISBN: 1-898486-22-0

*Printed and bound in Great Britain
by Biddles Ltd, Guildford and King's Lynn*

Contents

Books of Related Interest

For further information on these books contact:

Horizon Scientific Press
P.O. Box 1, Wymondham
Norfolk
NR18 0EH England

Tel: +44(0)1953-601106
Fax: +44(0)1953-603068
Email: mail@horizonpress.com
Internet: www.horizonpress.com

Our Web site has details of all our books including full chapter abstracts, book reviews, and ordering information:

www.horizonpress.com

Contributors

Jan Carlsson
Department of Odontology
Faculty of Medicine and Odontology
Umea University
Umea SE-901-85, Sweden

Richard Darveau
Department of Periodontics
School of Dentistry
University of Washington
Seattle, WA 98195, USA

Richard P. Ellen
University of Toronto
Faculty of Dentistry
Department of Graduate and
Postgraduate Studies
Toronto, ON Canada M5G 1G6

Daniel Grenier
Groupe de Recherche en Ecologie
Buccale
Faculté de Médecine Dentaire
Universite Laval, Cite Universitaire
Québec G1K 7P4, Canada

Ian R. Hamilton
Department of Oral Biology
Faculty of Dentistry
780 Bannatyne Avenue
University of Manitoba
Winnipeg R3E 0W2, Canada

Howard F. Jenkinson
Department of Oral and Dental
Science
University of Bristol Dental School
Bristol BS1 2LY, United Kingdom

Howard K. Kuramitsu
State University of New York
Departments of Oral Biology/
Microbiology
3435 Main Street
Buffalo, NY 14214-3092 USA

Richard J. Lamont
Department of Oral Biology
University of Washington
Seattle, WA 98195 USA

Philip D. Marsh
Centre for Applied Microbiology and
Research, Porton Down
Salisbury, SP4 0JG
and Leeds Dental Institute
Leeds LS2 9LU, United Kingdom

Denis Mayrand
Groupe de Recherche en Ecologie
Buccale
Faculté de Médecine Dentaire
Universite Laval, Cite Universitaire
Québec G1K 7P4, Canada

Foreword

As with other infectious diseases, the study of the microbial etiology of the major oral diseases, dental caries and periodontal diseases, has been significantly impacted by the introduction of molecular biological techniques. However, experience with these "reductionist" approaches have also made it abundantly clear that infectious disease outcomes do not depend solely on the properties of a single pathogen, factor, or gene but are also influenced by interactions of these organisms with other members of the microbiota as well as by host factors. The factors which determine bacterial ecology will be the primary focus of this monograph and they will be discussed in detail at the molecular level where possible.

We trust that the information provided will be of use to researchers in the field of oral microbial pathogenesis as well as for graduate level courses in oral microbiology, bacterial pathogenicity, or microbial ecology. We will have accomplished our goal if this monograph stimulates additional questions and the design of new approaches to test the premise that dental caries and periodontitis result primarily from ecological imbalances in the oral cavity.

Howard K. Kuramitsu and Richard P. Ellen

Acknowledgements

We thank Mr. Hugh Griffin of Horizon Press for his help and encouragement in producing this book. The coeditors also wish to acknowledge the patience and understanding of our respective wives, Kim and Judy, during the course of the planning and editing phases of preparation. RPE also wishes to thank the staff members of The Forsyth Institute, Boston, MA, for their cooperation and the use of the excellent library, office, and photography facilities. HKK also thanks Yiping Han for her critical comments on portions of the book.

From: *Oral Bacterial Ecology: The Molecular Basis*
ISBN 1-898486-22-0 ©2000 Horizon Scientific Press, Wymondham, U.K.

Introduction

MOLECULAR BASIS OF ORAL BACTERIAL ECOLOGY: IMPLICATIONS FOR PATHOGENICITY

Howard K. Kuramitsu and Richard P. Ellen

"The mouth has a variety of features which have enabled it to serve as a useful model for the discovery of basic principles of host-parasite interactions occurring in mucosal environments."

R.J. Gibbons (Kreshover Lecture, NIDR, 1988)

The rapid development and broad application of molecular biological techniques has led to profound advancements in our ability to address fundamental biological questions. The ability to isolate genes, determine their nucleotide sequences, amplify DNA fragments, characterize their gene products, construct defined mutants and fusions, as well as determine the environmental factors and regulators which control their expression has been an important impetus to such endeavors. In addition, the recent sequencing of whole genomes from a variety of microorganisms promises to increase our understanding of evolution. This latter approach will culminate before too long in the sequencing of the human genome. These advancements in genetics, of host and parasites alike, will likely suggest a variety of new strategies for treating both hereditary and infectious diseases.

Advances in our understanding of the functional aspects of organisms have also been made in parallel with progress initiated by the introduction of molecular biological techniques into biology. We are now entering an era of biological research where questions regarding complex interactions of organisms with each other and their environment can be examined at the molecular level (molecular ecology). This approach has already made it abundantly clear that the traditional approaches of examining individual organisms under laboratory conditions may not always reflect the relevant physiological state of these organisms in nature, especially in regard to their pathogenicity. For example, many microorganisms normally exist in complex communities called biofilms rather than individually as planktonic cells. An examination of the organisms in biofilms indicates that they are physiologically distinct from the same cells grown as laboratory suspensions. In this regard, it is of interest that some of the earliest studies on bacteria in a biofilm environment were

1

carried out with dental plaque (Gibbons and van Houte, 1973). As will be discussed in the following chapters, dental plaque represents a relevant and useful model system for examining the behavior of organisms in biofilms. This later insight is illustrative of a major thrust of this book: the principles and determinants of oral microbial ecology have direct application to microbial ecology at other anatomical sites. One of the most influential proponents of this concept was Ronald J. Gibbons, to whom this book is dedicated. This introductory chapter attempts to illustrate Ron's remarkable influence on current research in microbiology, especially in the context of themes that are raised in the chapters to follow.

It is clear that the colonization of a specific ecological site by an individual microorganism and sustenance of its progeny depends on its interaction with and adaptation to its immediate environment. Thus, the expression of an adhesin by an organism and the availability of a corresponding attachment receptor, the presence of antimicrobial effectors, the nutrient environment, the presence or absence of other symbiotic or antagonistic microorganisms, and the physical environment all may serve as determinants to define a microbial habitat. A corollary of this principle is that microorganisms are not found randomly in a particular environment. Ron and his colleagues at the Forsyth Dental Center were the first to test this fundamental principle by using organisms from the oral cavity (Gibbons and van Houte, 1975). Once this hypothesis was established it was then possible to begin to investigate the molecular basis for site-specific colonization, an endeavor which still occupies the interest of many microbiologists. These findings were also important in ultimately defining the characteristics of microorganisms that contribute to dental caries and periodontal diseases.

One important concept which Ron and his coworkers pioneered was that of a requirement for selective microbial adhesin-receptor interactions for host- and site-specific colonization (Gibbons, 1996). Although initially demonstrated for oral microorganisms (Gibbons and van Houte, 1971; Hillman *et al.*, 1970; Liljemark and Gibbons, 1971; van Houte *et al.*, 1970), Ron and his students extended and confirmed this basic concept for some pathogens which colonize extraoral environments (Ellen and Gibbons, 1972, 1974; Gibbons *et al.*, 1976; Gibbons, 1989). The ability to adhere is now regarded as a principal virulence property for pathogenic bacteria that infect surfaces exposed to bathing secretions. This has led to the identification and characterization of bacterial adhesins for numerous organisms and has now made it possible to design vaccines and specific peptide inhibitors of bacterial adhesion.

The opportunity for some bacteria to colonize a specific surface for which they have no innate affinity may also be fostered by the prior adhesion of other microorganisms that they have evolved to recognize. Such interactions (coaggregation) were initially demonstrated by Ron and his colleagues with oral bacteria and then extended to extraoral pathogens (Gibbons and Nygaard, 1970; Komiyama and Gibbons, 1984). These studies also indicated that an extracellular polymer of one organism, dextrans produced by oral streptococci for example, could mediate the selective attachment by the same or other species (Gibbons and Banghart, 1967). Interactions such as these are very important in biofilm formation, and the molecular basis for these properties is a major focus of several laboratories. Such studies now involve the isolation and sequencing of the genes that encode and regulate the expression of the adhesin molecules. Likewise, selective coadhesive interactions between pigmented bacteroides (those now classified as *Porphyromonas gingivalis*) and oral gram-positive bacteria originally demonstrated in vitro by Gibbons and coworkers (Slots and Gibbons, 1978) may help explain the transition of some developing oral biofilms into periodontopathic plaques.

The aggregation of different genera of oral bacteria, originally demonstrated by Gibbons and Nygaard (1970), might also promote cell signaling or facilitate gene transfer between these organisms in plaque. Recent investigations have shown that biofilms such as dental plaque have defined structures and that their formation and metabolism may be influenced by interbacterial signaling. Thus, an environmental cue picked up by one organism could be communicated to surrounding organisms of the same or different species. Aggregation may be a means of bringing organisms in close proximity for such signaling, for it is now known that population density, "quorum sensing", can initiate signaling cascades among bacteria. In addition, gene transfer may occur between different organisms in contact with each other via conjugation or transformation following the release of DNA by one member of the aggregate. The transfer of "pathogenicity island" genes from one oral bacterium to another has yet to be demonstrated but given the likelihood of such transfer in non-oral bacteria and emerging evidence for novel insertion sequences in oral bacterial genomes, it is a real possibility.

Another important concept which Ron introduced to the study of bacterial colonization was that some bacterial adhesins may bind to specific host molecules only following prior exposure of naturally sequestered cognate receptors (cryptitopes) (Gibbons and Hay, 1988; Gibbons *et al.*, 1990). Thus, the proteolytic activity of some oral bacteria like *P. gingivalis* may alter polypeptide receptors in the tooth pellicle or on epithelial cells, thereby exposing sites for their attachment or for that of other bacteria (Childs and Gibbons, 1990; Gibbons *et al.*, 1990). Further evidence for the importance of such a mechanism has been recently provided for the protease-enhanced attachment of the fimbriae of *P. gingivalis* to salivary proteins and to host cells.

The tooth surface's salivary pellicle receptors to which oral bacteria attached were a major interest of Ron and his colleagues at the Forsyth Dental Center. They developed the most widely used in vitro assay for bacterial adhesion to saliva-coated hydroxyapatite as a model for studying microbial attachment to teeth (Clark *et al.*, 1978), an assay that still serves as the standard. Over the years, they demonstrated that specific salivary components (proline-rich proteins [PRPs], blood-reactive glycoproteins, statherin) served as receptors for selective bacterial interactions (Gibbons *et al.*, 1988; Gibbons and Hay, 1988; Gibbons and Qureshi, 1978; Gibbons *et al.*, 1991). During the 1970's and early 1980's there was great interest in stereochemical recognition functions mediated by carbohydrate-binding proteins called lectins. The application of their quantitative bacterial adherence assay allowed Ron and his colleagues to generate adhesion isotherms to support his contention that the specific interactions between bacteria and salivary components were often lectin-mediated (Gibbons and Qureshi, 1979; Gibbons *et al.*, 1983, 1985). In many cases, bacterial surface proteins recognize sugar residues (e.g., sialic acid) in pellicle

3

glycoproteins. Some of these sugar residues may be part of cryptitopes and may require removal of the terminal oligosaccharides of the glycoproteins by bacterial glycosidases. Recognition of cryptitopes exposed due to enzymatic digestion or due to conformational change upon adsorption of salivary proteins to teeth (as Gibbons hypothesized for PRPs) may explain how some bacteria have evolved to bind these proteins in acquired pellicle unhindered by competitive inhibition by the same proteins free in saliva. Currently, attempts are underway to identify the binding domains of the adhesins from some of these bacteria and to develop specific inhibitors of such interactions.

Interactions between bacteria in a biofilm such as dental plaque can be either symbiotic or antagonistic. The Gibbons group was also the first to demonstrate that some oral streptococci express bacteriocins which inhibit the growth of other bacteria (Kelstrup and Gibbons, 1969). Prior to the advent of molecular biology, "fingerprinting" strains by bacteriocin typing, introduced to oral microbiology by Kelstrup and Gibbons (1970), was used as the standard method to trace bacterial transmission among hosts and intraoral sites. Antagonistic effects among bacterial strains might also help explain the association of certain groups of bacteria, but not others, in a particular plaque sample or region of dental plaque. The demonstration that bacteriocins specific for some strains of *S. mutans* can be isolated has suggested that it may be possible to genetically engineer autochthonous, nonpathogenic oral bacteria to secrete such bacteriocins to possibly reduce the incidence of dental caries. One such anti-*S. mutans* bacteriocin is presently being exploited to develop anticariogenic *S. mutans* strains for possible replacement therapy.

Enhancement of colonization of solid surfaces by exopolysaccharide synthesis is characteristic of bacteria in many different environments. The Gibbons group initially demonstrated the importance of glucan formation in adhesion and colonization of teeth by the cariogenic mutans streptococci (Gibbons and Banghart, 1967; Gibbons, 1968; Gibbons and Keyes, 1969). Their initial characterization of a "dextransucrase" activity from *S. mutans* led others to the demonstration that these organisms express three distinct glucosyltransferase (Gtf) enzymes whose combined action is responsible for colonization-promoting glucan formation. The genes coding for these enzymes were among the first to be isolated and characterized from oral bacteria and served as a paradigm for subsequent genetic characterization of oral bacterial pathogens. In addition, these enzymes and synthetic peptides based on their structure are currently the focus of experimental anti-caries treatments, either as vaccine candidates or as specific enzyme inhibitors, respectively.

One of the most important contributions of the Gibbons group to the study of bacterial ecology and its impact on infection and immunity was the demonstration that salivary immunoglobulin A can inhibit the attachment of bacteria to epithelial cells (Williams and Gibbons, 1972). This clearance mechanism probably plays a major role in determining the outcome of host-parasite interactions on surfaces bathed by secretions throughout the body, and enhancement of secretory immunity directed toward adhesins provides a basis for development of immunization strategies. Ron also proposed that the selective pressure of IgA responses may account for antigenic variation among strains of oral species (Bratthall and Gibbons, 1975; Gibbons and Howell, 1978; Gibbons and Qureshi, 1980), similar to the effect of intestinal antibodies on antigenic drift in *Vibrio cholerae* and other enteropathogens that were being studied at the time. His laboratory was also instrumental in exploring the nature of innate inhibition of bacterial adherence to mucosal receptors by antigenically similar salivary glycoproteins (Williams and Gibbons, 1975). The molecular basis for regulation of the indigenous microbiota by acquired and innate immunity components of secretions is currently being investigated in several laboratories using primarily molecular genetic approaches.

Following attachment of bacteria to oral surfaces, successful colonization is largely dependent upon the availability of nutrients for growth by the attached organisms. The Gibbons group was prominent among those that described the extensive sucrose hydrolysis by mutans streptococci which led to the identification of dextransucrase and intracellular "invertase" activities of the organisms and helped explain the unique relationship between sucrose, *S. mutans* strains, and dental caries (Gibbons and Banghart, 1967; Gibbons, 1972). Subsequent studies employing both biochemical and molecular genetic approaches have identified at least three distinct pathways for intracellular metabolism of sucrose by these acidogenic and acid tolerant organisms. In addition, Ron and his colleagues demonstrated the important findings that cariogenic streptococci could metabolize fructans in plaque as a carbohydrate source (DaCosta and Gibbons , 1968) and were capable of storing relatively large amounts of intracellular polysaccharide (Gibbons and Socransky, 1962). Ron's interest in intracellular polysaccharides actually goes back to his time as a rumenologist at the Maryland Agricultural Experimental Station (Gibbons *et al.*, 1955). Subsequent work from other laboratories has confirmed that this is an important virulence property of these organisms.

One of the primary goals of microbiologists working with pathogens is to identify their virulence properties. It was Ron and his colleagues who initially identified or characterized some of the most important virulence factors associated with the mutans streptococci: Gtfs, sucrose metabolizing enzymes, glucan-binding properties, and cell surface adhesins. In addition, Ron was directly involved with some of the earlier animal studies that demonstrated that the mutans streptococci could cause dental caries (Gibbons 1967, 1968; Gibbons and Keyes, 1969). Clearly, Ron Gibbons played a key role in establishing the properties that confer cariogenicity on the mutans streptococci, and he was both recognized and honored for these contributions.

Though perhaps overshadowed by his work on dental caries, it should be recognized that Ron also made seminal contributions to our knowledge of the microbiology of periodontitis. Together with his colleagues at the Forsyth Dental Center, Ron conducted the first mixed anaerobe infection studies in animals that implicated black-pigmented bacteroides in periodontitis (Macdonald *et.al.,* 1963; Socransky and Gibbons, 1965). This work led to the initial characterization of the nutrient requirements of these organisms, including the observation that menadione and hemin were required for their growth (Gibbons and Macdonald, 1960; Gibbons and Engle, 1964). It is now recognized that hemin can modulate the virulence of *P. gingivalis* by mechanisms which are presently under investigation at the molecular level. Early studies by Ron and coworkers also indicated that the species subsequently classified as *P. gingivalis* expressed collagenolytic activity (Gibbons and Macdonald, 1961) and derived energy from amino acid fermentation (Wahren and Gibbons,

1967). Since collagen breakdown is a hallmark of periodontitis, this might represent a significant virulence factor for these organisms, even though it is now recognized that most of the collagen degradation observed in inflamed periodontal tissue is probably mediated by host-derived matrix metalloproteinases. Interestingly, the enzyme(s) responsible for *P. gingivalis* collagenolytic activity are still a subject of some controversy.

Although many researchers in the field of oral microbiology are aware of the important contributions that Ron made to this area of biology, it is only by considering all of these together that one can really begin to appreciate the remarkable impact that one individual can have on an entire field. It is recognized that other laboratories which are not cited in this chapter also made seminal contributions to many of these areas and many are acknowledged appropriately in the following chapters. Nevertheless, it is generally recognized that Ron was a highly original leader in these areas of research. Although Ron himself did not employ molecular biology techniques in his work, we suspect that he would have been pleased to learn from this book how his intuitive and highly original ideas based on traditional microbiological and biochemical approaches have been confirmed and extended using these techniques. In a strong sense, the researchers whose contributions are discussed in the following chapters are advancing molecular microbial ecology following the pathways blazed by a remarkable scientist - Ronald J. Gibbons.

References

Note: Key references to information cited in this chapter from other than the Gibbons laboratory will be found in the following chapters.

Bratthall, D. and R.J. Gibbons. 1975. Antigenic variation of *Streptococcus mutans* colonizing gnotobiotic rats. Infect. Immun. 12: 1231-1236.

Childs, W.C. III and R.J. Gibbons. 1990. Selective modulation of bacterial attachment to oral epithelial cells by enzyme activities associated with poor oral hygiene. J. Periodontal Res. 25: 172-178.

Clark, W.B., L.L. Bamman, and R.J. Gibbons. 1978. Comparative estimates of bacterial affinities and adsorption sites on hydroxyapatite surfaces. Infect. Immun. 19: 846-853.

DaCosta, T. and R.J. Gibbons. 1968. Hydrolysis of levan by human plaque streptococci. Arch. Oral Biol. 13: 609-617.

Ellen, R.P. and R.J. Gibbons. 1972. M protein – associated adherence of *Streptococcus pyogenes* to epithelial surfaces: prerequisite for virulence. Infect. Immun. 5: 826-830.

Ellen, R.P. and R.J. Gibbons. 1974. Parameters affecting the adherence and tissue tropism of *Streptococcus pyogenes*. Infect. Immun. 9: 85-91.

Gibbons, R.J. 1967. Award for research in oral science. The significance of bacterial polysaccharides in dental caries. J. Dent. Res. 46: 1230.

Gibbons, R.J. 1968. Formation and significance of bacterial polysaccharides in caries etiology. Caries Res. 2: 164-171.

Gibbons, R.J. 1972. Presence of invertase-like enzymes and a sucrose permeation system in strains of *Streptococcus mutans*. Caries Res. 6: 122-131.

Gibbons, R.J. 1989. Bacterial adhesion to oral surfaces. A model for infectious diseases. J. Dent. Res. 68: 750-760.

Gibbons, R.J. 1996. Role of adhesion in microbial colonization of host tissues: a contribution of oral microbiology. J. Dent. Res. 75: 866-870.

Gibbons, R. J. and S.B. Banghart. 1967. Synthesis of extracellular dextran by cariogenic bacteria and its presence in human dental plaque. Arch. Oral Biol. 12: 11-23.

Gibbons, R.J., R.N. Doetsch, and J.C. Shaw. 1955. Further studies on polysaccharide production by bovine rumen bacteria. J. Dairy Sci. 38: 1147-1154.

Gibbons, R.J. and L.P. Engle. 1964. Vitamin K compounds in bacteria that are obligate anaerobes. Science 146: 1307-1309.

Gibbons, R.J., I. Etherden, and E.C. Moreno. 1983. Association of neuraminidase – sensitive receptors and and putative hydrophobic interactions with high-affinity binding sites for *Streptococcus sanguis* C5 in salivary pellicle. Infect. Immun. 42:

1006-1012.

Gibbons, R.J., I. Etherden, and E.C. Moreno. 1985. Contribution of stereochemical interactions in the adhesion of *Streptococcus sanguis* C5 to experimental pellicles. J. Dent. Res. 64: 96-101.

Gibbons, R.J. and D.I. Hay. 1988. Human salivary acidic proline-rich proteins and statherin promote the attachment of *Actinomyces viscosus* LY7 to apatitic surfaces. Infect. Immun. 56: 439-495.

Gibbons, R.J., D.I. Hay, W.C. Childs III, and G. Davis. 1990. Role of cryptic receptors (cryptitopes) in bacterial adhesion to oral surfaces. Arch. Oral Biol. 35: (suppl): 107S-114S.

Gibbons, R.J., D.I. Hay, J.O. Cisar, and W.B. Clark. 1988. Adsorbed salivary proline-rich protein 1 and statherin: receptors for type 1 fimbriae of *Actinomyces viscosus* T14V-J1 on apatitic surfaces. Infect. Immun. 56: 2990-2993.

Gibbons, R.J., D.I. Hay, and D. H. Schlesinger. 1991. Delineation of a segment of adsorbed salivary proline-rich proteins which promotes adhesion of *Streptococcus gordonii* to apatitic surfaces. Infect. Immun. 59: 2948-2954.

Gibbons, R.J. and T.H. Howell. 1978. Antigenic variation in a population of oral streptococci. Adv. Exp. Med. Biol. 107: 829-838.

Gibbons, R.J. and P.H. Keyes. 1969. Inhibition of insoluble dextran synthesis, plaque formation, and dental caries in hamsters by low molecular weight dextran. Arch. Oral Biol. 14: 721-724.

Gibbons, R.J. and J.B. Macdonald. 1960. Hemin and vitamin K compounds are required factors for the cultivation of certain strains of *Bacteroides melaninogenicus*. J. Bacteriol. 80: 164-170.

Gibbons, R.J. and J.B. Macdonald. 1961. Degradation of collagenous substrates by *Bacteroides melaninogenicus*. J. Bacteriol. 81: 614-621.

Gibbons, R.J. and M. Nygaard. 1970. Interbacterial aggregation of plaque bacteria. Arch. Oral Biol. 15: 1397-1400.

Gibbons, R.J. and S.S. Socransky. 1962. Intracellular polysaccharide storage by organisms in dental plaques. Its relations to dental caries and microbial ecology of oral cavity. Arch. Oral Biol. 7: 73-80.

Gibbons, R.J., D.M. Spinell, and Z. Skobe. 1976. Selective adherence as a determinant of the host tropisms of certain indigenous and pathogenic bacteria. Infect. Immun. 13: 238-246.

Gibbons, R.J. and J.V. Qureshi. 1978. Selective binding of blood group-reactive salivary mucins by *Streptococcus mutans* and other oral organisms. Infect. Immun. 22: 665-671.

Gibbons, R.J. and J.V. Qureshi. 1979. Inhibition of adsorption of *Streptococcus mutans* strains to saliva-coated hydroxyapatite by galactose and certain amines. Infect. Immun. 26: 1214-1217.

Gibbons, R.J. and J.V. Qureshi. 1980. Virulence-related physiologic changes and antigenic variation in populations of *Streptococcus mutans* colonizing gnotobiotic rats. Infect. Immun. 19: 1082-1091.

Gibbons, R.J. and J. van Houte. 1971. Selective bacterial adherence to oral epithelial surfaces and its role as an ecological determinant. Infect. Immun. 3: 567-573.

Gibbons, R.J. and J. van Houte. 1973. On the formation of dental plaques. J.

Periodontol. 44: 347-360.

Gibbons, R.J. and J. van Houte. 1975. Bacterial adherence in oral microbial ecology. Annu. Rev. Microbiol. 29: 19-44.

Hillman, J.D., J. van Houte, and R.J. Gibbons. 1970. Sorption of bacteria to human enamel powder. Arch. Oral Biol. 15: 899-903.

Kelstrup, J. and R.J. Gibbons. 1969. Bacteriocins from human and rodent streptococci. Arch. Oral Biol. 14: 251-258.

Kelstrup, J., S. Richmond, C. West, and R.J. Gibbons. 1970. Fingerprinting human oral streptococci by bacteriocin production and sensitivity. Arch. Oral Biol. 15: 1109-1116.

Komiyama, K. and R.J. Gibbons. 1984. Interbacterial adherence between *Actinomyces viscosus* and strains of *Streptococcus pyogenes, Streptococcus agalactiae*, and *Pseudomonas aeruginosa*. Infect. Immun. 44: 86-90.

Liljemark, W.F. and R.J. Gibbons. 1971. Ability of *Veillonella* and *Neisseria* species to attach to oral surfaces and their proportions present indigenously. Infect. Immun. 4: 264-268.

Macdonald, J.B., S.S. Socransky, and R.J. Gibbons. 1963. Aspects of the pathogenesis of mixed anaerobic infection of mucous membranes. J. Dent. Res. 42: 529-544.

Socransky, S.S. and R.J. Gibbons. 1965. Required role of *Bacteroides melaninogenicus* in mixed anaerobic infections. J. Infect. Dis. 115: 247-253.

Slots, J. and R.J. Gibbons. 1978. Attachment of *Bacteroides melaninogenicus* subsp. *asaccharolyticus* to oral surfaces and its possible role in colonization of the mouth and of periodontal pockets. Infect. Immun. 19: 254-264.

van Houte, J., R.J. Gibbons, and S.B. Banghart. 1970. Adherence as a determinant of the presence of *Streptococcus salivarius* and *Streptococcus sanguis* on the human tooth surface. Arch. Oral Biol. 15: 1025-1034.

Wahren, A. and R.J. Gibbons. 1970. Amino acid fermentation by *Bacteroides melaninogenicus*. Antonie van Leeuwenhoek 36: 149-159.

Williams, R.C. and R.J. Gibbons. 1972. Inhibition of bacterial adherence by secretory immunoglobulin A: a mechanism of antigen disposal. Science 177: 697-699.

Williams, R.C. and R.J. Gibbons. 1975. Inhibition of streptococcal attachment to receptors on human buccal epithelial cells by antigenically similar salivary glycoproteins. Infect. Immun. 11: 711-718.

From: *Oral Bacterial Ecology: The Molecular Basis*
ISBN 1-898486-22-0 ©2000 Horizon Scientific Press, Wymondham, U.K.

1

ORAL ECOLOGY AND ITS IMPACT ON ORAL MICROBIAL DIVERSITY

P. D. Marsh

Contents

Introduction

In order to introduce the concept of ecology to the microbiology of the oral cavity, the reader should consider the following question:

"Which of the following events is the odd one out?":
(1) The growth of algae in rivers and ponds following the leaching of nitrogenous fertilisers from farm land into water.
(2) The loss of marine life around shores following an oil spillage.
(3) The extinction of the dinosaurs due to climatic changes in the Cretaceous period.
(4) The development of caries and periodontal diseases in the mouth!

It is the contention of the authors of this volume that the processes underlying all of the above events are *similar*, and that it is possible to explain all of these four diverse biological events through an understanding of ecological principles. Indeed, a willingness to apply such principles can lead to several tangible benefits which include (1) a clearer understanding of the apparently complex relationship between the host and its resident oral microflora in health and disease, and (2) the identification of novel routes for disease prevention. A further benefit of an application of this ecological approach to understanding the cause of oral disease is that the principles can continue to be applied validly irrespective of subsequent changes to the nomenclature of the microflora or the discovery of "new" organisms. This is because emphasis is placed on determining the properties, and hence the function, of specific organisms in the disease process. This approach also seeks to relate microbiological observations to changes to the life-style and/or medical history of the patient.

Ecological Terminology

The microbial ecological terminology used in this chapter will be mainly as defined by Alexander (Alexander, 1971); thus, the site where micro-organisms grow is the **habitat**. The micro-organisms growing in a particular habitat constitute a **microbial community** made up of populations of individual species or less well-defined groups (**taxa**). Species found characteristically in a particular habitat are termed **autochthonous** micro-organisms. These multiply and persist at a site and contribute to the metabolism of a microbial community (with no distinction made regarding disease potential), and can be contrasted with **allochthonous** or exogenous organisms, which originate from elsewhere and are generally unable to colonize successfully unless the ecosystem is severely perturbed. Alternatively, the term **"resident microflora"** can be used to include any organism that is regularly isolated from a site; this term also has the benefit that no distinction concerning disease potential is made. The microbial community in a specific habitat together with the surroundings with which these organisms are associated is known as the **ecosystem**. In the terminology adopted here, the **niche** describes the *function* of an organism in a particular habitat. Thus, the niche is not the physical position of an organism but is its role within the community. This role is dictated by the biological properties of each microbial population. Species with identical functions in a particular habitat will compete for the same niche, while the co-existence of many species in a habitat is due to each population having a different role (niche) and thus avoiding competition.

Micro-organisms that have the potential to cause disease are termed **pathogens**. Those that cause disease only under exceptional circumstances are described as **opportunistic pathogens**, and can be distinguished from **true pathogens** which are consistently associated with a particular disease. It will be argued that oral diseases mediated by dental plaque are caused by resident organisms behaving opportunistically, and exploiting changes to the habitat or ecosystem.

Once established at a site, the composition of the resident microflora, and the proportions of the component populations, are relatively stable over time, despite regular exposure to modest environmental perturbations. The factors determining this stability (**microbial homeostasis**) will be discussed in detail later. Thus, a substantial disruption to the habitat needs to occur for the resident microflora at a site to be substantially disturbed. In the examples cited earlier, therefore, the fertiliser changes the nutritional status of the water, by providing additional nitrogen and permitting the outgrowth of previously minor components (algae) of the normal pond water communities (Codd, 1995). Eutrophication can be a further consequence because of the excessive oxygen utilisation by these microbes. Similarly, dust clouds following the impact of a meteorite (Alvarez *et al.*, 1980) reduce heat and light levels which, together with modern events such as soil spillages, are more extreme examples in which the environment is catastrophically affected, leading to disruption of key elements of the food chain (Raup, 1989).

It follows, therefore, that for disease to occur in the mouth, homeostatic mechanisms have to break down, thereby enabling potential disease-producing organisms to either colonise or to become predominant (Newman, 1990; Marsh, 1991). In this chapter, the resident microflora at distinct sites in the mouth will be described. Firstly, though, the ecological determinants that help define the composition and metabolism of these microbial communities will be considered, and factors capable of disrupting microbial homeostasis will be highlighted. Most published work has been focussed on understanding the role of dental plaque in disease, and this emphasis will also be reflected in this chapter.

The Resident Human Microflora

It has been estimated that the human body is made up of over 10^{14} cells of which only around 10% are mammalian (Sanders and Sanders, 1984). The remainder are the micro-organisms that comprise the resident microflora of the host. The composition of this microflora varies at distinct habitats, but is relatively consistent at each individual site among individuals. In other words, the composition of, for example, the skin microflora is relatively similar among people irrespective of age, gender, ethnic group, etc, but is consistently different from that of the mouth and gut (Figure 1) (Tannock, 1995). Thus, the skin is comprised almost entirely of Gram positive species; these are mainly aerobic (*Micrococcus* spp.) or facultatively anaerobic (*Corynebacterium* and *Staphylococcus* spp.) with a few obligately anaerobic species (*Propionibacterium* spp.) and yeasts (Marples, 1994). In contrast,

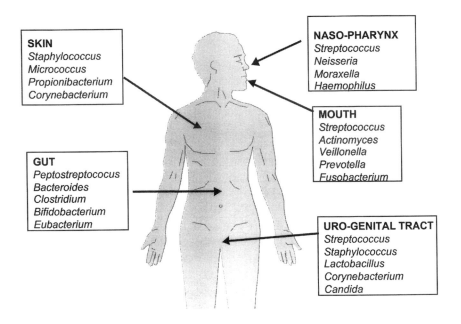

Figure 1. The distribution of the resident human microflora. The figure shows some of the predominant genera found at distinct sites.

the microfloras of the mouth and the lower digestive tract have high numbers of obligately anaerobic Gram-negative bacteria, although there are marked differences in the predominant genera and species at each habitat resulting in both sites having a microflora with a characteristic composition (Tannock, 1995). Indeed, following extensive studies comparing the microflora of dental plaque (in health and disease) and the gastro-intestinal tract, only 29 out of over 500 taxa found in the mouth were recovered from faecal samples, despite the continuous passage of these bacteria into the gut via saliva (Moore and Moore, 1994).

The principal bacterial genera isolated from the mouth are listed in Table 1; in addition, yeasts (predominantly *Candida*, but also *Rhodotorula* and *Saccharomyces* spp.), mycoplasmas and protozoa (e.g. *Entamoeba gingivalis*, *Trichomonas tenax* and *Giardia lamblia*) can be found in the healthy oral cavity. A number of viruses can also be detected in saliva (e.g. Herpes simplex, cytomegaloviruses, Hepatitis, Human Immunodeficiency Virus and Epstein-Barr virus). A fuller description of the resident oral microflora can be found elsewhere (Marsh and Martin, 1999).

In order to explain the continuous maintenance of microbial communities with a distinctive composition around the body, it has to be assumed that each of these habitats differs in terms of key ecological factors that enable certain populations to dominate at one site while rendering them non-competitive at others. Such factors can include the provision of appropriate receptors for attachment and colonisation, and essential nutrients and cofactors for growth, as well as an appropriate pH, redox

Table 1. Bacterial Genera Found in the Oral Cavity (adapted from Marsh and Martin, 1999; with permission)

GRAM-POSITIVE	GRAM-NEGATIVE
COCCI	
Abiotrophia	*Moraxella*
Enterococcus	*Neisseria*
Peptostreptococcus	*Veillonella*
Streptococcus	
Staphylococcus	
Stomatococcus	
RODS	
Actinomyces	*Actinobacillus*
Bifidobacterium	*(Bacteroides)**
Corynebacterium	*Campylobacter*
Eubacterium	*Cantonella*
Lactobacillus	*Capnocytophaga*
Propionibacterium	*Centipeda*
Pseudoramibacter	*Desulfovibrio*
Rothia	*Desulfobacter*
	Eikenella
	Fusobacterium
	Haemophilus
	Johnsonii
	Leptotrichia
	Porphyromonas
	Prevotella
	Selenomonas
	Simonsiella
	Treponema
	Wolinella

*The genus *Bacteroides* has been redefined. In time, the remaining oral bacteria still placed in this genus will be reclassified.

potential (Eh) and gaseous environment. For example, the density of the microflora on the skin varies according to moisture and lipid levels, and the highest population levels are found in the axilla. Moisture is probably the main limiting environmental determinant since occlusion of the forearm leads to an increase in the numbers of aerobic bacteria from 10^3 to 10^7 cells/cm^2, and a shift from a staphylococcal-dominated microflora to one with enhanced numbers of coryneforms (Aly and Maibach, 1981). This situation can be contrasted to the conditions in the oral cavity, the surfaces of which are bathed in saliva, and yet in which staphylococci, micrococci and corynebacteria are isolated only occasionally (Marsh and Martin, 1999). The environmental determinants that define the mouth as a microbial habitat will be described shortly in a subsequent section. It is important to note, however, that the example of the skin microflora described above clearly shows that a substantial

Table 2. Detrimental Effects Associated with the Absence or Suppression of the Resident Microflora at a Site

Resident microflora	Consequence
Absent[1]	thin intestinal walls poorly developed villi poor nutrient adsorption vitamin deficiencies reduced host defences caecum enlargement
Suppressed[2]	overgrowth by drug-resistant organisms colonisation by exogenous species

[1]data based on germ-free animal studies
[2]data based on the effects of antibiotics on the human microflora
Adapted from Marsh (1989).

change in the local environment is capable of causing a major perturbation in the balance of the resident microflora at a site.

The resident microflora at any site does not have merely a passive relationship with the host, but contributes directly and indirectly to the normal development of the physiology, nutrition and defence systems of the organism (Table 2) (Rosebury, 1962; Grubb *et al.*, 1989; Marsh, 1989). Knowledge of this area has come from two sources: (1) observations of the physiology of germ-free animals, and (2) studies of the consequences following suppression of the natural microflora of man and animals by antimicrobial agents. The gut of germ-free animals is poorly-developed, but when these animals are colonised by components of the natural resident microflora, many of these anatomical and physiological deficiencies reverse (Rosebury, 1962; Sanders and Sanders, 1984). A range of nutrient deficiencies in absorption or metabolism of vitamins has also been reported in humans on long-term antibiotic treatment. Antibiotics can also lead to the rapid suppression of the resident oral microflora and result in overgrowth by drug-resistant, but previously minor components of the microflora, or colonisation by exogenously-acquired (and often pathogenic) micro-organisms (Lacey *et al.*, 1983; Sanders and Sanders, 1984; Woodman *et al.*, 1985). These observations confirm that the resident microflora also acts directly as an important component of the host defences by being a significant barrier to exogenous (allochthonous) populations. This barrier effect is termed "**colonisation resistance**" (van der Waaij *et al.*, 1971), and has been well-characterised in other habitats, such as the gut. Now, many food products are being designed as **probiotics** and contain micro-organisms such as lactobacilli and/or bifidobacteria to enhance or restore the colonisation resistance properties of the digestive tract (Fuller, 1992) when the microflora of this site has been perturbed. The mechanisms involved in colonisation resistance are listed in Table 3. Studies are underway to deliberately colonise oral surfaces in infants with strains that would prevent subsequent colonisation by natural strains of, for example, mutans streptococci that could cause disease (**replacement

---

Table 3. Properties of the Resident Microflora that Contribute to "Colonisation Resistance"

Competition for receptor sites for adhesion
Competition for essential endogenous nutrients and co-factors
Creation of micro-environments that prevent the growth of exogenous species
Production of inhibitory factors (e.g. bacteriocins, H_2O_2, acids, etc)

---

therapy) (Hillman, 1999). In general, therefore, the microflora of a site lives in harmony with the host and, indeed, all parties benefit from the association. Studies are just beginning in attempts to unravel the molecular mechanisms involved in regulating these host-microbe interactions. It has been proposed recently that this harmonious relationship is a result of complex molecular signalling between members of the resident microflora and host cells so as to actively avoid the unnecessary activation of the host inflammatory response (Henderson and Wilson, 1998).

The microbial colonization of all environmentally-exposed surfaces of the body (both external and internal) begins at birth. Such surfaces are exposed to a wide range of micro-organisms derived from the environment and from other persons. Each surface, however, because of its physical and biological properties, is suitable for colonization by only a proportion of these microbes. Some of the key ecological factors that define the mouth as a potential microbial habitat will now be described.

The Mouth as a Microbial Habitat

The mouth is continuously bathed with saliva, and this has a profound influence on the ecology of the mouth (Scannapieco, 1994). The mean pH of saliva is between pH 6.75-7.25, which favours the growth of many micro-organisms, and the ionic composition of saliva promotes its buffering properties and its ability to remineralise enamel. The major organic constituents of saliva are proteins and glycoproteins, such as mucin. These influence the oral microflora by:
(a) adsorbing to the tooth surface to form a conditioning film (the acquired pellicle) to which micro-organisms can attach (see Chapter 3),
(b) acting as primary sources of nutrients (carbohydrates and proteins) which foster the growth of the resident microflora in the absence of a significant pH fall (see Chapter 2),
(c) aggregating micro-organisms and thereby facilitating their clearance from the mouth by swallowing, and
(d) inhibiting the growth of some exogenous micro-organisms, via their role as components of the host defences.

The natural flow of saliva, and mastication, will detach micro-organisms not firmly attached to an oral surface. Although saliva contains between 10^8-10^9 viable microbes/ ml, these organisms are all derived from the teeth and oral mucosa, especially the tongue. The rate of swallowing ensures that these organisms are unable to maintain themselves in saliva by cell division. Thus, adherence to a surface is a critical ecological determinant if micro-organisms are to persist in the mouth (Gibbons,

Table 4. Distinct Microbial Habitats Within the Mouth

Habitat	Comment
Lips, cheek, palate	biomass restricted by desquamation; different surfaces have specialised host cell types
Tongue	highly papillated surface; acts as a reservoir for anaerobes
Teeth	non-shedding surface enabling large masses of microbes to accumulate (e.g. biofilms such as dental plaque). Teeth have distinct surfaces for microbial colonisation; each surface (e.g. fissures, smooth surfaces, approximal, gingival crevice) will support a distinct microflora because of their intrinsic biological properties.

1989; Ellen and Burne, 1996). Adherence involves specific inter-molecular interactions between adhesins on the surface of the micro-organism and receptors in the acquired pellicle (conditioning film) on the host surface, and is described in detail in Chapter 3. Oral organisms are not distributed randomly but usually display **tissue tropisms**, i.e. they selectively attach and grow on certain surfaces (Gibbons, 1989). The composition of the microflora of these surfaces will be described later in this Chapter.

The mouth is not a homogeneous environment for microbial colonisation. Distinct habitats exist, for example, the mucosal surfaces (such as the lips, cheek, palate, and tongue) and teeth (Table 4) which, because of their biological features, support the growth of a distinctive microbial community. The properties of some of these habitats will change during the life of an individual. For example, during the first few months of life the mouth consists only of mucosal surfaces for microbial colonisation. With the development of the primary dentition, hard non-shedding surfaces appear providing a unique surface in the body for microbial colonisation. The primary dentition is usually complete by the age of 3 years, and the permanent teeth appear between 6-12 years of age. Inevitably, local environmental conditions will vary during these periods of change, which will in turn influence the composition and metabolism of the resident microbial community at a site. The eruption of teeth also generates another habitat, the gingival crevice (where the tooth emerges from the gums; Figure 2), and an additional major nutrient source for that site (gingival crevicular fluid, GCF).

Ecological conditions within the mouth will also be affected by extraction of teeth, the insertion of prostheses such as dentures, and any dental treatment including scaling, polishing and restorations. Transient fluctuations in the stability of the oral

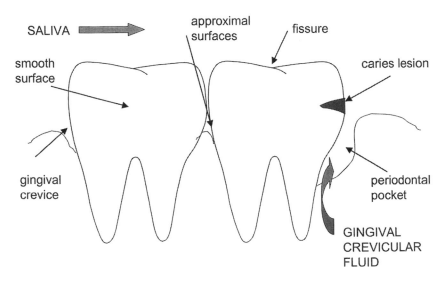

Figure 2. The main habitats associated with the tooth surface in health and disease.

ecosystem may be induced by the frequency and type of food ingested, periods of antibiotic therapy, and variations in the composition and rate of flow of saliva. For example, a side-effect of medication can be a reduction in saliva flow, and this can predispose a site to caries. In old age, the activity of the host defences can wane, and this might explain the increased isolation of staphylococci and enterobacteria from the oral cavity of the elderly (Percival *et al.*, 1991; Marsh *et al.*, 1992).

The mouth is the only normally-accessible site in the body that has hard, non-shedding surfaces for microbial colonization. Teeth allow the accumulation of large masses of micro-organisms (predominantly bacteria) and their extracellular products (collectively, this is termed dental plaque), especially at stagnant or retentive sites. In contrast, elsewhere in the body, desquamation ensures that the microbial load is relatively light on most mucosal surfaces. The mouth has some specialised epithelial surfaces which contribute to the diversity of the microflora. The papillary structure of the dorsum of the tongue provides refuge for many micro-organisms which would otherwise be removed by mastication and the flow of saliva. Such sites on the tongue can also have a low redox potential, which facilitates the growth of obligately anaerobic bacteria. Indeed, the tongue can act as a reservoir for some of the Gram-negative anaerobes that are implicated in the aetiology of periodontal diseases (Chapter 6). The mouth also contains keratinised (as in the palate) as well as non-keratinised stratified squamous epithelium which may affect the intra-oral distribution of micro-organisms by displaying distinct receptors for microbial attachment.

The ecological complexity of the mouth is increased markedly by the range of habitats associated with the tooth surface. Teeth have distinct surfaces, each of which are optimal for colonization and growth by different populations of micro-organisms. Again, this is due to the physical nature of the particular surface and the resulting biological properties of the area (Table 5). The areas between adjacent teeth (approximal) and in the gingival crevice afford protection from the normal removal forces, such as mastication, salivary flow and oral hygiene practices. Both sites have a low redox potential (Eh) and, in addition, the gingival crevice region is bathed in the nutritionally-rich GCF, particularly during inflammation, and so these areas are able to support a more diverse community including higher proportions of obligately anaerobic bacteria. Smooth surfaces are more exposed to the prevailing environmental conditions and, consequently, can be colonized only by a limited number of bacterial species adapted to such extreme conditions. Pits and fissures of the biting (occlusal) surfaces of the teeth also offer protection from some of these environmental factors and, in addition, are also susceptible to food impaction. Such protected areas are associated with the largest microbial communities and, in general, the most disease.

The formation of dental plaque involves an ordered colonization (microbial succession) by a range of bacteria. The early colonisers interact with, and adhere to, saliva-coated enamel (Gibbons, 1989), while later colonisers bind to already attached species (coaggregation) (Kolenbrander, *et al.*, 1993). Plaque is an example of a biofilm (Novak, 1997; Newman and Wilson, 1999), and, while it is found naturally in health, it is also associated with dental caries and periodontal disease (Marsh, 1994). In disease, there is a shift in the composition of the plaque microflora away from the species that predominate in health (see Chapters 5 and 6) (Marsh and Martin, 1999). Bacteria growing on a surface as a biofilm display an altered phenotype, and are also more resistant to antimicrobial agents (for reviews, see Costerton, *et al.*, 1987, and Wilson, 1996); the formation and properties of dental plaque will be described in more detail later in this Chapter.

The relationship between the environment and the microbial community is not unidirectional. Although the properties of the environment dictate which micro-organisms can occupy a given site, the metabolism of the microbial community can modify the physical and chemical properties of their surroundings. Thus, the environmental conditions change during the formation of dental plaque. For example, the metabolism of early colonisers will deplete oxygen and produce carbon dioxide and hydrogen, which lowers the redox potential and provides environmental conditions suitable for anaerobes. The local habitat on the tooth will also vary in health and disease (Figure 2). As caries progresses, the advancing front of the lesion penetrates the dentine. The nutritional sources will change and local conditions may become acidic and more anaerobic due to the accumulation of products of bacterial metabolism. Similarly, in disease, the gingival crevice develops into a periodontal pocket and the production of GCF is increased. These new environments will result in the selection of the microbial community most suitably adapted to the prevailing conditions (Table 5).

The health of the mouth is dependent on the integrity of the mucosa (and enamel) which acts as a physical barrier preventing penetration by micro-organisms or antigens. The host has a number of additional defence mechanisms which play an important role in maintaining the integrity of these oral surfaces (Table 6). These defences are divided into non-specific (or innate) and specific (or immune) factors. The former, unlike antibodies, do not require prior exposure to an organism or antigen

Table 5. Factors Affecting the Growth of Micro-organisms in the Oral Cavity

Factor	Range	Comment
Temperature	35-36°C	Increases during inflammation, e.g. in periodontal disease
Oxygen	0-21%	Oxygen abundant at mucosal surfaces; gradients exist in biofilms such as plaque
Redox potential (Eh)	+200 to <-200mV	Gradients exist in biofilms such as plaque; lowest value in periodontal pockets
pH	6.75-7.25	Plaque pH falls during sugar metabolism; sub-gingival plaque pH rises during inflammation
Nutrients	endogenous	peptides, proteins and glycoproteins in saliva and GCF support growth of resident microflora
	exogenous	sucrose facilitates overgrowth by acidogenic and aciduric species in plaque

for activity and so provide a continuous, broad spectrum of protection. An alternative terminology is for the non-specific and specific factors to be termed innate immunity and adaptive immunity, respectively (see Chapter 4).

Several anti-bacterial factors are present in saliva (Table 6) (Scannapieco, 1994) which are important in controlling the bacterial and fungal colonization of the mouth, and include lysozyme, lactoferrin, and the sialoperoxidase system. Antibodies have been detected, with secretory IgA (sIgA) being the predominant class of immunoglobulin; IgG and IgM are also present but in lower concentrations. A major role of sIgA is to reduce or prevent microbial colonisation. Antimicrobial peptides

Table 6. Specific and Non-Specific Host Defence Factors of the Mouth (Marsh and Martin, 1999)

Defence Factor	Main Function
Non-Specific:	
Saliva flow	Physical removal of micro-organisms
Mucin/agglutinins	Physical removal of micro-organisms
Lysozyme-protease-anion	Cell lysis
Lactoferrin	Iron sequestration
Apo-lactoferrin	Cell killing
Sialoperoxidase system	Hypothiocyanite production (neutral pH) Hypocyanous acid production (low pH)
Antibacterial peptides (e.g. histatins, cystatins, defensins)	Antibacterial and antifungal activity; bacterial protease inhibition
Specific:	
Intra-epithelial lymphocytes and Langerhans cells	Cellular barrier to penetrating bacteria and/or antigens
sIgA	Prevents microbial adhesion and metabolism
IgG, IgA, IgM	Prevent microbial adhesion; opsonins; complement activators
Complement	Activates neutrophils
Neutrophils/macrophages	Phagocytosis

are present, and include histidine-rich polypeptides (histatins), and cystatins, and these may control the levels of yeasts. Some histatins can also inhibit coaggregation between certain oral bacteria (Murakami *et al.*, 1991).

Serum components can reach the mouth by the flow of a serum-like gingival crevicular fluid (GCF) through the junctional epithelium of the gingiva (Mukherjee, 1985). The flow of GCF is relatively slow at healthy sites, but increases during the inflammatory responses associated with periodontal diseases (see Chapters 4 and 6). GCF can influence the ecology of the site by:
(a) removing non-adherent microbial cells,
(b) introducing additional components of the host defences, and
(c) acting as a novel source of nutrients for the resident micro-organisms.

In contrast to saliva, IgG is the predominant immunoglobulin in GCF; IgM and IgA are also present, as is complement. GCF contains large numbers of neutrophils, as well as a minor number of lymphocytes and monocytes. The neutrophils in GCF are viable and can phagocytose bacteria within the crevice. The resident microflora has evolved mechanisms by which members can evade the host defences; these will be discussed in Chapter 4.

Many bacteria from sub-gingival plaque are proteolytic and interact synergistically to degrade host proteins and glycoproteins in GCF to provide peptides, amino acids and carbohydrates for growth (van der Hoeven, 1998). Essential co-factors, including haemin for black-pigmented anaerobes, can also be obtained from the degradation of haeme-containing molecules such as transferrin, haemopexin, haemoglobin and haptoglobin (see Chapter 2) (Carlsson *et al.*, 1984).

Factors Affecting the Growth of Micro-organisms in the Oral Cavity

Key environmental factors which influence the growth of micro-organisms in the mouth are listed in Table 5, and additional details will be provided below.

Temperature
The human mouth is kept at a relatively constant temperature (35-36°C) which provides stable conditions suitable for the growth of a wide range of micro-organisms, although transient temperature shocks will occur following the intake of hot and cold beverages and food. Periodontal pockets with active disease have a higher mean temperature (up to 39°C) compared with healthy sites (Fedi and Killoy, 1992), and the temperature in a pocket tends to increase with increasing depth (Mukherjee, 1981). Such changes in temperature can affect gene expression in periodontal pathogens, such as *Porphyromonas gingivalis*. A rise in temperature to 39°C down-regulated expression of the some of the major proteases of this micro-organism (Percival *et al.*, 1999) as well as the gene coding for the major sub-unit protein of fimbriae which mediates attachment of the bacterium to host cells (Amano *et al.*, 1994; Xie *et al.*, 1997). In contrast, an increase in temperature up-regulated synthesis of superoxide dismutase, which is involved in the neutralisation of toxic oxygen metabolites (Amano *et al.*, 1994). Temperature has been shown to vary between different sub-gingival sites, even within the same individual, and this can influence

the proportions of certain bacterial species, such as the putative periodontal pathogens *P. gingivalis, Bacteroides forsythus* and *Campylobacter rectus* (Maiden *et al.*, 1998).

Redox Potential/Anaerobiosis

Despite the fact that the mouth is overtly aerobic, it is perhaps surprising that the oral microflora comprises few, if any, truly aerobic (oxygen-requiring) species (Table 1). The majority of organisms are either facultatively anaerobic (i.e. can grow in the presence or absence of oxygen) or obligately anaerobic (i.e. require reduced conditions, and oxygen can be toxic to these organisms). In addition, there are some capnophilic (CO_2-requiring) and micro-aerophilic species (requiring low concentrations of oxygen for growth). Members of the oral microflora produce a range of enzymes to protect themselves from oxidative stress (Bobo *et al.*, 1973; Marquis, 1995; Bowden and Hamilton, 1998), and strict anaerobes can also be protected by close association (physical and metabolic) with oxygen-consuming species (Bradshaw *et al.*, 1996, 1998).

The redox potential has been measured during plaque development on a clean enamel surface. As stated earlier, the development of plaque is associated with a specific succession of micro-organisms. Early colonisers utilise O_2 and produce CO_2; later colonisers may produce H_2 and other reducing agents such as sulphur-containing compounds and volatile fermentation products. An initial Eh of plaque of over +200 mV (highly oxidized) can fall to as low as -141 mV (highly reduced) after 7 days (Kenney and Ash, 1969). Thus, as the Eh is gradually lowered, sites become suitable for the survival and growth of a changing pattern of organisms, particularly anaerobes. Periodontal pockets are more reduced (mean value -48 mV) than healthy gingival crevices in the same individuals (mean value +73 mV) (Kenney and Ash, 1969). It is likely that the Eh falls still lower since highly anaerobic organisms, such as oral spirochaetes and sulphate-reducing bacteria, can be isolated (Table 1) (Marsh and Martin, 1999). Approximal areas will also have a low Eh.

The oxygen tension has been measured at sub-gingival sites. The pO_2 was lowest in deep pockets (mean value = 12 mm Hg) (Loesche *et al.*, 1983; Mettraux *et al.*, 1984). Gradients of O_2 concentration and Eh will develop in microbial communities over short distances, particularly in a thick biofilm such as plaque, due to bacterial metabolism, and this may influence bacterial behaviour. For example, the activity of intracellular glycolytic enzymes and the pattern of fermentation products of oral streptococci varies depending on the degree of anaerobiosis (Yamada *et al.*, 1985).

pH

Many micro-organisms require a pH around neutrality for growth, and are sensitive to extremes of acid or alkali. The pH of most surfaces of the mouth is regulated by saliva (mean pH = 6.75-7.25) thereby providing an optimum pH for microbial growth at most sites (Edgar and Higham, 1996). Bacterial population shifts within the plaque microflora can occur following fluctuations in environmental pH. After sugar consumption, the pH in plaque can fall rapidly to below pH 5.0 by the production of acids (predominantly lactic acid) by bacterial metabolism; the pH then recovers slowly to base-line values (Jensen and Schachtele, 1983). Many of the predominant plaque bacteria that are associated with healthy sites can tolerate brief conditions of low pH, but are inhibited or killed by more frequent or prolonged exposures to acidic conditions (Harper and Loesche, 1984; Bradshaw *et al.*, 1989), such as occurs in subjects who commonly consume sugar-containing snacks or drinks between meals. This can result in the enhanced growth of, or colonization by, acid-tolerant (**aciduric**) species, especially mutans streptococci and *Lactobacillus* species, which are normally absent or only minor components in dental plaque at healthy sites. Such a change in the bacterial composition of plaque predisposes a surface to dental caries (see Chapter 5). The acid tolerance of these bacteria is achieved by the possession of particular metabolic strategies and the induction of a specific set of stress response proteins (Bowden and Hamilton, 1998).

In contrast, the pH of the gingival crevice in health is approximately 6.9, and this can rise during the host inflammatory response in periodontal disease to between pH 7.2-7.4, with a few patients having pockets with a mean pH of around 7.8 (Eggert *et al.*, 1991). The rise in pH is probably due to the bacterial production of ammonia from urea and from the deamination of amino acids. Such a relatively modest rise in pH could still perturb the balance of the resident microflora of the gingival crevice. Laboratory studies of mixed cultures of three black-pigmented, Gram-negative anaerobic bacteria have shown that a rise in pH from pH 7.0 to 7.5 led to the periodontal pathogen *P. gingivalis* increasing from <1% of the microbial community to predominate within the culture (Marsh *et al.*, 1993). The activity of some of the major proteases of periodontal pathogens is also optimal around pH 7.5-8.0 (McDermid *et al.*, 1988).

Nutrients

Populations within a microbial community are dependent solely on the habitat for the nutrients essential for their growth. Therefore, the association of an organism with a particular habitat, including highly fastidious species, is direct evidence that all of the necessary growth-requiring nutrients are present.

(i) *Endogenous Nutrients*

The persistence and diversity of the resident oral microflora is due primarily to the metabolism of the endogenous nutrients provided by the host, rather than by exogenous factors in the diet (Chapter 2) (Beckers and van der Hoeven, 1982; Beighton and Hayday, 1986). The main source of endogenous nutrients is saliva, which contains amino acids, peptides, proteins and glycoproteins (which also act as a source of sugars and amino-sugars), vitamins and gases (Scannapieco, 1994). Moreover, the gingival crevice is supplied with GCF which, in addition to delivering components of the host defences, contains potential sources of novel nutrients, such as albumin and other host proteins and glycoproteins, including haeme-containing molecules, that are not found in saliva (Cimasoni, 1983; Mukherjee, 1985). The difference in the source of endogenous nutrients is one of the reasons for the variation in the microflora of the gingival crevice compared with other oral sites, particularly on teeth.

Evidence for the importance of endogenous nutrients has also come from the observation that a diverse microbial community persists in the mouth of humans and animals fed by intubation (stomach tube) (Littleton *et al.*, 1967). Also, the oral microflora of animals with dietary habits ranging from insectivores and herbivores to carnivores is broadly similar at the genus level (Dent, 1979; Dent and Marsh, 1981). Plaque bacteria produce glycosidases which release carbohydrates from the oligosaccharide side chains of salivary mucins (Beighton *et al.*, 1986). Similarly, organisms isolated from the gingival crevice and periodontal pocket can degrade host proteins and glycoproteins including albumin, transferrin, haemoglobin, and immunoglobulins (Carlsson *et al.*, 1984; Sundqvist *et al.*, 1985). Oral micro-organisms interact synergistically as a consortium to break down these endogenous nutrients. No single species has the full enzyme complement to totally metabolise these nutrients. Individual organisms possess different but overlapping patterns of enzyme activity, so that they co-operate and interact with species with complementary degradative profiles to achieve complete degradation (van der Hoeven, 1998).

(ii) Exogenous (dietary) Nutrients
Superimposed upon these endogenous nutrients is the complex array of foodstuffs ingested periodically in the diet. Despite the complexity of the diet, fermentable carbohydrates are one of the few components that have been found to influence markedly the ecology of the mouth. Such carbohydrates can be broken down to acids while, additionally, sucrose can be converted by bacterial enzymes (glucosyltransferases, GTF, and fructosyltransferases, FTF) into two main classes of polymer (glucans and fructans) which can be used to consolidate attachment or act as extracellular nutrient storage compounds, respectively (Walker and Jacques, 1987; Kuramitsu, 1993; Russell, 1994). Excess carbohydrate is converted to intracellular polysaccharide storage compounds. The frequent consumption of dietary carbohydrates is associated with an increase in the proportions of mutans streptococci and lactobacilli in plaque, a decrease in levels of acid-sensitive species (e.g. *Streptococcus sanguis, S. gordonii*), and a shift to a homo-fermentative pattern of metabolism, with lactate as the predominant fermentation product (Loesche, 1986). Laboratory studies suggest that it is the repeated low pH generated from sugar metabolism rather than the availability of excess carbohydrate *per se* that is responsible for the perturbations to the microflora (Bradshaw *et al.*, 1989).

Dairy products (milk, cheese) have some influence on the ecology of the mouth. The ingestion of milk or milk products can protect the teeth of animals against caries (Reynolds and del Rio, 1984; Bowen *et al.*, 1991). This may be due to the buffering capacity of milk proteins or due to decarboxylation of amino acids after proteolysis since several bacterial species can metabolise casein. Milk proteins (casein) and casein derivatives can also adsorb on to the tooth surface and reduce the adhesion of mutans streptococci (Schüpbach *et al.*, 1996). The milk glycoprotein, kappa-casein, can inhibit GTF adsorption into the pellicle and reduce enzyme activity, thereby suppressing glucan formation (Vacca-Smith and Bowen, 1995).

Cheese has been shown to increase salivary flow rates in animals and to rapidly elevate plaque pH changes in humans following a sucrose rinse. Xylitol, a sugar substitute that cannot be metabolised by oral bacteria and which has been added to some confectionery, can be inhibitory to the growth of *Streptococcus mutans*. Lower levels of this species are found in plaque and saliva of those who frequently consume confectionery containing this alternative sweetener (Isokangas *et al.*, 1991). The role of nutrition as an ecological determinant will be discussed in greater detail in Chapter 2.

Host Genetics and Social Behaviour

Studies of periodontal disease suggest that gender and race can affect the microflora and possibly also influence disease susceptibility. In an adult periodontitis group, *P. gingivalis* and *Peptostreptococcus anaerobius* were associated more with black subjects whereas *Fusobacterium nucleatum* was found more commonly in white individuals (Schenkein *et al.*, 1993). The reasons for this are unknown, but may reflect some variation in the local immune response. No significant differences in microflora were found between males and females in these groups, although the presence of female hormones in GCF has been found by some groups (but not by others) to enhance the growth of some putative periodontal pathogens. A wide selection of strains of *Actinobacillus actinomycetemcomitans* (a pathogen in juvenile periodontitis; see Chapter 6) have been isolated from different geographical areas and screened for their genetic relatedness. Some strains were found to over-produce a particular virulence factor (a leukotoxin), and these were all isolated from individuals who could be traced to North West Africa (Haubek *et al.*, 1997).

The sub-gingival microflora of twins has also been compared (Moore *et al.*, 1993). The microflora of twin children living together was more similar than that of unrelated children of the same age. Further analysis showed that the microflora of identical twins was more similar than that of fraternal twins, suggesting some genetic influence.

Social traits can affect the composition of the resident oral microflora. As stated earlier, the frequent intake of fermentable substrates in the diet leads to the selection of acid-tolerating species (and an increased risk of dental caries) (Loesche, 1986), while smoking is a major risk factor for periodontal diseases, and regular tobacco smokers have increased levels and prevalence of potential pathogens such as *P. gingivalis* and *A. actinomycetemcomitans* (Zambon *et al.*, 1996).

Acquisition of the Resident Oral Microflora

The foetus in the womb is normally sterile. Acquisition of the resident microflora depends on the successive transmission of micro-organisms to the site of potential colonization. In the mouth, this is by passive transfer from the mother, from organisms present in milk, water (and eventually food), and the general environment, as well as from the saliva of individuals in close proximity to the baby (Davey and Rogers, 1984; Berkowitz and Jones, 1985; Greenstein and Lamster, 1997; Asikainen and Chen, 1999). Acquisition of micro-organisms from the birth canal itself may be of only limited significance. Studies on the transmission of yeasts and lactobacilli from the vagina of the mother suggest that these organisms do not initially become established as part of the resident oral microflora of the newborn, although they may be present transiently. Saliva is probably the main vehicle for transmission. Bacteriocin-typing and genotyping of strains has enabled the transfer of *S. salivarius*, mutans streptococci and Gram-negative species (e.g. *P. gingivalis* and *A. actinomycetemcomitans*) from mother to child to be followed (**vertical transmission**;

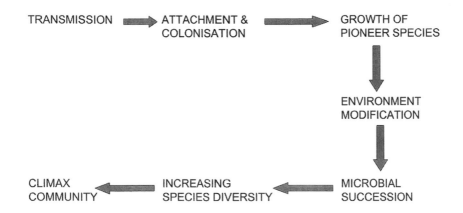

Figure 3. Ecological stages in the establishment of microbial communities in the mouth.

Greenstein and Lamster, 1997; Asikainen and Chen, 1999). Generally, similar clonal types of species are found within family groups, while different patterns are usually observed between such groups. For example, the genotypes of mutans streptococci found in children appeared identical to those of their mothers in 71% of 34 infant-mother pairs examined (Li and Caufield, 1995). No evidence of father-infant (or father-mother) transmission of mutans streptococci was observed, although **horizontal transmission** of some periodontal pathogens, such as *P. gingivalis,* may occur between spouses (for reviews, see Greenstein and Lamster, 1997; Asikainen and Chen, 1999).

Pioneer Community and Ecological Succession.
The mouth is highly selective for micro-organisms even during the first few days of life. Only a few of the species common to the oral cavity of adults, and even less of the large number of bacteria found in the environment, are able to colonize the mouth of the newborn. The first micro-organisms to colonize are termed **pioneer species**, and collectively they make up the pioneer microbial community (Figure 3). These pioneer species continue to grow and colonize until environmental resistance is encountered. This can be due to several limiting forces which act as barriers to further development, and include both physical (desquamation, and shear forces from chewing and saliva flow) and chemical (nutrient requirements, Eh, pH, and the antibacterial properties of saliva) factors.

One genus or species usually predominates during the development of the pioneer community. In the mouth, the predominant organisms are streptococci and in particular, *S. salivarius, S. mitis* and *S. oralis* (Smith *et al.*, 1993; Pearce *et al.*, 1995). Many of the pioneer species possess IgA$_1$ protease activity (Cole *et al.*, 1994), which may confer an advantage by enabling producer organisms to evade the effects of this key host defence factor that coats most oral surfaces (Table 6). With time, the metabolic activity of the pioneer community modifies the environment, thereby providing conditions suitable for colonization by a succession of other populations. This may be by:

(a) changing the local Eh or pH,

(b) modifying or exposing new receptors for attachment ("cryptitopes"; Gibbons *et al.*, 1990 - see Chapter 3), or

(c) generating novel nutrients, for example, as end products of metabolism (lactate, succinate, etc.) or as breakdown products (peptides, haëmin, etc) which can be used as primary nutrients by other organisms as part of a food chain.

Microbial succession involves the progressive development of a pioneer community (containing few species) through several stages in which the number of types of organism increases. Eventually a stable situation is reached with a high species diversity (**climax community**; Figure 3). Succession is associated with a change from a site possessing few niches (i.e. functions) to one with a multitude of potential niches. A climax community reflects a highly dynamic state and must not be regarded as a static situation. A change in the environment could lead to a reaction from the microflora resulting in an altered composition and metabolic activity.

The oral cavity of the newborn contains only epithelial surfaces for colonization. The pioneer populations consist of mainly aerobic and facultatively anaerobic species. In a study of forty full-term babies, a range of streptococcal species was recovered during the first three days of life, and *S. oralis, S. mitis* biovar 1 and *S. salivarius* were the numerically dominant species (Table 7) (Pearce *et al.*, 1995). Indeed, *S. salivarius* has been isolated from the mouth of infants as early as 18 hours after birth. The diversity of the streptococcal microflora increased with time, and all babies were colonised by more than one species of *Streptococcus* by one month of age. At this age, *S. salivarius* and *S. mitis* biovar 1 were isolated most commonly and predominated (Table 7).

The diversity of the pioneer oral community increases during the first few months of life, and several species of Gram-negative anaerobes appear. In a study of edentulous infants with a mean age of 3 months (range: 1-7 months), *Prevotella melaninogenica* was the most frequently isolated anaerobe, being recovered from

Table 7. Streptococcal Species Isolated from the Mucosal Surfaces of Babies (Pearce *et al.*, 1995)

| *Streptococcus* | Percentage viable count | | |
| | Age | | |
	1-3 days	2 weeks	1 month
S. oralis	41	24	20
S. mitis biovar 1	30	28	30
S. mitis biovar 2	4	1	1
S. salivarius	10	30	28
S. sanguis	4	3	2
S. anginosus	3	5	5
S. gordonii	1	2	4

76% of infants (Table 8) (Könönen *et al.*, 1992). Other commonly isolated bacteria were *F. nucleatum*, *Veillonella* spp., and non-pigmented *Prevotella* spp. In contrast, *Capnocytophaga* spp. (including *C. gracilis*), *Prevotella loescheii* and *Prevotella intermedia* were recovered from 4-23% of infants, while *Eikenella corrodens* and *Wolinella succinogenes* were only found in a single mouth. The number of different anaerobes in the same mouth varied from 0-7 species (Könönen *et al.*, 1992).

The same infants were followed longitudinally during the eruption of the primary dentition (Könönen *et al.*, 1994). Gram-negative anaerobic bacteria were isolated more commonly, and a greater diversity of species were recovered from around the gingival margin of the newly erupted teeth (mean age of the infants = 32 months) (Table 8) (Könönen *et al.*, 1994). Also, mutans streptococci and *S. sanguis* appear in the mouth following tooth eruption (Carlsson *et al.*, 1970; Berkowitz *et al.*, 1975). These findings confirm that a change in the environment, such as the eruption of teeth, has a significant ecological impact on the resident microflora.

During the first year of life, members of the genera *Neisseria*, *Veillonella*, *Actinomyces*, *Lactobacillus*, and *Rothia* are commonly isolated, particularly after tooth eruption (McCarthy *et al.*, 1965). Some of the genera (*Porphyromonas* and *Actinobacillus*) associated with the aetiology of periodontal disease have been cultivated from the plaque of infants aged around 12 months, albeit infrequently and in low numbers (Milnes *et al.*, 1993).

The acquisition of some bacteria may occur optimally only at certain ages. Studies of the transmission of mutans streptococci to children have identified a specific **"window of infectivity"** at 19-31 months (median age = 26 months) (Caufield *et al.*, 1993). This opens up the possibility of targeting preventive strategies

Table 8. The Effect of Tooth Eruption on the Composition of the Oral Microflora in Young Children (Könönen *et al.*, 1992, 1994)

Bacterium	Percentage isolation frequency	
	Mean Age	
	3 months	32 months
Prevotella melaninogenica	76	100
non-pigmented *Prevotella*	62	100
P. loescheii	14	90
P. intermedia	10	67
P. denticola	ND[*]	71
Fusobacterium nucleatum	67	100
Fusobacterium spp.	ND	71
Selenomonas spp.	ND	43
Capnocytophaga spp.	19	100
Leptotrichia spp.	24	71
Campylobacter spp.	5	43
Eikenella corrodens	5	57
Veillonella spp.	63	63

[*]ND = not detected

over this critical period to reduce the likelihood of colonisation in the infant. Indeed, reducing the carriage of mutans streptococci in mothers can reduce transmission of these potentially cariogenic bacteria to their offspring, and delay the onset of caries (Köhler *et al.*, 1983, 1984).

Allogenic and Autogenic Succession.
The development of a climax community at an oral site can involve examples of both allogenic and autogenic succession. In allogenic succession, factors of non-microbial origin are responsible for an altered pattern of community development. Species such as mutans streptococci and *S. sanguis* only appear in the mouth once teeth have erupted (Carlsson *et al.*, 1970; Berkowitz *et al.*, 1975; Smith *et al.*, 1993). In the cases in which they have been isolated from the edentulous (toothless) mouth, these species are associated with the insertion of artificial devices such as acrylic obturators in children with cleft palate.

The increase in number and diversity of obligate anaerobes once teeth are present is an example of autogenic succession in which community development is influenced by microbial factors, such as the lowering of the redox potential and consumption of oxygen in plaque by pioneer species. Other examples of autogenic succession include the development of food chains and foodwebs, whereby the metabolic end product of one organism becomes a primary nutrient source for a second:

	primary **feeder**		**secondary** **feeder**	
complex substrate	$\rightarrow$	**product**	$\rightarrow$	**simpler product**

A further example is the exposure of new receptors on host macromolecules for bacterial adhesion ("cryptitopes"; Gibbons *et al.*, 1990; see Chapter 3).

Clonal Turnover and Succession
The application of molecular techniques (e.g. ribotyping) can differentiate among strains within a species on the basis of genetic variation, thereby allowing specific **clonal types** to be recognised. Relatively few clones are found within species of pathogenic bacteria, and a limited number of these may be responsible for the majority of infections (Kilian, 1998). In contrast, species that comprise the resident human microflora display large numbers of clones; this may be a strategy to help such species evade the host defences (see Chapter 4). Clones of some species appear to persist for long periods at a site whereas others appear to be transient, and undergo replacement by distinct clones. For example, clonal replacement appears to maintain *S. mitis* biovar 1 in the mouth of neonates; 93 clonal types were detected among 101 strains of *S. mitis* colonising 40 infants over a one month period (Fitzsimmons *et al.*, 1996). The clonal types that could be isolated were found to vary at different sampling times, suggesting that individual clones did not persist and were replaced by new clones. In a further study of the clonal diversity of *S. mitis* biovar 1, limited sharing of genotypes was found among three members of a particular family under study, and each individual carried between 6-13 types (Hohwy and Kilian, 1995).

Differences were also found between isolates recovered from the pharyngeal and buccal mucosa of the same individual. The reasons for, and the mechanisms involved in, the persistence of certain clones of some species but the continual turnover of clones of other species is not yet understood. Wide variations in the expression of carbohydrate and protein antigens were found among the different genotypes of *S. mitis* biovar 1 from the family group, suggesting that this "turnover" might play a role as an "immune-evasion" mechanism for this species.

Distribution of the Resident Oral Microflora

The populations making up the resident microbial community of the oral cavity are not found with equal frequency throughout the mouth. This is not due to random distribution, but is a consequence of a dynamic selection process which depends on the physical and biological properties of particular sites. The climax communities from distinct sites in the oral cavity will be compared in the following sections. Little information is available concerning mucosal surfaces; in contrast, because of its association with caries and periodontal diseases, most attention has been focused on the microbial composition of dental plaque.

Lips and Palate

There have been relatively few studies of the microbial community of the lips and palate. The lips form the border between the oral and skin microflora, which as discussed earlier consists predominantly of staphylococci, micrococci and Gram-positive rods (e.g. *Corynebacterium* and *Propionibacterium* spp.) (Marples, 1994). A study of this transition area would be of great interest in that it might indicate how the microflora at this site copes with the constant pressure of "contamination" from two contrasting ecosystems. Facultatively anaerobic streptococci probably comprise a large proportion of the microflora on the lips. *Veillonella* and *Neisseria* have been found, but only at low levels (Marsh and Martin, 1999). *S. vestibularis* is recovered from the muccobuccal fold between the lower lip and the gingiva; on occasions, black-pigmented anaerobes (including *P. gingivalis*), *A. actinomycetemcomitans* and fusobacteria have been detected using nucleic acid probes (Socransky *et al.*, 1999). *Candida albicans* can colonize damaged lip mucosal surfaces in the corners of the mouth ("angular cheilitis") (Marsh and Martin, 1999).

The microflora of the normal palate can show large variations between subjects, not only in the total colony forming units removed (which may reflect differences in the area sampled and the success in removing organisms) but also in the proportions of the individual species. The majority of the bacteria are Gram positive (streptococci and *Actinomyces* spp.); *Veillonella* and Gram-negative anaerobic rods are also regularly recovered but at lower levels (Table 9). Haemophili have also been found when the appropriate selective media have been used, and also low levels of *P. gingivalis* and *A. actinomycetemcomitans* have been detected using probe technology (Socransky *et al.*, 1999). *Candida* spp. are not regularly isolated from the normal palate although their prevalence does increase if dentures are worn. The mucosa of the palate can become infected with *C. albicans* (denture stomatitis) (Verran, 1998).

Table 9. Predominant Microflora of the Healthy Human Palatal Mucosa

Micro-organism	Percentage of the total cultivable microflora	Percentage isolation frequency
Streptococcus	52	100
Actinomyces	15	100
Lactobacillus	1	87
Neisseria	2	93
Veillonella	1	100
Prevotella	4	100
Candida	+*	7

*present, but in numbers too low to count.
Adapted from Marsh and Martin (1999).

Cheek

The predominant populations of bacteria isolated from the cheek (buccal mucosa) are shown in Table 10, while Table 11 gives the relative proportions of streptococci at this site as reported in a separate study. Streptococci are most numerous, especially *S. mitis*. Conflicting data have been obtained for *S. sanguis*; this may be due to differences in sampling procedures and identification methods, wide subject-to-subject variation or it might reflect a highly localized distribution of this species on the cheek surface. As with other mucosal surfaces, obligate anaerobes are not regularly isolated and when present they do not constitute a large percentage of the microflora. Haemophili are commonly isolated in relatively-high numbers. Spirochaetes and other motile organisms have been observed occasionally (by phase contrast microscopy) attached to the buccal mucosa. *Simonsiella* spp. are isolated primarily from the cheek cells of humans and animals. These organisms have a unique cellular morphology being composed of unusually large, multicellular filaments in groups or multiples of eight cells.

Tongue

The dorsum of the tongue with its highly papillated surface has a large surface area and supports, therefore, a higher bacterial density than that of other oral mucosal surfaces. Cultural studies have demonstrated the presence of a relatively diverse microflora with several obligately anaerobic species present. The relative proportions of the resident microflora are shown in Table 10, with the most prevalent facultatively anaerobic streptococci listed in Table 11 (Frandsen *et al.*, 1991). Again, streptococci are the most numerous group of bacteria (approximately 40% of the total cultivable microflora) with *S. salivarius*- and *S. mitis*-group organisms predominating (Smith *et al.*, 1993). Anaerobic streptococci (mainly *Peptostreptococcus* spp.) have also been isolated in some studies while *Stomatococcus mucilagenosus* (strains produce an extracellular slime) is found almost exclusively on the tongue. Other major groups of bacteria (and their proportions) include *Veillonella* spp. (16%), Gram-positive rods (16%) of which *A. naeslundii* and *A. odontolyticus* are common, and haemophili

Table 10. Proportions of Some Bacterial Populations at Different Sites in the Normal Oral Cavity.

	Percentage viable count			
Bacterium	Saliva	Cheek	Tongue dorsum	Supragingival plaque
S. sanguis	1	6	1	7
S. salivarius	3	3	6	2
S. oralis/S. mitis	21	29	33	23
mutans streptococci	4	3	3	5
A. naeslundii	2	1	5	5
A. odontolyticus	2	1	7	13
Haemophilus spp	4	7	15	7
Capnocytophaga spp	<1	<1	1	<1
Fusobacterium spp	1	<1	<1	1
Black-pigmented Anaerobes	<1	<1	1	+*
Simonsiella	-	+*	-	-
Stomatococcus	-	-	+*	-

*detected on occasions

(15%). Both pigmenting (*P. intermedia, P. melaninogenica*) and non-pigmenting anaerobes can be recovered from the tongue and this site is regarded as a potential reservoir (along with the tonsils) for some of the organisms implicated in periodontal diseases (van der Velden, *et al.*, 1986; Socransky *et al.*, 1999). Other organisms, including lactobacilli, yeasts, fusobacteria, spirochaetes and other motile bacteria, have been found in low numbers (<1% of the total microflora) on the tongue.

Table 11. Relative Proportions of Streptococci at Different Sites in the Healthy Mouth (Frandsen *et al.*, 1991)

Species	Percentage Streptococcal Count				
	Cheek	Tongue	Pharynx	Supra-gingival plaque	Sub-gingival plaque
S. sanguis	49	-*	2	13	2
S. gordonii	2	+#	4	9	+
S. oralis	+	-	+	5	+
S. mitis 1	22	13	21	10	5
S. mitis 2	+	52	17	6	2
S. salivarius	2	17	30	+	-
S. vestibularis	2	+	4	6	-
S. anginosus	4	12	14	16	85
S. mutans	-	-	-	1	-

Results are pooled from several studies, and are expressed as percentages of the total streptococcal count
-*not detected.   #recovered on occasions

Similar findings have been obtained from a comprehensive study of the anterior dorsal surface of the tongue in infants (aged 8-13 months). Streptococci accounted for 52% of the microflora, and *S. salivarius* and *S. mitis* were the predominant species (Table 12) (Milnes *et al.*, 1993). High proportions of *Neisseria* (20%) were also found, together with lower levels of *Actinomyces* (5%) and occasional Gram-negative species including haemophili, fusobacteria, *Prevotella*, *Capnocytophaga*, and *Actinobacillus*; *Stomatococcus mucilagenosus* was recovered from almost half of the samples. *Neisseria* have been shown in laboratory studies to be able to protect obligate anaerobes at aerated sites by the rapid consumption of oxygen (Bradshaw *et al.*, 1996), thereby creating sharp gradients over relatively short distances.

Oral malodour is associated with the microflora of the tongue. A recent study found an increase in bacterial load, especially in Gram-negative anaerobes (including *Porphyromonas*, *Prevotella* and *Fusobacterium* spp.) on the tongue of subjects with high odour (Hartley *et al.*, 1996). Malodour production is strongly implicated with

Table 12. The Predominant Microflora of the Tongue from Pre-School Children (Milnes *et al.*, 1993)

Bacterium	Mean proportion (%)	Isolation frequency (%)
S. anginosus	4.7	42
S. oralis	3.8	30
S. mitis	11.8	75
S. mutans	1.0	8
S. sobrinus	0.5	2
S. salivarius	22.3	94
S. sanguis	7.6	58
Total streptococci	51.7	
A. naeslundii	4.2	46
A. odontolyticus	1.1	17
R. dentocariosa	0.9	21
C. matruchotii	0.1	4
Lactobacillus spp.	0.3	6
Total Gram-positive rods	6.6	13
Neisseria spp.	20.2	>90
Veillonella spp.	6.3	73
Total Gram-negative cocci	26.5	
Prevotella spp.	0.4	15
Fusobacterium spp.	0.6	25
Leptotrichia spp.	0.2	13
Haemophilus spp.	0.6	19
Actinobacillus spp.	0.1	4
Capnocytophaga spp.	0.1	6
Stomatococcus spp.	5.5	46
Aerobic Gram-negative rods	2.3	40
Anaerobic Gram-negative rods	1.8	40
Yeasts	1.0	4

Data are from 9 children, aged 8-13 months.

high proteolytic activity and the production of volatile sulphur compounds. The predominant sulphur compounds include hydrogen sulphide, H_2S, and methyl mercaptan, CH_3SH, with smaller concentrations of dimethyl sulphide, $(CH_3)_2S$, and dimethyl disulphide, $(CH_3S)_2$.

The tongue may also act as a reservoir for exogenous pathogens. Mucoid *Pseudomonas aeruginosa* strains were isolated from the tongue (and cheek) of cystic fibrosis patients; these sites could act as a reservoir of this organism promoting cross-contamination of other sites and surfaces (Lindemann *et al.*, 1985).

Teeth

The microbial community associated with teeth is referred to as dental plaque. Its composition varies at each tooth surface due to the local environmental conditions. For these reasons, plaque is described on the basis of the sampling site by terms such as smooth surface, approximal, fissure, or gingival crevice plaque. Similarly, samples taken from above the level of the gum margin are given the general name "supra-gingival plaque", while those from below the gum margin are described as "sub-gingival plaque". As teeth are non-shedding surfaces, the highest concentrations of micro-organisms are found in stagnant sites which afford protection from removal forces. Dental plaque is an example of a biofilm; bacteria growing in biofilms can display novel properties, including an enhanced resistance to antimicrobial agents. The properties of plaque as a biofilm are discussed later. Compared to other surfaces in the mouth, dental plaque contains high proportions of Gram positive (mainly *Actinomyces* spp.) and anaerobic Gram negative rods and filaments (*Prevotella, Porphyromonas, Fusobacterium* spp.) (Table 10). Mutans streptococci and some members of the *S. mitis-* and *S. anginosus*-groups are found in highest numbers on teeth (Table 11) (Frandsen *et al.*, 1991), and these organisms have a strong affinity for hard surfaces. Indeed, these species do not usually appear in the mouth until after tooth eruption (Carlsson *et al.*, 1970; Berkowitz *et al.*, 1975).

The composition of plaque on different surfaces will now be described. Most studies have employed conventional cultural approaches, and over 300 different taxa have been cultivated from dental plaque (though not all from an individual sample!). However, because of the labour-intensive and time-consuming nature of such studies, molecular and immunological techniques are now being used more commonly. Molecular studies, in which 16S rRNA sequences are amplified and compared to those present in data-bases, have led to the recognition of a whole range of new species and taxa, most of which cannot, as yet, be cultivated in the laboratory (Wade, 1996, 1999). Such approaches also suggest that the mouth may act as a reservoir of organisms capable of causing disease elsewhere in the body. For example, in a random sample of 40 healthy people, 14 showed evidence of DNA in their saliva of the unculturable aetiological agent of Whipple's disease, *Tropheryma whippelii* (Street *et al.*, 1999). The future application of molecular techniques to samples of dental plaque will undoubtedly lead to further discoveries and an increase in the recognised diversity of the oral microflora.

Table 13. The Predominant Cultivable Microflora of 10 Occlusal Fissures in Adults (Theilade *et al.*, 1982)

Bacterium	Median percentage of total cultivable microflora	Range	Percentage isolation frequency
Streptococcus	45	8-86	100
Staphylococcus	9	0-23	80
Actinomyces	18	0-46	80
Propionibacterium	1	0-8	50
Eubacterium	0	0-27	10
Lactobacillus	0	0-29	20
Veillonella	3	0-44	60
Individual species:			
mutans streptococci	25	0-86	70
S. sanguis-group	1	0-15	50
S. oralis-group	0	0-13	30
S. anginosus-group	0	0-3	10
A. naeslundii	3	0-44	70
L. casei	0	0-10	10
L. plantarum	0	0-29	10

Fissure Plaque

The microbiology of fissure plaque has been determined using either 'artificial fissures' implanted in occlusal surfaces of pre-existing restorations, or by sampling 'natural' fissures. The microflora is mainly Gram-positive and is dominated by streptococci, especially extracellular polysaccharide-producing species (Theilade *et al.*, 1974). In one study, no obligately anaerobic Gram-negative rods were found, while others have recovered anaerobes including *Veillonella*, and *Propionibacterium* species, occasionally and in low numbers (Table 13) (Theilade *et al.*, 1982). *Neisseria* and *Haemophilus* spp. have also been isolated on occasions. The total anaerobic microflora can vary markedly between individual fissures (e.g. 1 x 10^6 to 33 x 10^6 CFU per fissure), suggesting major differences in the ecology at each site. The factors that determine the final composition of the microflora in fissures are not known, but the influence of saliva at this site must be of great significance, and food may become impacted. The simpler community found in fissures compared to other enamel surfaces probably reflects a more severe environment.

Approximal Plaque

The main organisms isolated from a study of approximal plaque are shown in Table 14 (Bowden *et al.*, 1975). Although streptococci are present in high numbers, these sites are frequently dominated by Gram-positive rods, particularly *Actinomyces* spp. The more reduced nature of this site compared to that of fissures can be gauged

Table 14. The Predominant Cultivable Microflora of Approximal Plaque (Bowden *et al.*, 1975)

Bacterium	Mean percentage of total cultivable microflora	Range	Percentage isolation frequency
Streptococcus	23	0.4-70	100
Gram-positive rods (predominantly			
Actinomyces)	42	4-81	100
Gram-negative rods (predominantly			
Prevotella)	8	0-66	93
Neisseria	2	0-44	76
Veillonella	13	0-59	93
Fusobacterium	0.4	0-5	55
Lactobacillus	0.5	0 2	24
Rothia	0.4	0-6	36
Individual species			
mutans streptococci	2	0-23	66
S. sanguis	6	0-64	86
S. salivarius	1	0-7	54
S. anginosus-group	0.5	0-33	45
A. israelii	17	0-78	12
A. naeslundii	19	0-74	97

from the higher recovery of obligately anaerobic organisms although spirochaetes are not commonly found. Again, the range and percentage isolation frequency of most bacteria is high, suggesting that each site represents a distinct habitat which should be looked at in isolation with regard to the relationship between the resident microflora and the clinical state of the enamel.

The variability of plaque composition was highlighted in a study of several small samples taken from different sites around the contact area of teeth extracted for orthodontic purposes (Table 15) (Babaahmady *et al.*, 1997). The recovery and proportions of different groups of bacteria varied according to the location of the sample site around the contact area. The isolation frequency of *S. mutans* and *S. sobrinus* was higher at sub-sites from below the contact area, and this is also the most caries-prone site. The reasons for this are not known at present but will be intimitely related to the biological properties of mutans streptococci, e.g. in terms of their ability to tolerate low pH or to produce intracellular or extracellular polysaccharides. Similarly, *A. naeslundii* and *A. odontolyticus* were found more commonly below the contact area, while *Neisseria*, *S. sanguis* and *S. mitis* biovar 1 were recovered more frequently at sub-sites away from, and to the side of, the contact area. Such variations emphasize the need for accurate sampling of discrete sites when attempting to correlate the composition of plaque with disease.

Table 15. Distribution of selected bacteria in approximal plaque at three sub-sites away from (A) to the side of (S), and below (B) the contact area (Babaahmady *et al.*, 1997)

Bacterium	Away		Side		Below	
	%*	IF†	%	IF	%	IF
S. mutans	9.9	43	7.7	62	13.3	86
S. sobrinus	1.9	5	3.9	20	4.9	33
S. sanguis	4.4	57	8.3	67	7.8	62
S. gordonii	2.1	3	1.6	33	2.6	39
S. mitis 1	13.2	67	7.6	76	8.4	67
S. mitis 2	2.7	33	0.7	29	1.5	33
S. oralis	8.7	57	3.2	48	7.6	62
S. anginosus-group	5.0	38	3.7	43	2.9	43
S. salivarius	3.6	33	1.2	38	2.0	33
A. naeslundii	20.6	76	20.1	86	18.3	91
A. odontolyticus	4.1	38	7.9	33	5.8	57
A. israelii	4.3	48	11.6	67	4.7	62
Lactobacillus spp.	0.4	14	0.2	14	<0.1	10
Neisseria spp.	5.9	48	2.3	42	0.2	29
Veillonella spp.	3.3	29	2.3	38	10.7	76
Fusobacterium spp.	0.1	10	0.7	19	0.4	14

*% percentage viable count.
†percentage isolation frequency.

Gingival Crevice Plaque

An obviously distinct ecological climate is found in the gingival crevice. This is reflected in the higher species diversity of the bacterial community at this site although the total numbers of bacteria can be low (10^3-10^6 CFU/crevice). In contrast to the microflora of fissures and approximal surfaces, higher levels of obligately anaerobic bacteria can be found, many of which are Gram-negative (Table 16) (van Palenstein Helderman, 1975; Slots, 1977). Indeed, spirochaetes and anaerobic streptococci are isolated almost exclusively from this site. The ecology of the crevice is influenced by the anatomy of the site and the flow and properties of GCF. Many organisms that are asaccharolytic but proteolytic are found in the gingival crevice; they derive their energy from the hydrolysis of host proteins and peptides and from the catabolism of amino acids. In disease, the gingival crevice enlarges to become a periodontal pocket (Figure 2) and the flow of GCF increases. The diversity of the microflora increases still further with many fastidious and obligately anaerobic bacteria present (Moore and Moore, 1994); these will be described in more detail in Chapter 6. Among the genera and species associated with the healthy gingival crevice are members of the *S. mitis*-group and *S. anginosus*-group (Table 11) (Frandsen *et al.*, 1991); in addition, Gram-positive rods such as *Actinomyces meyeri*, *A. odontolyticus*, *A. naeslundii*, *Actinomyces georgiae*, *Capnocytophaga ochracea* and *Rothia dentocariosa* can also be found. The most commonly isolated black-pigmented anaerobe in the healthy gingival crevice is *P. melaninogenica* while *Prevotella nigrescens* has also been recovered on occasions; *P. gingivalis* is rarely isolated from healthy sites. Fusobacteria are among the commonest anaerobes found in the healthy gingival crevice.

Table 16. The Predominant Cultivable Microflora of the Healthy Gingival Crevice (Slots, 1977)

Bacterium	Percent viable count		Percentage isolation frequency
	Median	Range	
Gram-positive facultatively anaerobic cocci (predominantly *Streptococcus*)	40	2-73	100
Gram-positive obligately anaerobic cocci (predominantly *Peptostreptococcus*)	1	0-6	14
Gram-positive facultatively anaerobic rods (predominantly *Actinomyces*)	35	10-63	100
Gram-positive anaerobic rods	10	0-37	86
Gam-negative facultatively anaerobic cocci (predominantly *Neisseria*)	0.3	0-2	14
Gram-negative obligately anaerobic cocci (predominantly *Veillonella*)	2	0-5	57
Gram-negative facultatively anaerobic rods	ND*	ND	ND
Gram-negative obligately anaerobic rods	13	8-20	100

Samples were taken from the gingival crevice of seven adults humans.
*ND, not detected.

Denture Plaque

The microflora of denture plaque from healthy sites (i.e. with no sign of denture stomatitis) is highly variable (Theilade *et al.*, 1983) as can be deduced from the wide ranges in viable counts obtained for individual bacteria shown in Table 17. Clear differences are also apparent between the tissue-facing and the exposed surfaces of the denture. In the relatively stagnant area on the tissue-facing surface, plaque tends to be more acidogenic, thereby favouring streptococci (especially mutans streptococci) and sometimes *Candida* spp. (Verran, 1998). In edentulous subjects, dentures become the primary habitat for mutans streptococci and members of the *S. sanguis*-group. It has been claimed that the microflora of denture plaque overlying healthy palatal mucosa is similar to that of fissure plaque in that streptococci, actinomyces and sometimes lactobacilli predominate. Denture plaque can harbour obligate anaerobes including *A. israelii* and low proportions of Gram-negative rods. Interestingly, *S. aureus* was regularly isolated in one study of denture plaque (Table 17). This species is also found commonly in the mucosa of patients with denture stomatitis. Since dentures can spend time out of the mouth, sometimes in less than hygienic environments, a range of micro-organisms not normally associated with the mouth can sometimes be isolated, including *Streptococcus pneumoniae*, *Haemophilus influenzae* and *Neisseria meningitidis,* as well as members of the *Enterobacteriaceae* (Verran, 1998).

Table 17. The Predominant Cultivable Microflora of Denture Plaque.

Micro-organism	Percentage viable count		Percentage isolation frequency
	Median	Range	
Streptococcus	41	0-81	88
mutans streptococci	<1	0-48	50
S. sanguis-group	1	0-4	63
S. oralis-group	2	0-30	75
S. anginosus-group	2	0-51	63
S. salivarius	0	0-41	38
Staphylococcus	8	1-13	100
S. aureus	6	0-13	88
"S. epidermidis"	0	0-7	13
Gram-positive rods	33	1-74	100
Actinomyces	21	0-54	88
A. israelii	3	0-47	63
A. naeslundii	3	0-48	63
A. odontolyticus	1	0-17	63
Lactobacillus	0	0-48	25
Propionibacterium	<1	0-5	50
Veillonella	8	3-20	100
Gram-negative rods	0	0-6	38
Yeasts	0.002	0-0.5	63

Dental Plaque from Animals

There is interest in the microbial composition of dental plaque from animals for two main reasons: (a) to study the influence of widely different diets and life-styles on the microflora, and (b) to determine the similarity between the microflora of an animal with that of humans to ascertain their relevance as a model of human oral diseases. At the genus level, the plaque microflora is similar among animals representing such diverse dietary groups as insectivores, herbivores and carnivores (Dent, 1979; Dent and Marsh, 1981). Thus, *Actinomyces, Streptococcus, Neisseria, Veillonella*, and *Fusobacterium* are widely distributed in both zoo and non-zoo primates and other animals, and can, therefore, be genuinely considered as members of the resident plaque microflora. This again emphasises (a) the significance of endogenous nutrients in maintaining the stability and diversity of the resident microflora, and (b) the specificity of the interaction between this microflora and oral surfaces. Following recent taxonomic studies, differences between isolates from man and animals have emerged at the species level. For example, *Streptococcus rattus* and *Streptococcus macacae* are isolated exclusively from rodents and primates, respectively, whereas other species of mutans streptococci are found in humans.

Dental Plaque as a Microbial Biofilm

Studies of a range of distinct ecosystems have shown that the vast majority of micro-organisms exist in Nature associated with a surface. The term **"biofilm"** is used to describe communities of micro-organisms attached to a surface; such microbes are usually spatially organised into a three-dimensional structure and are enclosed in a matrix of extracellular material derived both from the cells themselves and from the environment. For extensive reviews on the properties of biofilms, see Costerton *et al.*, 1987, 1994, 1995. If (a) biofilm microbes were simply planktonic (liquid-phase) cells that had adhered to a surface, and (b) the properties of microbial communities were merely the sum of those of the constitutive populations, then interest in such issues would have had limited scientific impact. However, research over recent years has revealed that cells growing as biofilms have unique properties, some of which have clinical significance (Table 18). For example, biofilms can be 100-1,000 times more resistant to antimicrobial agents than the same cells growing in liquid culture (Nickel *et al.*, 1985; Wright *et al.*, 1997); some of the proposed mechanisms for this are listed in Table 18 (Costerton *et al.*, 1987; Gilbert *et al*, 1997). Likewise, growth in a microbial community results in bacterial species displaying (a) a broader habitat range, (b) enhanced resistance to environmental stresses, and (c) an increased metabolic efficiency and versatility than when individual species are grown in isolation (Caldwell *et al.*, 1997).

Originally, biofilms were considered to be dense, compressed accumulations of cells. A major recent advance in the study of biofilms has come from the application of novel techniques, such as confocal scanning laser microscopy (CSLM), which enable biofilms to be studied *in situ* without any processing of samples which could distort their structure (Stoodley *et al.*, 1999). Optical thin sections can be generated throughout the depth of the biofilm, and these can be combined using imaging software to generate three-dimensional images. Such studies have shown that plaque might have a more open architecture than hitherto thought, with channels passing through its depth (Wood *et al.*, 2000). In the future, the location of specific organisms might be visualized in plaque using immunological or nucleic acid probes, while other molecular probes could indicate viability and metabolic activity of cells. CSLM can also be used in combination with "reporter gene technology" to identify genes that are expressed only within a biofilm (Stoodley *et al.*, 1999). This technology

---

Table 18. General Properties of a Biofilm

Protection from host defences and predators
Protection from desiccation
Protection from antimicrobial agents
- Surface-associated phenotype*
- Slow growth rate
- Poor penetration
- Inactivation/neutralisation
Novel gene expression and phenotype*.
Persistence in flowing systems.
Spatial and environmental heterogeneity.
Spatial organisation facilitating metabolic interactions.
Elevated concentration of nutrients.

---

* one consequence of altered gene expression can also be an increased resistance to antimicrobial agents.

41

involves the insertion of a marker into the bacterial chromosome downstream of a promoter so that a recognisable "signal" (e.g. fluorescence) is produced when the gene is activated.

The properties of an organism may be affected in a biofilm in two ways. The attachment of cells to a surface may cause a direct effect (Bradshaw, 1995), perhaps by triggering "touch sensors" on the cell surface. Adhesion to surfaces has been shown to specifically induce the expression of certain genes, e.g. those involved in exopolysaccharide synthesis by *Pseudomonas aeruginosa* (Boyd and Chakrabarty, 1995) and by mutans streptococci (Burne *et al.*, 1997; Burne, 1998). In addition, the growth environment within the biofilm may be significantly different from planktonic culture, and this may result in an altered pattern of gene expression as an "indirect" effect of growth in a biofilm (Bradshaw, 1995). Organisms may be growing more slowly, e.g. due to a particular nutrient limitation or an unfavourable pH, and this may result in altered gene expression, and hence an altered phenotype.

Dental Plaque Formation

Plaque formation can be divided into several arbitrary stages, although it should be appreciated that biofilm formation is a dynamic process, and that the attachment, growth, removal and re-attachment of bacteria is continuous. These stages are:

1. the adsorption of host and bacterial polymers to the tooth to form a surface conditioning film (the acquired pellicle) (Lie, 1977; Al-Hashimi and Levine, 1989),

2. the transport of micro-organisms to the polymer-coated tooth surface; generally, this is a passive process facilitated by the flow of oral secretions as very few early colonisers are motile,

3. long-range physico-chemical interactions between the microbial cell surface and the pellicle-coated tooth (Busscher *et al.*, 1992; Busscher and van der Mei, 1997). The interplay of van der Waals attractive forces and electrostatic repulsion due to the interaction of charged molecules on the bacterial cell surface and in the conditioning film produces a weak area of net attraction that facilitates reversible adhesion (see Chapter 3),

4. short-range interactions involving specific, stereo-chemical interactions between adhesins on the microbial surface and receptors in the acquired pellicle; these interactions usually result in irreversible adhesion (Ellen and Burne, 1996; see Chapter 3),

5. the coaggregation of micro-organisms to already attached cells; for extensive reviews, see Kolenbrander and London, 1992; Kolenbrander, 1993); this stage results in an increased diversity of the plaque microflora. Thus, many late colonisers, such as obligately anaerobic bacteria, bind not to the acquired pellicle but to already attached species such as streptococci and *Actinomyces* spp. (Kolenbrander and London, 1993). In addition, coaggregation can provide the mechanism that facilitates a range of physical and metabolic synergistic interactions.

6. the multiplication of the attached organisms to produce confluent growth and a biofilm (Marsh, 1995; Marsh and Bradshaw, 1995). Biofilms provide the physical opportunities for close range cell-to-cell interactions, such as those involved in cell density-dependent signalling (quorum sensing; Bloomquist *et al.*, 1996; Liljemark *et al.*, 1997) and possibly also for gene transfer, although there is only limited evidence

at present for this occurring in plaque. For example, a 55 kDa antigen has a restricted distribution in clinical strains of *P. gingivalis*, and there is some evidence that the gene encoding this protein was acquired via horizontal gene transfer (Hanley *et al.*, 1999). There is also evidence of horizontal gene transfer of penicillin binding proteins from *S. mitis* to *S. pneumoniae*, and this may occur in the naso-pharyngeal region. A soluble, low molecular weight compound, termed START, has been identified in cell-free supernatants of oral streptococci such as *S. gordonii*, that can stimulate bacterial growth (Liljemark *et al.*, 1997). Extracellular polymer synthesis also occurs, especially from the metabolism of sucrose, and the resultant glucans and fructans contribute to the plaque matrix, and will help determine the architecture of the biofilm (Walker and Jacques, 1987; Kuramitsu, 1993; Russell, 1994). Glucosyltransferases (GTF) can also be found in an active form in the acquired pellicle, where they can synthesise polymer that can act as receptors for glucan-binding proteins on colonising streptococci (Chapter 3) (Vacca-Smith *et al.*, 1996). In some species of streptococci, multiple gene products are involved in glucan production, although similar products can be made by a single gene in other species (Table 19) (Kuramitsu, 1993; Russell, 1994). The precise regulation of these genes and how the different products contribute to polymer production has yet to be fully elucidated. Evidence is also appearing that

Table 19.   Streptococcal Glucosyltransferase Genes, Enzymes and their Products

Streptococcal Species	Gene	Class of Enzyme	Primer Dependency	Glucan Structure	Glucan Properties
S. mutans	*gtfB*	GTF-I	No	73% α-(1-3)- / 27% a-(1-6)-	Insoluble
	gtfC	GTF-I	No	66% α-(1-3)- / 34% a-(1-6)-	Insoluble
	gtfD	GTF-S	Yes	70% α-(1-6)-[†]	Soluble
S. sobrinus	*gftI*	GTF-I	Yes	80-90%[•] α-(1-3)-	Insoluble
	Not Cloned	GTF-S	Yes	70% α-(1-6)-[†]	Soluble
	gtfS	GTF-S	No	100% α-(-16)-[tt]	Soluble
	gtfT	GTF-S	No	92% α-(1-6)-[†]	Soluble
S. salivarius	*gtfJ*	GTF-I	Yes	90% α-(1-3)-	Insoluble
	gtfK	GTF-S	Yes[#]	100% α-(1-6)-	Soluble
	gtfL	GTF-I	No	50% α-(1-6)- / 50% α-(1-3)-	Insoluble
	gtfM	GTF-S	No	95% α-(1-6)-	Soluble
S. gordonii	*gtfG*	——[‡]	No	50-83%[•] α-(1-6)-[†]	Insoluble/Soluble

Both *Streptococcus sobrinus* and *Streptococcus salivarius* may possess more than the four *gtf* genes listed above.
[#] GtfK is primer-stimulated
[†] Contains α-(1-3)-branch points
[tt] Short chain linear oligoisomaltosaccharides
[•] Strain dependent
[‡] Both insoluble and soluble products are formed depending on the strain and conditions prevailing *in vitro*.

suggests that the structure of glucans synthesised by enzymes adsorbed to the surface is different to that produced by the same enzyme in solution (Kopec *et al.*, 1997).

Recent studies of the structure of smooth surface plaque using CSLM have shown it to have a relatively open architecture; channels are present that can traverse the depth of the biofilm (Wood *et al.*, 2000), in an analogous manner to that seen in environmental biofilms (Costerton *et al.*, 1994). It remains to be determined whether a similar open architecture will develop at more stagnant sites where compaction may occur. These channels may be filled with bacterial extracellular polysaccharides (EPS). High levels of EPS may lead to an enhanced cariogenic challenge at the tooth surface because, in deep biofilms, sugars will be able to penetrate further while the significant buffering effect of bacteria will be reduced, thereby producing a more pronounced pH fall at the plaque-enamel interface (Dibdin and Shellis, 1988). Indeed, demineralisation of enamel slices was maximal when the artificial plaque consisted of 95% EPS and only 5% bacteria (Zero *et al.*, 1986).

Bacteria growing as a biofilm are more resistant to antimicrobial agents (Wilson, 1996; Gilbert *et al.*, 1997). Chlorhexidine has been incorporated into several types of products for use in the mouth. Older biofilms (three days) of *S. sanguis* were less susceptible to chlorhexidine than younger (one day) biofilms; the biofilm killing concentration for the former being 200 μg/ml compared with 50 μg/ml for the latter (Millward and Wilson, 1989). Diverse mixed cultures of oral bacteria when grown as a biofilm were unaffected when exposed to concentrations of chlorhexidine equivalent to MIC values of the component species (Kinniment *et al.*, 1996). Ten-fold higher concentrations were needed to demonstrate some effectiveness, but even at these levels, some species were unaffected. Similarly, *P. gingivalis* growing as a biofilm could tolerate concentrations of metronidazole 160 times the MIC for planktonic cells (Wright *et al.*, 1997).

Bacteria are also able to detach themselves from a surface, or from within the biofilm, so as to be able to colonize elsewhere. Recently, a protein-releasing enzyme synthesized by *S. mutans* was shown to be able to liberate surface proteins from its own cell surface and thereby detach itself from a mono-species biofilm (Lee *et al.*, 1996). Similarly, a protease produced by *Prevotella loescheii* can hydrolyse its own fimbrial-associated adhesin which is responsible for coaggregation with *S. oralis* as well as binding to host molecules such as fibrin (Cavedon and London, 1993). This protease may control the detachment of this species from its coaggregating partner, thereby providing the cell with a specific mechanism by which it can colonise other sites.

The final outcome of these phases is the development of a complex, multi-species, spatially- and functionally-organised biofilm. The diversity of potential mechanisms for adherence together with the molecular heterogeneity of the microbial and host surface probably means that biofilm formation involves multiple interactions.

Plaque may also act as a reservoir for true pathogens. It has been reported that oral hygiene is poor among patients in intensive care, and that dental plaque from these patients contains large numbers of potential respiratory pathogens (Scannapieco and Mylotte, 1996). Aspiration of these pathogens into the lower respiratory tract would increase the likelihood of serious lung infection, especially in immuno-

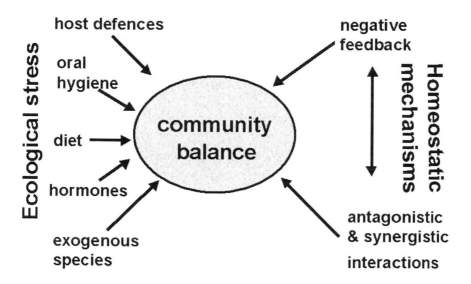

Figure 4. Factors involved in microbial homeostasis.

compromised or elderly people. *Helicobacter pylori* has also been detected in dental plaque on occasions, and this organism is strongly associated with chronic gastritis and peptic ulcers, and is a risk factor for gastric cancer (Madinier *et al.*, 1997; Oshowa *et al.*, 1998). *H. pylori* is not a normal bacterial inhabitant of the mouth, and its presence is probably associated with gastro-oesophageal reflux. Its intermittent persistence in the mouth may aid its transmission from person-to-person. This pathogen may be retained in dental plaque by selective adherence to already attached bacteria, namely *Fusobacterium* spp., by coaggregation (see Chapter 3) (Andersen *et al.*, 1998).

Microbial Homeostasis in Dental Plaque

In spite of its microbial diversity, once established at a site, the composition of dental plaque is characterized by a remarkable degree of stability or balance among the component species (Marsh, 1989). This stability is maintained in spite of the host defences, and despite the regular exposure of the plaque community to a variety of modest environmental stresses. These stresses include diet, the repeated challenge by exogenous species, the regular use of dentifrices and mouthwashes containing antimicrobial agents, and changes in saliva flow and hormone levels (Figure 4). The ability to maintain community stability in a variable environment has been termed **microbial homeostasis** (Alexander, 1971; Marsh, 1989). This stability stems not from any metabolic indifference among the components of the microflora but results rather from a balance of dynamic microbial interactions, including both synergism (such as proto-cooperation and commensalism) and antagonism. When the environment is perturbed, self-regulatory mechanisms (homeostatic reactions) come into force to restore the original balance. An essential component of such mechanisms is negative feedback, whereby a potential change in the level of one or more organisms results in a response by others to oppose or neutralize such a change. There is a tendency for homeostasis to be greater in microbial communities with a higher species diversity. Some examples of microbial interactions in plaque that may contribute to this homeostasis are described below.

Microbial Interactions in Dental Plaque
In a biofilm such as dental plaque, microorganisms are in close proximity with one another and this facilitates interactions. These interactions can be beneficial to one or more of the interacting populations, while others can be antagonistic (Table 20). Microbial metabolism within plaque will produce gradients in factors affecting the growth of other species. These would include the depletion of essential nutrients with the simultaneous accumulation of toxic or inhibitory by-products or of metabolic end products that can be used by other bacteria (Figure 5), or the consumption of potentially growth-limiting factors such as oxygen enabling the growth of obligate anaerobes. These gradients lead to the development of vertical and horizontal stratifications within the plaque biofilm. Some of this environmental heterogeneity has recently been visualised in experimental biofilms using a combination of fluorescent life-time imaging to measure pH with two photon excitation microscopy (Vroom *et al.*, 1999). After a sucrose pulse, pH gradients were detected in both the lateral and axial planes with, for example, small zones of low pH being surrounded by areas of much higher pH. Such environmental heterogeneity has two important consequences for microbial interactions. Organisms with widely differing requirements can grow, and it ensures the co-existence of species that would be incompatible with one another in a homogeneous habitat. In this way, plaque as a microbial habitat makes a major contribution to the microbial diversity of the mouth.

Bacterial growth in most natural ecosystems is limited by the availability of essential nutrients and co-factors, and there is now considerable evidence that the growth of the resident oral microflora is dependent on the utilisation of endogenous nutrients (van der Hoeven, 1998). Competition for nutrients will be one of the primary ecological determinants in dictating the prevalence of a particular species in dental plaque. Individual species possess different but overlapping patterns of enzymes (proteases and glycosidases), so that the concerted action of several species is necessary for the complete degradation of complex, host molecules (van der Hoeven, 1998). The growth of some organisms will be dependent on others removing, for example, the terminal sugar from the oligosaccharide side chain of the glycoprotein

Table 20. Microbial Interactions in Dental Plaque

Beneficial	Antagonistic
Enzyme complementation	Bacteriocins
Food chains/food webs	Hydrogen peroxide
Coaggregation	Organic acids
Inactivation of inhibitors	Low pH
Subversion of host defences	Nutrient competition

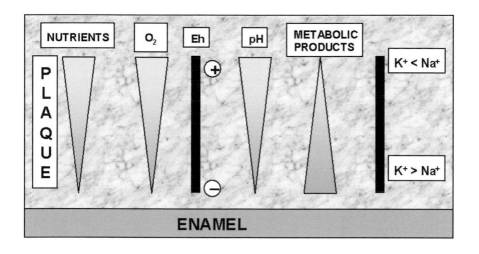

Figure 5. Schematic representation of gradients in dental plaque (adapted from Marsh and Martin, 1999).

(Bradshaw *et al.*, 1994). Such nutritional interdependencies are common, and this is a key factor behind the observed microbial diversity seen at other sites in the human body, such as the gut (Gibson and Macfarlane, 1995).

An example of microbial co-operation in the breakdown of host macromolecules has been provided by laboratory studies of the enrichment of subgingival bacteria on human serum (used to mimic GCF) (ter Steeg *et al.*, 1987, 1988; ter Steeg and van der Hoeven, 1989). Prolonged growth of subgingival plaque in a chemostat led to the selection of consortia of bacteria with different metabolic capabilities (ter Steeg *et al.*, 1987, 1988). Shifts in the microbial composition of the consortia occurred at different stages of glycoprotein breakdown. Initially, carbohydrate side-chains were removed by organisms with complementary glycosidase activities, including *S. oralis, Eubacterium saburreum* and *Prevotella* spp. This was followed by the hydrolysis of the protein core by, for example, *P. intermedia, P. oralis, F. nucleatum*, and to a lesser extent, *Eubacterium* spp.; some amino acid fermentation occurred and the remaining carbohydrate side-chains were metabolised leading to the emergence of *Veillonella* spp. A final phase was characterized by progressive protein degradation and extensive amino acid fermentation; the predominant species included *Peptostreptococcus micros* and *Eubacterium brachy* (ter Steeg and van der Hoeven, 1989). Significantly, individual species grew only poorly in pure culture in serum. A consequence of these interactions involving enzyme complementation is that different species avoid direct competition for individual nutrients, and hence are able to co-exist. This type of interaction is an example of proto-cooperation or mutualism, whereby there is benefit to all participants that are involved in the interaction.

Bacterial polymers are also targets for degradation. Extracellular polysaccharides synthesized by many plaque bacteria can be metabolized by other bacteria in the absence of exogenous (dietary) carbohydrates. The fructan of *S. salivarius* and other streptococci, and the glycogen-like polymer of *Neisseria*, are particularly labile, and only low levels of fructan can be detected in plaque *in vivo*. In addition, mutans streptococci, members of the *S. mitis*-group, *S. salivarius, A. israelii, Capnocytophaga* spp., and *Fusobacterium* spp. possess exo- and/or endo-hydrolytic activity and metabolize streptococcal glucans (Schachtele *et al.*, 1976). The metabolism of these polymers is usually cited as an example of commensalism (an interaction beneficial to one organism but with a neutral effect on the other), although polymer metabolism might lead to antagonism (inhibition of one or more of the interacting species) if glucan-mediated adhesion is affected (Schachtele *et al.*, 1975).

Many other types of nutritional interactions are known to exist among plaque micro-organisms whereby the products of metabolism of one organism (primary feeder) is the main source of nutrients for another (secondary feeder). The best described interaction of this type is the utilization by *Veillonella* spp. of lactate produced from the metabolism of dietary carbohydrates by a range of other species, but particularly *Streptococcus* and *Actinomyces* species (Mikx and van der Hoeven, 1975). *Veillonella* spp. could reduce the cariogenic potential of other plaque bacteria by converting lactate into propionic and acetic acids, which are weaker acids. Fewer caries lesions were obtained in rats inoculated with either *S. mutans* or *S. sanguis* and *Veillonella* than in animals infected with either streptococcus alone (Mikx *et al.*, 1972). Strains of *Neisseria, Corynebacterium*, and *Eubacterium* are also able to metabolize lactate.

Similarly, a mutually beneficial interaction between *S. sanguis* and *Campylobacter rectus* has been described whereby the anaerobe scavenges inhibitory oxygen, or possibly hydrogen peroxide produced by the streptococcus, while *S. sanguis* provides *C. rectus* with formate following the fermentation of glucose under carbohydrate-limiting conditions (Ohta *et al.*, 1990). These and other interactions will be described in more detail in Chapter 2. Thus, a complex array of interbacterial nutritional interactions can take place in plaque, with the growth of some species being dependent on the metabolism of other organisms. Indeed the diversity of the plaque microflora is due, in part, to (a) the development of such food chains and food webs (Figure 6) and to (b) the lack of a single nutrient limiting the growth of all bacterial species. Bacterial survive by adopting, where possible, alternative metabolic strategies in order to avoid direct competition. Another beneficial interaction among plaque bacteria is coaggregation (Kolenbrander and London, 1992) which can aid colonisation of surfaces and also facilitate metabolic interactions between mutually-dependent strains.

Antagonism is also a major contributing factor in the determination of the composition of microbial ecosystems such as dental plaque. The production of antagonistic compounds (Table 20) can give an organism a competitive advantage when interacting with other microbes. One of the most common types of antagonistic compounds produced are bacteriocins or bacteriocin-like substances. Bacteriocins are relatively low molecular weight polypeptides that are coded for by a plasmid; the producer strains are resistant to the action of the bacteriocins they produce. Bacteriocins are produced by most species of oral streptococci (e.g. mutacin by *S. mutans*, Alaluusua *et al.*, 1991, and sanguicin by *S. sanguis*, Fujimura and Nakamura,

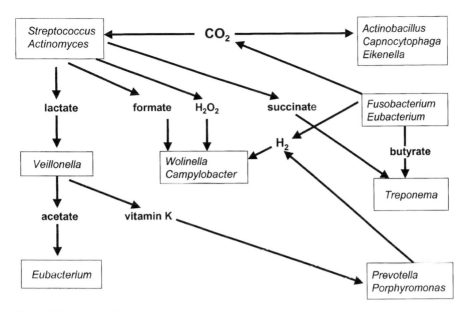

Figure 6. Examples of food webs among oral bacteria.

1979), as well as by *Corynebacterium matruchotii*, black-pigmented anaerobes, and *A. actinomycetemcomitans*. In contrast, *Actinomyces* species are not generally bacteriocinogenic. Although bacteriocins are usually limited in their spectrum of activity, many of the streptococcal bacteriocins are broad spectrum, inhibiting species belonging to several Gram-positive and Gram-negative bacteria. A mutant of *S. mutans* that over-produced a bacteriocin was found to colonise human volunteers better than the parent strain (Hillman *et al.*, 1987). In addition, bacteriocin-production can prevent colonisation by allochthonous species. Some *S. salivarius* strains can produce an inhibitor (enocin or salivaricin) with activity against Lancefield Group A streptococci (Sanders and Sanders, 1982). Enocin-producing strains may prevent colonization of the mouth by this pathogen in a manner similar to that proposed for streptococci in the pharynx.

Other inhibitory factors produced by plaque bacteria include organic acids, hydrogen peroxide, and enzymes (Table 20). The production of hydrogen peroxide by members of the *S. mitis*-group has been proposed as a mechanism whereby the numbers of periodontopathic bacteria are reduced in plaque to levels at which they are incapable of initiating disease (Hillman *et al.*, 1985). Perhaps significantly, some periodontal pathogens (e.g. *A. actinomycetemcomitans*) are able to produce factors inhibitory to oral streptococci. Certain types of periodontal disease (see Chapter 6) might result, therefore, from an ecological imbalance between dynamically-interacting groups of bacteria. The low pH generated from carbohydrate metabolism is also inhibitory to many plaque species, particularly Gram-negative organisms and some acid-sensitive streptococci such as *S. gordonii* (Bradshaw *et al.*, 1989).

The production of antagonistic factors will not necessarily lead to the complete exclusion of sensitive species. The presence of distinct micro-habitats within a biofilm such as plaque enables bacteria to survive that would be incompatible with one another in a homogeneous environment. Also where there is competition for nutrients, the production of inhibitory factors might be a mechanism whereby less-competitive species can persist (negative feed-back).

Dental Plaque as a Microbial Community

A microbial community has been defined as a group of interacting bacteria growing in association with one another, often on a surface. Growth in a microbial community allows component species to display a broader habitat range, an enhanced resistance to environmental perturbation, and an increased metabolic efficiency than when growing alone (Caldwell *et al.*, 1997). Evidence that plaque functions as a true microbial community has come from:

(a) modulation of local environmental conditions (pH, oxygen tension, redox potential, etc), for example, enabling the growth of obligate anaerobes in an overtly aerobic habitat, and of pH-sensitive bacteria during periods of low pH (Costerton *et al.*, 1994),

(b) the synergistic catabolism of complex host macromolecules so that substrates can be utilised that would be recalcitrant to degradation by individual species (Homer and Beighton, 1992; ter Steeg and van der Hoeven, 1989), and

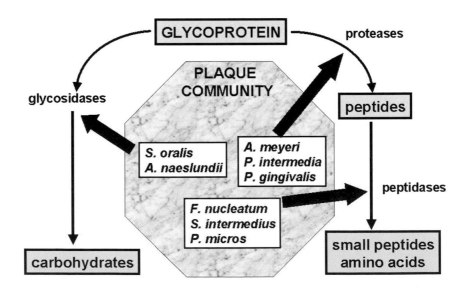

Figure 7. The catabolism of complex host substrates involving the concerted and sequential action of communities of oral bacteria.

(c) efficient nutrient and energy cycling via cross-feeding and food webs (van der Hoeven, 1998).

Great metabolic diversity exists within the plaque microflora, ranging from organisms that can catalyse the initial splitting of complex host polymers into smaller units, to those such as sulphate reducing bacteria and methanogens that gain energy from the utilisation of simple end products of metabolism (Marsh and Bradshaw, 1998; van der Hoeven, 1998). These organisms are likely to be spatially organised, perhaps mediated by coaggregation, to generate metabolically-favourable consortia of interdependent species. In such a microbial community, the metabolic efficiency of the whole is greater than that of the sum of the individual species since substrate utilisation involves both the concerted and sequential catabolism of these complex molecules (Figure 7). One reason for the inability to recover fastidious bacteria on laboratory media is our insistence on trying to grow these organisms as pure cultures rather than as part of the community in which they have evolved.

Growth of microbial communities as a biofilm confers additional benefits since cells are protected from the host defences, antimicrobial agents and from other hostile factors (Costerton *et al.*, 1987, 1994). In addition, the closely coupled physical and metabolic interactions leave few niches unfilled, thereby reducing the likelihood of colonisation by exogenous microbes, and this contributes to the natural microbial stability of the flora of plaque (microbial homeostasis).

Despite the microbial diversity of dental plaque, homeostasis does break down on occasions. The main causes for this can be divided into either (a) deficiencies in the immune response, or (b) other (non-immune) factors (Table 21). The remainder of this chapter will focus on recent studies which have shown how homeostasis breaks down in plaque, leading to caries and periodontal diseases. These studies have provided new insights into how plaque bacteria may cause disease, and this in turn has identified new opportunities for controlling disease.

Ecological Perturbations in Dental Plaque and their Relationship to Disease

Numerous cross-sectional and longitudinal epidemiological studies have compared the microflora of plaque at sites with disease (caries and periodontal diseases) with that at normal healthy surfaces. The results from such studies will be described in Chapters 5 and 6, but, in general, caries is associated with increases in the proportions

---

Table 21. Factors Responsible for the Breakdown of Microbial Homeostasis (Marsh and Martin, 1999).

Immunological Factors	Non-immunological Factors
sIgA-deficiency	Xerostomia
Neutrophil dysfunction	Antibiotics
Chemotherapy-induced	Dietary carbohydrates/low pH
myelosuppression	Increased GCF flow
Infection-induced	Oral contraceptives
myelosuppression (e.g. AIDS)	

of acidogenic and aciduric bacteria, especially mutans streptococci and lactobacilli (Loesche, 1986; Marsh and Martin, 1999) whereas higher levels of proteolytic (and often Gram-negative) anaerobic bacteria (e.g. *P. gingivalis* and *A. actinomyctemcomitans*) are recovered from periodontal pockets (Moore and Moore, 1994). This has led to the aetiology of these diseases being considered in the same manner as many medical infections, i.e. with a relatively specific microbial aetiology. This has provided impetus to studies attempting to prevent these common dental diseases using strategies similar to those adopted for many classical pathogens, e.g. using vaccines or antimicrobial agents. However, a single species of micro-organism is generally responsible for these medical diseases, and these pathogens often express distinct virulence factors, including toxin production or the ability to invade cells and spread within the host tissues; in addition, these pathogens cause disease at sites that are usually sterile or not usually colonised by that organism. In contrast, plaque-mediated diseases occur at sites with a pre-existing, complex and normal microflora, and the implicated organisms rarely possess particularly specific pathogenic determinants. Thus, for example, a key feature of cariogenic bacteria is their ability to rapidly transport and catabolise dietary sugars to acid and to polysaccharides, and continue their metabolism even under the conditions of low pH that they have generated (Bowden and Hamilton, 1998). Likewise, although some of the anaerobes isolated from periodontal pockets can cause tissue damage directly (by the production of enzymes such as collagenases and hyaluronidase), most of the damage to the host is probably due to an "uncontrolled" inflammatory response to the microbial insult, partly due to the proteolytic inactivation of host regulatory molecules, as well as to other host cellular effects that deregulate connective tissue homeostasis (Darveau *et al.*, 1997; Birkedal-Hansen, 1998; see also Chapters 5 and 6).

The origin of these plaque pathogens has been the subject of much investigation. For many years, the putative pathogens were rarely cultured from plaque from healthy sites, and so they were considered by many to be acquired exogenously from other people, as occurs with most other pathogenic bacteria (e.g. *Mycobacterium tuberculosis, Salmonella typhi*). More recently, the application of highly sensitive immunological (e.g. ELISA; Gmür and Guggenheim, 1994; Di Murro *et al.*, 1997) or molecular (e.g. oligonucleotide probe or PCR; Kisby *et al.*, 1989; Greenstein and Lamster, 1997; Asikainen and Chen, 1999; Socransky *et al.*, 1999) techniques to plaque has led to the more frequent detection of low levels of most dental "pathogens" at a range of healthy sites, even in young children. This suggests that plaque-mediated diseases result from imbalances in the resident microflora resulting from an enrichment within the microbial community of these "pathogens" (Newman, 1990; Marsh, 1991, 1994). Nevertheless, in either situation (i.e. natural low levels of "pathogens" or low levels of exogenously acquired "pathogens") the cariogenic or periodontopathic bacteria would be in the minority and would be non-competitive with the rest of the resident microflora, and probably at levels too low to be clinically significant (Figure 8). For disease to occur, these species would have to achieve an appropriate level of numerical dominance, and outcompete the already established microflora. This would require substantial changes in local environmental conditions and the imposition of strong selective pressures that would favour the growth of these organisms (Marsh, 1998).

Studies of a range of habitats have provided clues as to the type of factors capable of disrupting the intrinsic homeostasis that exists within microbial communities. A common feature is a change in the nutrient status of a site, for example, following the introduction of a novel substrate. Thus, again using the example cited earlier, nitrogenous fertilisers that are washed off farm land into surface water result in an overgrowth by algae, with secondary effects due to the concomitant consumption of dissolved oxygen in the water (Codd, 1995). Other effects can result from a chemical or physical change to the habitat, such as acidification of soil and lakes due to environmental pollution (acid rain) and the introduction of implants such as catheters, respectively.

Analogous effects can occur in the mouth. Data from both laboratory models and clinical studies suggest that changes to the local oral environment can perturb the microbial balance found in health and select for organisms associated with plaque-mediated diseases. Rinsing with low pH buffers led to the enrichment in plaque of mutans streptococci in human volunteers (Svanberg, 1980), while a tightly controlled mixed culture study showed conclusively that it was the low pH generated from the catabolism of dietary carbohydrates rather than the availability of fermentable sugars *per se* that selects for acidogenic and cariogenic bacteria such as *S. mutans* and lactobacilli (Bradshaw *et al.*, 1989). Similarly, laboratory studies have indicated that an increased flow of GCF leads to the enrichment of proteolytic species associated with the aetiology of periodontal diseases by providing essential nutrients including haeme-containing molecules (ter Steeg *et al.*, 1987, 1988; ter Steeg and

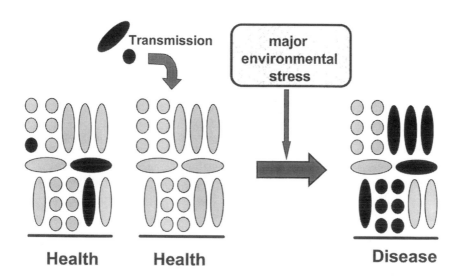

Figure 8. Relationship of oral pathogens to disease in dental plaque. Pathogens (cells in black) may be acquired from other individuals or be present in low numbers at sub-clinical levels in dental plaque. In either situation, a major environmental change would have to occur for these pathogens to out-compete the numerically dominant members of the resident microflora (cells shown in light grey). (Adapted from Marsh, 1998.)

van der Hoeven, 1989); such species also become more competitive under the alkaline conditions associated with sites undergoing active proteolysis (Marsh *et al.*, 1993).

The concept that caries and periodontal diseases arise as a result of environmental perturbations to the habitat has been encapsulated in the "ecological plaque hypothesis" (Marsh, 1991, 1994, 1998; Marsh and Bradshaw, 1997). Key features of this hypothesis are that (a) the selection of "pathogenic" bacteria is directly coupled to changes in the environment, and (b) diseases need not have a specific aetiology; any species with relevant traits can contribute to the disease process. Thus, the significance to disease of newly-discovered species can be predicted on the basis of their physiological characteristics. For example, bacteria associated with dental caries can display a continuum from those that are slightly acidogenic, and hence only make a minor contribution, to species that are both highly acidogenic and also aciduric. Mutans streptococci are the organisms that are best adapted to the cariogenic environment (high sugar/low pH) but such properties are not unique to these bacteria. Strains of other species, such as members of the *S. mitis*-group, also share some of these properties, and therefore, will contribute to the rate of demineralisation of enamel (Sansone *et al.*, 1993). The role in disease of any subsequently discovered novel bacterium could be gauged by an assessment of its acidogenic/aciduric properties. Following on this line of argument, another key element of the ecological plaque hypothesis is the fact that disease could be prevented not only by targeting the putative pathogens directly, e.g. by antimicrobial or anti-adhesive strategies, but also by interfering with the selection pressures responsible for their enrichment (Marsh 1991, 1994, 1998; Marsh and Bradshaw, 1997). In the case of dental caries, the regular conditions of sugar/low pH challenge could be reduced by the use of sugar substitutes, the stimulation of saliva flow after meals by chewing gum and the use of products containing antimicrobial agents, many of which at low concentrations (as occurs often in the mouth) can interfere with glycolysis (Marsh, 1993; Marsh and Bradshaw, 1997). In periodontal diseases, attempts could be made to alter the local environment by reducing the flow of GCF, or the site could be made less anaerobic by the use of redox agents (Wilson *et al.*, 1992; Gibson *et al.*, 1994; Marsh and Bradshaw, 1997).

Commonly, when a clinician is faced with plaque-mediated disease, only the symptoms are treated. The appearance of disease should alert the clinician to identify the causal factor(s) driving this local ecological catastrophe in plaque, and deal with both the **cause** and the **effect** of the disease. Examples of potential causal factors include an inappropriate diet, smoking, and the long term use of medications that, as a side-effect, reduce the flow of saliva or suppress the activity of components of the specific host defences. Thus, an appreciation of the ecology of the oral cavity will enable the enlightened clinician to take a more holistic approach and take into account the nutrition, physiology, host defences and general well-being of the patient, as these will affect the balance and activity of the resident oral microflora. Future developments in oral care will recognise these inter-relationships and use multiple strategies to maintain homeostasis (and hence a favourable ecology) in plaque.

Concluding Remarks

The use of rigorous isolation techniques, coupled with improvements in microbial taxonomy, has led to the recognition of a highly diverse microflora that inhabits the various surfaces of the normal mouth. This diversity is particularly apparent in dental plaque where the spatial heterogeneity of biofilms ensures that species can co-exist despite possessing conflicting nutritional and atmospheric requirements.

The recent application of molecular approaches to microbial identification has facilitated the identification of species and taxa not previously described, and which cannot as yet be cultivated in the laboratory. Such taxa are being found commonly in healthy individuals, implying that the resident oral flora has an even greater complexity than hitherto imagined. The continued use and refinement of these powerful techniques will be necessary to fully describe these microbial communities. A likely outcome of these studies will be the recognition that the mouth in general, and dental plaque in particular, can act as a significant reservoir for many medically-important pathogens. Such findings may open up further opportunities for the control of these opportunistic pathogens.

Relatively little is known of the true architecture of dental plaque, of the physical location of particular species within these biofilms, nor how the biofilm "mode of life" affects microbial gene expression. The availability of new techniques such as confocal microscopy, species-specific probes, reporter genes, differential display of mRNA profiles, and DNA chip technology will help resolve many of these fundamental issues. Another important area of study, which is just beginning, is to understand the "cross-talk" that may occur both among the component species of the resident microflora, but also between these communities and the host tissues. Advances in knowledge in these areas offer exciting opportunities for the control and prevention of dental diseases in the future, while still retaining the beneficial properties afforded to the host by a resident microflora.

Acknowledgement

The author gratefully acknowledges the assistance of Dr. D.J. Bradshaw, CAMR, Salisbury, with the manuscript, and the provision of the data for Table 19 by Dr. N. Jacques, IDR, Sydney.

References

Alaluusua, S., T. Takei, T. Ooshima, and S. Hamada. 1991. Mutacin activity of strains isolated from children with varying levels of mutans streptococci and caries. Archives of Oral Biology. 36: 251-255.

Alexander, M. 1971. Microbial Ecology. John Wiley, New York.

Al-Hashimi, I., and M. J. Levine. 1989. Characterization of *in vivo* salivary-derived enamel pellicle. Archives of Oral Biology. 34: 289-295.

Alvarez, L. W., W. Alvarez, F. Asaro, and H. V. Michel. 1980. Extraterrestial cause for the Cretaceous-Tertiary extinction. Science. 208: 1095-1108.

Aly, R., and H. Maibach. 1981. Microbial interactions on skin., p. 29-39. *In* H. Maibach and R. Aly (ed.), Skin microbiology: relevance to clinical infection. Springer Verlag, New York.

Amano, A., A. Sharma, H. T. Sojar, H.K. Kuramitsu, and R.J. Genco. 1994. Effects of temperature stress on expression of fimbriae and superoxide dismutase by *Porphyromonas gingivalis*. Infection and Immunity. 62: 4682-4685.

Andersen, R. N., N. Ganeshkumar, and P. E. Kolenbrander. 1998. *Helicobacter pylori* adheres selectively to *Fusobacterium* spp. Oral Microbiology and Immunology. 13: 51-54.

Asikainen, S., and C. Chen. 1999. Oral ecology and person-to-person transmission of *Actinobacillus actinomycetemcomitans* and *Porphyromonas gingivalis*. Periodontology 2000. 20: 65-81.

Babaahmady, K. G., P. D. Marsh, S. J. Challacombe, and H. N. Newman. 1997. Variations in the predominant cultivable microflora of dental plaque at defined sub-sites on approximal tooth surfaces in children. Archives of Oral Biology. 42: 101-111.

Beckers, H. J. A., and J. S. van der Hoeven. 1982. Growth rates of *Actinomyces viscosus* and *Streptococcus mutans* during early colonisation of tooth surfaces in gnotobiotic rats. Infection and Immunity. 35: 583-587.

Beighton, D., and H. Hayday. 1986. The influence of the availability of dietary food on the growth of streptococci on the molar teeth of monkeys (*Macaca fascicularis*). Archives of Oral Biology. 31: 449-454.

Beighton, D., K. Smith, and H. Hayday. 1986. The growth of bacteria and the production of exoglycosidic enzymes in the dental plaque of macaque monkeys. Archives of Oral Biology. 31: 829-835.

Berkowitz, R. J., and P. Jones. 1985. Mouth-to-mouth transmission of the bacterium *Streptococcus mutans* between mother and child. Archives of Oral Biology. 30: 377-379.

Berkowitz, R. J., H. V. Jordan, and G. White. 1975. The early establishment of *Streptococcus mutans* in the mouths of infants. Archives of Oral Biology. 20: 171-174.

Birkedal-Hansen, H. 1998. Links between microbial colonization, inflammatory response, and tissue destruction., p. 170-178. *In* B. Guggenheim and S. Shapiro (ed.), Oral biology at the turn of the century. Misconceptions, truths, challenges and prospects. Karger, Basel.

Bloomquist, C. G., B. E. Reilly, and W. F. Liljemark. 1996. Adherence, accumulation, and cell division of a natural adherent bacterial population. Journal of Bacteriology. 178: 1172-1177.

Bobo, R. A., E. J. Newton, L. F. Jones, L. H. Farmer, and J. J. Farmer, 3d. 1973. Nursery outbreak of *Pseudomonas aeruginosa*: epidemiological conclusions from five different typing methods. Applied Microbiology. 25: 414-420.

Bowden, G. H., J. M. Hardie, and G. L. Slack. 1975. Microbial variations in approximal dental plaque. Caries Research. 9: 253-277.

Bowden, G. H. W., and I. R. Hamilton. 1998. Survival of oral bacteria. Critical Reviews in Oral Biology and Medicine. 9: 54-85.

Bowen, W. H., S. K. Pearson, B. C. van Wuyckhuyse, and L. A. Tabak. 1991. Influence of milk, lactose-reduced milk, and lactose on caries in desalivated rats. Caries Research. 25: 283-286.

Boyd, A., and A. M. Chakrabarty. 1995. *Pseudomonas aeruginosa* biofilms: role of the alginate exopolysaccharide. Journal of Industrial Microbiology. 15: 162-168.

Bradshaw, D. J. 1995. Metabolic responses in biofilms. Microbial Ecology in Health and Disease. 8: 313-316.

Bradshaw, D. J., K. A. Homer, P. D. Marsh, and D. Beighton. 1994. Metabolic cooperation in oral microbial communities during growth on mucin. Microbiology. 140: 3407-3412.

Bradshaw, D. J., P. D. Marsh, C. Allison, and K. M. Schilling. 1996. Effect of oxygen, inoculum composition and flow rate on development of mixed culture oral biofilms. Microbiology. 142: 623-629.

Bradshaw, D. J., P. D. Marsh, G. K. Watson, and C. Allison. 1998. Role of *Fusobacterium nucleatum* and coaggregation in anaerobe survival in planktonic and biofilm oral microbial communities during aeration. Infection and Immunity. 66: 4729-4732.

Bradshaw, D. J., A. S. McKee, and P. D. Marsh. 1989. Effects of carbohydrate pulses and pH on population shifts within oral microbial communities *in vitro*. Journal of Dental Research. 68: 1298-1302

Burne, R. A. 1998. Regulation of gene expression in adherent populations of oral streptococci., p. 41-53. *In* D. J. LeBlanc, M. S. Lanz, and L. M. Switalski (ed.), Microbial pathogenesis: Current and emerging issues. Indiana University, Indianapolis.

Burne, R. A., Y.-Y. M. Chen, and J. E. C. Penders. 1997. Analysis of gene expression in *Streptococcus mutans* in biofilms *in vitro*. Advances in Dental Research. 11: 100-109.

Busscher, H. J., M. M. Cowan, and H. C. van der Mei. 1992. On the relative importance of specific and non-specific approaches to oral microbial adhesion. FEMS Microbiology Reviews. 88: 199-210.

Busscher, H. J., and H. C. van der Mei. 1997. Physico-chemical interactions in initial microbial adhesion and relevance for biofilm formation. Advances in Dental Research. 11: 24-32.

Caldwell, D. E., G. M. Wolfaardt, D. R. Korber, and J. R. Lawrence. 1997. Do bacterial communities transcend Darwinism?, p. 105-191. *In* J. G. Jones (ed.), Advances in Microbial Ecology, vol. 15. Plenum, New York.

Carlsson, J., H. Grahnen, G. Johnsson, and S. Wikner. 1970. Establishment of *Streptococcus sanguis* in the mouths of infants. Archives of Oral Biology. 15: 1143-1148.

Carlsson, J., B. F. Herrmann, J. F. Hofling, and G. J. Sundqvist. 1984. Degradation of albumin, haemopexin, haptoglobin and transferrin by black-pigmented *Bacteroides* species. Journal of Medical Microbiology. 18: 39-46.

Caufield, P. W., G. R. Cutter, and A. P. Dasanayake. 1993. Initial acquisition of mutans streptococci by infants: evidence for a discrete window of infectivity. Journal of Dental Research. 72: 37-45.

Cavedon, K., and J. London. 1993. Adhesin degradation: a possible function for a *Prevotella loescheii* protease? Oral Microbiology and Immunology. 8: 283-287.

Choi, B. K., C. Wyss, and U. B. Göbel. 1996. Phylogenetic analysis of pathogen-related oral spirochaetes. Journal of Clinical Microbiology. 34: 1922-1925.

Cimasoni, G. 1983. Crevicular Fluid Updated. S. Karger, Basel.

Codd, G. A. 1995. Cyanobacterial toxins: occurrence, properties and biological significance. Water Science and Technology. 32: 149-156.

Cole, M. F., M. Evans, S. Fitzsimmons, J. Johnson, C. Pearce, M. J. Sheridan, R. Wientzen, and G. Bowden. 1994. Pioneer oral streptococci produce immunoglobulin A1 protease. Infection and Immunity. 62: 2165-2168.

Costerton, J. W., K. J. Cheng, G. G. Geesey, T. I. Ladd, J. C. Nickel, M. Dasgupta, and T. J. Marrie. 1987. Bacterial biofilms in nature and disease. Annual Reviews of Microbiology. 41: 435-464.

Costerton, J. W., Z. Lewandowski, D. E. Caldwell, D. R. Korber, and H. M. Lappin-Scott. 1995. Microbial biofilms. Annual Reviews of Microbiology. 49: 711-745.

Costerton, J. W., Z. Lewandowski, D. DeBeer, D. E. Caldwell, D. R. Korber, and G. James. 1994. Biofilms, the customized microniche. Journal of Bacteriology. 176: 2137-2142.

Darveau, R. P., A. Tanner, and R. C. Page. 1997. The microbial challenge in periodontitis. Periodontology 2000. 14: 12-32.

Davey, A. L., and A. H. Rogers. 1984. Multiple types of the bacterium *Streptococcus mutans* in the human mouth and their intra-family transmission. Archives of Oral Biology. 29: 453-460.

Dent, V. 1979. The bacteriology of dental plaque from a variety of zoo-maintained mammalian species. Archives of Oral Biology. 24: 277-282.

Dent, V. E., and P. D. Marsh. 1981. Evidence for a "basic plaque" microbial community of animals. Archives of Oral Biology. 26: 171-179.

Di Murro, C., M. Paolantonio, V. Pedrazzoli, D. E. Lopatin, and M. Cattabriga. 1997. Occurrence of *Porphyromonas gingivalis*, *Bacteroides forsythus*, and *Treponema denticola* in periodontally healthy and diseased subjects as determined by an ELISA technique. Journal of Periodontology. 68: 18-23.

Dibdin, G. H., and R. P. Shellis. 1988. Physical and biochemical studies of *Streptococcus mutans* sediments suggest new factors linking the cariogenicity of plaque with its extracellular polysaccharide content. Journal of Dental Research. 67: 890-895.

Edgar, M., and S. M. Higham. 1996. Saliva and the control of plaque pH., p. 81-94. *In* W. M. Edgar and D. M. O'Mullane (ed.), Saliva and oral health, Second Edition. British Dental Journal, London.

Eggert, F. M., L. Drewell, J. A. Bigelow, J. E. Speck, and M. Goldner. 1991. The pH of gingival crevices and periodontal pockets in children, teenagers and adults. Archives of Oral Biology. 36: 233-238.

Ellen, R. P., and R. A. Burne. 1996. Conceptual advances in research on the adhesion of bacteria to oral surfaces., p. 201-248. *In* M. Fletcher (ed.), Bacterial adhesion, molecular and ecological diversity. John Wiley & Sons Inc, New York.

Fedi, P. F. J., and W. J. Killoy. 1992. Temperature differences at periodontal sites in health and disease. Journal of Periodontology. 63: 24-27.

Fitzsimmons, S., M. Evans, C. Pearce, M. J. Sheridan, R. Wientzen, G. Bowden, and M. F. Cole. 1996. Clonal diversity of *Streptococcus mitis* biovar 1 isolates from the oral cavity of human neonates. Clinical Diagnostic Laboratory and Immunology. 3: 517-522.

Frandsen, E. V., V. Pedrazzoli, and M. Kilian. 1991. Ecology of viridans streptococci in the oral cavity and pharynx. Oral Microbiology and Immunology. 6: 129-133.

Fujimura, S., and T. Nakamura. 1979. Sanguicin, a bacteriocin of oral *Streptococcus sanguis*. Antimicrobial Agents and Chemotherapy. 16: 262-265.

Fuller, R. (ed.). 1992. Probiotics. The scientific basis. Chapman & Hall, London.

Gibbons, R. J. 1989. Bacterial adhesion to oral tissues: a model for infectious diseases. Journal of Dental Research. 68: 750-760.

Gibbons, R. J., D. I. Hay, W. C. Childs III, and G. Davis. 1990. Role of cryptic receptors (cryptitopes) in bacterial adhesion to oral surfaces. Archives of Oral Biology. 35(Suppl.): 107S-114S.

Gibson, G. R., and G. T. Macfarlane (ed.). 1995. Human colonic bacteria. Role in nutrition, physiology and pathology. CRC Press, Boca Raton.

Gibson, M. T., D. Mangat, G. Gagliano, M. Wilson, J. Fletcher, J. Bulman, and H. N. Newman. 1994. Evaluation of the efficacy of a redox agent in the treatment of chronic periodontitis. Journal of Clinical Periodontology. 21: 690-700.

Gilbert, P., J. Das, and I. Foley. 1997. Biofilm susceptibility to antimicrobials. Advances in Dental Research. 11: 160-167.

Gmür, R., and B. Guggenheim. 1994. Interdental supragingival plaque - a natural habitat of *Actinobacillus actinomycetemcomitans*, *Bacteroides forsythus*, *Campylobacter rectus*, and *Prevotella nigrescens*. Journal of Dental Research. 73: 1421-1428.

Greenstein, G., and I. Lamster. 1997. Bacterial transmission in periodontal diseases: a critical review. Journal of Periodontology. 68: 421-431.

Grubb, R., T. Midtvedt, and E. Norin (ed.). 1989. The regulatory and protective role of the normal microflora. Macmillan Press Ltd, Basingstoke.

Hakenbeck, R., A. Konog, I. Kern, M. van der Linden, W. Keck, D. Billot-Klein, R. Legrand, B. Schoot, and L. Gutmann. 1998. Acquisition of five high-Mr penicillin-binding protein variants during transfer of high-level beta-lactam resistance from

Streptococcus mitis to *Streptococcus pneumoniae*. Journal of Bacteriology. 180: 1831-1840.

Hanley, S. A., J. Aduse-Opoku, and M. A. Curtis. 1999. A 55-kilodalton immunodominant antigen of *Porphyromonas gingivalis* W50 has arisen via horizontal gene transfer. Infection and Immunity. 67: 1157-1171.

Harper, D. S., and W. J. Loesche. 1984. Growth and acid tolerance of human dental plaque bacteria. Archives of Oral Biology. 29: 843-848.

Harper-Owen, R., D. Dymock, V. Booth, A. J. Weightman, and W. G. Wade. 1999. Detection of unculturable bacteria in periodontal health and disease by PCR. Journal of Clinical Microbiology. 37: 1469-1473.

Hartley, M. G., M. A. EL-Maaytah, C. McKenzie, and J. Greenman. 1996. The tongue microbiota of low odour and malodorous individuals. Microbial Ecology in Health and Disease. 9: 215-223.

Haubek, D., J. M. Direnzo, E. M. B. Tinoco, J. Westergaard, N. J. Lopez, C.-P. Chung, K. Poulsen, and M. Kilian. 1997. Racial tropism of a highly toxic clone of *Actinobacillus actinomycetemcomitans* associated with juvenile periodontitis. Journal of Clinical Microbiology. 35: 3037-3042.

Henderson, B., and M. Wilson. 1998. Commensal communism and the oral cavity. Journal of Dental Research. 77: 1674-1683.

Hillman, J. D. 1999. Replacement therapy for dental caries. p. 587-599. *In* H. N. Newman and M. Wilson (ed.), Dental plaque revisited: oral biofilms in health and disease. BioLine, Cardiff.

Hillman, J. D., A. L. Dzuback, and S. W. Andrews. 1987. Colonization of the human oral cavity by a *Streptococcus mutans* mutant producing increased bacteriocin. Journal of Dental Research. 66: 1092-1094.

Hillman, J. D., S. S. Socransky, and M. Shivers. 1985. The relationships between streptococcal species and periodontopathic bacteria in human dental plaque. Archives of Oral Biology. 30: 791-795.

Hohwy, J., and M. Kilian. 1995. Clonal diversity of the *Streptococcus mitis* biovar 1 population in the human oral cavity and pharynx. Oral Microbiology and Immunology. 10: 19-25.

Homer, K. A., and D. Beighton. 1992. Synergistic degradation of transferrin by mutans streptococci in association with other dental plaque bacteria. Microbial Ecology in Health and Disease. 5: 111-116.

Isokangas, P., J. Tenovuo, E. Söderling, H. Männistö, and K. K. Mäkinen. 1991. Dental caries and mutans streptococci in the proximal areas of molars affected by the habitual use of xylitol chewing gum. Caries Research. 25: 444-448.

Jensen, M. E., and C. F. Schachtele. 1983. The acidogenic potential of reference foods and snacks at interproximal sites. Journal of Dental Research. 62: 889-892.

Kenney, E. B., and M. Ash. 1969. Oxidation-reduction potential of developing plaque, periodontal pockets and gingival sulci. Journal of Periodontology. 40: 630-633.

Kilian, M. 1998. Clonal basis of bacterial virulence., p. 131-142. *In* B. Guggenheim and S. Shapiro (ed.), Oral biology at the turn of the century. Misconceptions, truths, challenges and prospects. Karger, Basel.

Kinniment, S. L., J. W. T. Wimpenny, D. Adams, and P. D. Marsh. 1996. The effect of chlorhexidine on defined, mixed culture oral biofilms grown in a novel model system. Journal of Applied Bacteriology. 81: 120-125.

Kisby, L. E., E. D. Savitt, C. K. French, and W. J. Peros. 1989. DNA probe detection of key periodontal pathogens in juveniles. Journal of Pedodontics. 13: 222-230.

Köhler, B., I. Andréen, and B. Jonsson. 1984. The effect of caries-preventive measures in mothers on dental caries and the oral presence of the bacteria *Streptococcus mutans* and lactobacilli in their children. Archives of Oral Biology. 29: 879-883.

Köhler, B., D. Bratthall, and B. Krasse. 1983. Preventive measures in mothers influence the establishment of the bacterium *Streptococcus mutans* in their infants. Archives of Oral Biology. 1983: 225-231.

Kolenbrander, P. E. 1993. Coaggregation of human oral bacteria: potential role in the accretion of dental plaque. Journal of Applied Bacteriology. 74 Suppl: 79S-86S.

Kolenbrander, P. E., N. Ganeshkumar, F. J. Cassels, and C. V. Hughes. 1993. Coaggregation: specific adherence among human oral plaque bacteria. Federation of American Societies for Experimental Biology Journal. 7: 406-413.

Kolenbrander, P. E., and J. London. 1993. Adhere today, here tomorrow: oral bacterial adherence. Journal of Bacteriology. 175: 3247-3252.

Kolenbrander, P. E., and J. London. 1992. Ecological significance of coaggregation among oral bacteria. Advances in Microbial Ecology. 12: 183-217.

Könönen, E., S. Asikainen, and H. Jousimies-Somer. 1992. The early colonisation of gram-negative anaerobic bacteria in edentulous infants. Oral Microbiology and Immunology. 7: 28-31.

Könönen, E., S. Asikainen, M. Saarela, J. Karjalainen, and H. Jousimies-Somer. 1994. The oral Gram-negative anaerobic microflora in young children: longitudinal changes from edentulous to dentate mouth. Oral Microbiology and Immunology. 9: 136-141.

Kopec, L. K., A. M. Vacca-Smith, and W. H. Bowen. 1997. Structural aspects of glucans formed in solution and on the surface of hydroxyapatite. Glycobiology. 7: 929-934.

Kuramitsu, H. K. 1993. Virulence factors of mutans streptococci - Role of molecular genetics. Critical Reviews in Oral Biology and Medicine. 4: 159-176.

Lacey, R. W., V. L. Lord, G. L. Howson, D. E. A. Luxton, and I. S. Trotter. 1983. Double-blind study to compare the selection of antibiotic resistance by amoxycillin or cephradine in the commensal flora. Lancet. ii: 529-532.

Lee, S. F., Y. H. Li, and G. W. H. Bowden. 1996. Detachment of *Streptococcus mutans* biofilm cells by an endogenous enzyme activity. Infection and Immunity. 64: 1035-1038.

Li, Y., and P. W. Caufield. 1995. The fidelity of initial acquisition of mutans streptococci by infants from their mothers. Journal of Dental Research. 74: 681-685.

Lie, T. 1977. Scanning and transmission electron microscopy study of pellicle morphogenesis. Scandinavian Journal of Dental Research. 85: 217-231.

Liljemark, W. F., C. G. Bloomquist, B. E. Reilly, C. J. Bernards, D. W. Townsend, A. T. Pennock, and J. L. LeMoine. 1997. Growth dynamics in a natural biofilm and its impact on oral disease management. Advances in Dental Research. 11: 14-23.

Lindemann, R. A., M. G. Newman, A. K. Kaufman, and T. V. Le. 1985. Oral colonization and susceptibility testing of *Pseudomonas aeruginosa* oral isolates from cystic fibrosis patients. Journal of Dental Research. 64: 54-57.

Littleton, N. W., R. M. McCabe, and C. H. Carter. 1967. Studies of oral health in persons nourished by stomach tube. II. Acidogenic properties and selected bacterial components of plaque material. Archives of Oral Biology. 12: 601-609.

Loesche, W. J. 1986. Role of *Streptococcus mutans* in human dental decay. Microbiological Reviews. 50: 353-380.

Loesche, W. J., F. Gusberti, G. Mettraux, T. Higgins, and S. Syed. 1983. Relationship between oxygen tension and subgingival bacterial flora in untreated human periodontal pockets. Infection and Immunity. 42: 659-667.

Madinier, I. M., T. M. Fosse, and R. A. Monteil. 1997. Oral carriage of *Helicobacter pylori*: A review. Journal of Periodontology. 68: 2-6.

Maiden, M. F., A. C. Tanner, P. J. Macuch, L. Murray, and R. L. J. Kent. 1998. Subgingival temperature and microbiota in initial periodontitis. Journal of Clinical Periodontology. 25: 786-793.

Marples, R. 1994. The normal flora of human skin is a biofilm, p. 127-129. *In* J. Wimpenny, Nichols, W., Stickler, D., Lappin-Scott, H. (ed.), Bacterial biofilms and their control in medicine and industry. BioLine, Cardiff.

Marquis, R. E. 1995. Oxygen metabolism, oxidative stress and acid-base physiology of dental plaque biofilms. Journal of Industrial Microbiology. 15: 198-207.

Marsh, P. D. 1989. Host defenses and microbial homeostasis: role of microbial interactions. Journal of Dental Research. 68(Special Issue): 1567-1575.

Marsh, P. D. 1991. Sugar, fluoride, pH and microbial homeostasis in dental plaque. Proceedings of the Finnish Dental Society. 87: 515-525.

Marsh, P. D. 1993. Antimicrobial strategies in the prevention of dental caries. Caries Research. 27 Suppl 1: 72-76.

Marsh, P. D. 1994. Microbial ecology of dental plaque and its significance in health and disease. Advances in Dental Research. 8: 263-271.

Marsh, P. D. 1995. Dental plaque, p. 282-300. *In* H. M. Lappin-Scott and J. W. Costerton (ed.), Microbial biofilms. Cambridge University Press, Cambridge.

Marsh, P. D. 1998. The control of oral biofilms: new approaches for the future., p. 22-31. *In* B. Guggenheim and S. Shapiro (ed.), Oral biology at the turn of the century. Misconceptions, truths, challenges and prospects. Karger, Basel.

Marsh, P. D., and D. J. Bradshaw. 1995. Dental plaque as a biofilm. Journal of Industrial Microbiology. 15: 169-175.

Marsh, P. D., and D. J. Bradshaw. 1997. Physiological approaches to plaque control. Advances in Dental Research. 11: 176-185.

Marsh, P. D., and D. J. Bradshaw. 1998. Microbial community aspects in dental plaque, p. 43-55. *In* H. J. Busscher and L. V. Evans (ed.), Oral Biofilms and Plaque Control. Harwood Academic Publishers, Reading.

Marsh, P. D., and M. V. Martin. 1999. Oral microbiology. Fourth edition. Wright, Bristol.

Marsh, P. D., A. S. McKee, and A. S. McDermid. 1993. Continuous culture studies., p. 105-123. *In* H. N. Shah, D. Mayrand, and R. J. Genco (ed.), Biology of the species *Porphyromonas gingivalis*. CRC Press, Boca Raton.

Marsh, P. D., R. S. Percival, and S. J. Challacombe. 1992. The influence of denture wearing and age on the oral microflora. Journal of Dental Research. 71: 1374-1381.

McCarthy, C., M. L. Snyder, and R. B. Parker. 1965. The indigenous oral flora of man. I. The newborn to the 1-year-old infant. Archives of Oral Biology. 10: 61-70.

McDermid, A. S., A. S. McKee, and P. D. Marsh. 1988. Effect of environmental pH on enzyme activity and growth of *Bacteroides gingivalis* W50. Infection and Immunity. 56: 1096-1100.

Mettraux, G. R., F. A. Gusberti, and H. Graf. 1984. Oxygen tension (pO_2) in untreated human periodontal pockets. Journal of Periodontology. 55: 516-521.

Mikx, F. H. M., and J. S. van der Hoeven. 1975. Symbiosis of *Streptococcus mutans* and *Veillonella alcalescens* in mixed continuous culture. Archives of Oral Biology. 20: 407-410.

Mikx, F. H. M., J. S. van der Hoeven, K. G. König, A. J. M. Plasschaert, and B. Guggenheim. 1972. Establishment of defined microbial ecosystems in germ-free rats. I. the effect of the interaction of *Streptococcus mutans* or *Streptococcus sanguis* with *Veillonella alcalescens* on plaque formation and caries activity. Caries Research. 6: 211-223.

Millward, T. A., and M. Wilson. 1989. The effect of chlorhexidine on *Streptococcus sanguis* biofilms. Microbios. 58: 155-164.

Milnes, A. R., G. H. Bowden, D. Gates, and R. Tate. 1993. Normal microbiota on the teeth of preschool children. Microbial Ecology in Health and Disease. 6: 213-227.

Milnes, A. R., G. H. Bowden, D. Gates, and R. Tate. 1993. Predominant cultivable microorganisms on the tongue of pre-school children. Microbial Ecology in Health and Disease. 6: 229-235.

Moore, W. E. C., J. A. Burmeister, C. N. Brooks, R.R. Ranney, K.H. Hinkelmann,

R.M. Schieken, and L.V. Moore. 1993. Investigation of the influences of puberty, genetics and environment on the composition of subgingival periodontal floras. Infection and Immunity. 61: 2891-2898.

Moore, W. E. C., and L. V. H. Moore. 1994. The bacteria of periodontal diseases. Periodontology 2000. 5: 66-77.

Mukherjee, S. 1981. The temperature of the periodontal pockets. Journal of Clinical Periodontology. 8: 17-20.

Mukherjee, S. 1985. The role of crevicular iron in periodontal disease. Journal of Periodontology. 56 (Suppl): 22-27.

Murakami, Y., H. Nagata, A. Amano, M. Takagaki, S. Shizukishi, A. Tsunemitsu, and S. Aimato. 1991. Inhibitory effects of human salivary histatins and lysozyme on coaggregation between *Porphyromonas gingivalis* and *Streptococcus mitis*. Infection and Immunity. 59: 3284-3286.

Newman, H. N. 1990. Plaque and chronic inflammatory disease. A question of ecology. Journal of Clinical Periodontology. 17: 533-541.

Newman, H.N. and Wilson, M. (eds.). 1999. Dental plaque revisited. BioLine, Cardiff.

Nickel, J. C., J. B. Wright, I. Ruseska, T. J. Marrie, C. Whitfield, and J. W. Costerton. 1985. Antibiotic resistance of *Pseudomonas aeruginosa* colonizing a urinary catheter *in vitro*. European Journal of Clinical Microbiology. 4: 213-218.

Novak, M. J. (ed.). 1997. Biofilms on oral surfaces: Implications for health and disease., vol. 11. Advances in Dental Research.

Ohta, H., J. C. Gottschal, K. Fukui, and K. Kato. 1990. Interrelationships between *Wolinella recta* and *Streptococcus sanguis* in mixed continuous cultures. Microbial Ecology in Health and Disease. 3: 237-244.

Oshowa, A., D. Gillam, A. Botha, M. Tunio, J. Holton, P. Boulos, and M. Hobsley. 1998. *Helicobacter pylori*: the mouth, stomach and gut axis. Annals of Periodontology. 3: 276-280.

Pearce, C., G. H. Bowden, M. Evans, S. P. Fitzsimmons, J. Johnson, M. J. Sheridan, R. Wientzen, and M. F. Cole. 1995. Identification of pioneer viridans streptococci in the oral cavity of human neonates. Journal of Medical Microbiology. 42: 67-72.

Percival, R. S., S. J. Challacombe, and P. D. Marsh. 1991. Age-related microbiological changes in the salivary and plaque microflora of healthy adults. Journal of Medical Microbiology. 35: 5-11.

Percival, R. S., P. D. Marsh, D. A. Devine, M. Rangarajan, J. Aduse-Opoku, P. Shepherd, and M. A. Curtis. 1999. Effect of temperature on growth, hemagglutination, and protease activity of *Porphyromonas gingivalis*. Infection and Immunity. 67: 1917-1921.

Raup, D. A. 1989. The case for extraterrestial causes of extinction. Philosophical Transactions of the Royal Society of London B Biological Science. 325: 421-431.

Reynolds, E. C., and A. del Rio. 1984. Effect of casein and whey-protein solutions on caries experience and feeding patterns of the rat. Archives of Oral Biology. 29: 927-933.

Rosebury, T. 1962. Micro-organisms indigenous to man. McGraw-Hill, New York.

Russell, R. R. B. 1994. The application of molecular genetics to the microbiology of dental caries. Caries Research. 28: 69-82.

Sanders, C. C., and W. E. Sanders. 1982. Enocin: an antibiotic produced by *Streptococcus salivarius* that may contribute to protection against infections due to group A streptococci. Journal of Infectious Disease. 146: 683-690.

Sanders, W. E., and C. C. Sanders. 1984. Modification of normal flora by antibiotics: effects on individuals and the environment, p. 217-241. *In* R. K. Koot and M. A. Sande (ed.), New dimensions in Antimicrobial Chemotherapy. Churchill Livingstone, New York.

Sansone, C., J. van Houte, K. Joshipura, R. Kent, and H. C. Margolis. 1993. The association of mutans streptococci and non-mutans streptococci capable of acidogenesis at a low pH with dental caries on enamel and root surfaces. Journal of Dental Research. 72: 508-516.

Scannapieco, F. A. 1994. Saliva-bacterium interactions in oral microbial ecology. Critical Reviews in Oral Biology and Medicine. 5: 203-248.

Scannapieco, F. A., and J. M. Mylotte. 1996. Relationships between periodontal disease and bacterial pneumonia. Journal of Periodontology. 67: 1114-1122.

Schachtele, C. F., S. K. Harlander, D. W. Fuller, P. K. Zolinder, and W.-L. S. Leung. 1976. Bacterial interference with sucrose-dependent adhesion of oral streptococci., p. 401-412. *In* H. M. Stiles, W. J. Loesche, and T. C. O'Brien (ed.), Proceedings, Microbial Aspects of Dental Caries, vol. 2. Information Retrieval, New York.

Schachtele, C. F., R. H. Staat, and S. K. Harlander. 1975. Dextranases from oral bacteria: inhibition of water-insoluble glucan production and adherence to smooth surfaces by *Streptococcus mutans*. Infection and Immunity. 12: 309-317.

Schenkein, H. A., J. A. Burmeister, and T. E. Koertge. 1993. The influence of race and gender on periodontal microflora. Journal of Periodontology. 64: 292-296.

Schüpbach, P., J. R. Neeser, M. Golliard, M. Rouvet, and B. Guggenheim. 1996. Incorporation of caseinoglycomacropeptide and caseinophosphopeptide into the salivary pellicle inhibits adherence of mutans streptococci. Journal of Dental Research. 75: 1779-1788.

Slots, J. 1977. Microflora in the healthy gingival sulcus in man. Scandinavian Journal of Dental Research. 85: 247-254.

Slots, J., and R. J. Gibbons. 1978. Attachment of *Bacteroides melaninogenicus* subsp. *asaccharolyticus* to oral surfaces and its possible role in colonization of the mouth and of periodontal pockets. Infection and Immunity. 19: 254-264.

Smith, D. J., J. M. Anderson, W. F. King, J. van Houte, and M. A. Taubman. 1993. Oral streptococcal colonization of infants. Oral Microbiology and Immunology. 8: 1-4.

Socransky, S. S., A. D. Haffajee, L. A. Ximenez-Fyvie, M. Feres, and D. Mager. 1999. Ecological considerations in the treatment of *Actinobacillus actinomycetemcomitans* and *Porphyromonas gingivalis* periodontal infections. Periodontology 2000. 20: 341-362.

Street, S., H. D. Donoghue, and G. H. Neild. 1999. *Tropheryma whippelii* DNA in saliva of healthy people. Lancet. 354: 1178-1179.

Stoodley, P., J. D. Boyle, D. DeBeer, and H. M. Lappin-Scott. 1999. Evolving perspectives of biofilm structure. Biofouling. 14: 75-90.

Sundqvist, G., J. Carlsson, B. Herrmann, and A. Tarnvik. 1985. Degradation of human immunoglobulins G and M and complement factors C3 and C5 by black-pigmented *Bacteroides*. Journal of Medical Microbiology. 19: 85-94.

Svanberg, M. 1980. *Streptococcus mutans* in plaque after mouth rinsing with buffers of varying pH values. Scandinavian Journal of Dental Research. 88: 76-78.

Tannock, G. 1995. Normal Microflora. An introduction to microbes inhabiting the human body. Chapman & Hall, London.

ter Steeg, P. F., and J. S. van der Hoeven. 1989. Development of periodontal microflora on human serum. Microbial Ecology in Health and Disease. 2: 1-10.

ter Steeg, P. F., J. S. van der Hoeven, M. H. de Jong, P. J. J. van Munster, and M. J. H. Jansen. 1987. Enrichment of subgingival microflora on human serum leading to accumulation of *Bacteroides* species, peptostreptococci and fusobacteria. Antonie van Leeuwenhoek Journal of Microbiology. 53: 261-272.

ter Steeg, P. F., J. S. van der Hoeven, M. H. de Jong, P. J. J. van Munster, and M. J. H. Jansen. 1988. Modelling the gingival pocket by enrichment of subgingival microflora in human serum in chemostats. Microbial Ecology in Health and Disease. 1: 73-84.

Theilade, E., E. Budtz-Jorgensen, and J. Theilade. 1983. Predominant cultivable microflora of plaque on removable dentures in patients with healthy oral mucosa. Archives of Oral Biology. 28: 675-680.

Theilade, E., O. Fejerskov, T. Karring, and J. Theilade. 1982. Predominant cultivable microflora of human dental fissure plaque. Infection and Immunity. 36: 977-982.

Theilade, E., O. Fejerskov, W. Prachyabrued, and M. Kilian. 1974. Microbiologic study on developing plaque in human fissures. Scandinavian Journal of Dental Research. 82: 420-427.

Vacca-Smith, A. M., and W. H. Bowen. 1995. The effect of milk and kappa casein on streptococcal glucosyltransferase. Caries Research. 29: 498-506.

Vacca-Smith, A. M., A. R. Venkitaraman, K. M. Schilling, and W. H. Bowen. 1996. Characterization of glucosyltransferase of human saliva adsorbed onto hydroxyapatite surfaces. Caries Research. 30: 354-360.

van der Hoeven, J. S. 1998. The ecology of dental plaque: the role of nutrients in the control of the oral microflora., p. 57-82. *In* H. J. Busscher and L. V. Evans (ed.), Oral biofilms and plaque control. Harwood, Amsterdam.

van der Velden, U., A. J. van Winkelhoff, F. Abbas, and J. de Graaff. 1986. The habitat of periodontopathic bacteria. Journal of Clinical Periodontology. 13: 243-248.

van der Waaij, D., J. M. Berghuis de Vries, and J. E. C. Lekker-Kerk van der Wees. 1971. Colonisation resistance of the digestive tract in conventional and antibiotic-treated mice. Journal of Hygiene. 69: 405-411.

van Houte, J., J. Russo, and K. S. Prostak. 1989. Increased pH-lowering ability of *Streptococcus mutans* cell masses associated with extracellular glucan-rich matrix material and the mechanisms involved. Journal of Dental Research. 68: 451-459.

van Palenstein Helderman, W. H. 1975. Total viable count and differential count of *Vibrio (Campylobacter) sputorum*, *Fusobacterium nucleatum*, *Selenomonas sputigena*, *Bacteroides ochraceus* and *Veillonella* in the inflamed and non inflamed gingival crevice. Journal of Periodontal Research. 10: 230-241.

Verran, J. 1998. Denture plaque, denture stomatitis and the adhesion of *Candida albicans* to inert materials, p. 175-191. *In* H. J. Busscher and L. V. Evans (ed.), Oral biofilms and plaque control. Harwood, Amsterdam.

Vroom, J. M., K. J. de Grauw, H. C. Gerritsen, D. J. Bradshaw, P. D. Marsh, G. K. Watson, C. Allison, and J. J. Birmingham. 1999. Depth penetration and detection of pH gradients in biofilms using two-photon excitation microscopy. Applied and Environmental Microbiology. 65: 3502-3511.

Wade, W. G. 1996. The role of *Eubacterium* species in periodontal disease and other oral infections. Microbial Ecology in Health and Disease. 9: 367-370.

Wade, W. 1999. Unculturable bacteria in oral biofilms, p. 313-322. *In* H. N. Newman and M. Wilson (ed.), Dental plaque revisited: oral biofilms in health and disease. BioLine, Cardiff.

Walker, G. J., and N. A. Jacques. 1987. Polysaccharides of oral streptococci, p. 39-68. *In* A. Reizer and A. Peterkofsky (ed.), Sugar Transport and Metabolism in Gram-positive Bacteria. Ellis Horwood, Chichester.

Wilson, M. 1996. Susceptibility of oral bacterial biofilms to antimicrobial agents. Journal of Medical Microbiology. 44: 79-87.

Wilson, M., M. Gibson, D. Strahan, and W. Harvey. 1992. A preliminary evaluation of the use of a redox agent in the treatment of chronic periodontitis. Journal of Periodontal Research. 27: 522-527.

Wood, S. R., J. Kirkham, P. D. Marsh, R. C. Shore, B. Nattress, and C. Robinson. 2000. Architecture of intact natural human plaque biofilms studied by confocal laser scanning microscopy. Journal of Dental Research. 79: 21-27.

Woodman, A. J., J. Vidic, H. N. Newman, and P. D. Marsh. 1985. Effect of repeated high dose prophylaxis with amoxycillin on the resident oral flora of adult volunteers. Journal of Medical Microbiology. 19: 15-23.

Wright, T. L., R. P. Ellen, J. M. Lacroix, S. Sinnadurai, and M. W. Mittelman. 1997. Effects of metronidazole on *Porphyromonas gingivalis* biofilms. Journal of Periodontal Research. 32: 473-477.

Xie, H., S. Cai, and R. J. Lamont. 1997. Environmental regulation of fimbrial gene expression in *Porphyromonas gingivalis*. Infection and Immunity. 65: 2265-2271.

Yamada, T., S. Takahashi-Abbe, and K. Abbe. 1985. Effects of oxygen on pyruvate formate-lyase *in situ* and sugar metabolism of *Streptococcus mutans* and *Streptococcus sanguis*. Infection and Immunity. 47: 129-134.

Zambon, J. J., S. G. Grossi, E. E. Machtei, A. W. Ho, R. Dunford, and R. J. Genco. 1996. Cigarette smoking increases the risk for subgingival infection with periodontal pathogens. Journal of Periodontology. 67: 1050-1054.

Zero, D. T., J. van Houte, and J. Russo. 1986. The intra-oral effect on enamel demineralization of extracellular matrix material synthesized from sucrose by *Streptococcus mutans*. Journal of Dental Research. 65: 918-923.

From: *Oral Bacterial Ecology: The Molecular Basis*
ISBN 1-898486-22-0 ©2000 Horizon Scientific Press, Wymondham, U.K.

2

GROWTH AND NUTRITION AS ECOLOGICAL FACTORS

Jan Carlsson

Contents

Introduction

Six billion human individuals inhabit our planet, and this is also the number of new microbial cells that may be produced in one to two hours in the mouth of each individual. The latter estimate refers to a study in which water, a glucose solution, or a peptone-yeast-extract-glucose solution were siphoned into the mouths of six individuals for 4 hours; all the while the subjects were chewing a piece of paraffin wax (Carlsson and Johansson, 1973). The fluid accumulated in the mouth was continuously collected. At various times after the start of each experiment the number of viable microorganisms in the fluid was determined. When water was pumped into the mouth, the microbial density of the fluid decreased during the first hour and then remained at a steady level throughout the rest of the experiment (Figure 1). The number of microorganisms in the fluid during the last two hours of the experiment was considered to mirror the yield of new microorganisms in the mouth. The glucose solution did not increase the yield of microorganisms, but the peptone-yeast-extract-glucose solution did. On our planet most people live in poverty struggling for survival. This is also the case with the microbial cells of the mouth. In the present chapter, how microorganisms manage to survive and grow in the harsh environment of the oral cavity will be discussed .

Microbial Growth

Kinetics of Microbial Growth

Microbial growth kinetics, *i.e.,* the relationship between the specific growth rate of a microbial population and the nutrient concentration, has had a central role in the evaluation of microbial physiology and ecology. For many applications the classical Monod equation has been used (Monod, 1949).

$$\mu = \mu_{max} \ \frac{A}{K_A + A}$$

where the specific growth rate (μ) is related to the concentration of a single growth-controlling nutrient (A) via two parameters, the maximum specific growth rate (μ_{max}), and the nutrient affinity constant (K_A). This constant is numerically the nutrient concentration at which $\mu = \mu_{max}/2$. Over the years many refined theoretical models of microbial growth kinetics have been suggested (Button, 1985, 1998; Kovárová-Kovar and Egli, 1998; Smith and Pippin, 1998). They are rarely supported by experimental data and usually imply that microbial populations grow in suspension with the nutrients freely accessible. These are conditions that seldom, if ever, occur in nature. Today it is realized that it may never be possible to describe the kinetics of growth with a single set of constants (Kovárová-Kovar and Egli, 1998). In most models the enormous microbial capacity for adaptation has been overlooked. For an understanding of microbial ecology, the interest should focus on the time scale of the adaptive responses and the molecular and genetic events that regulate or mediate these processes.

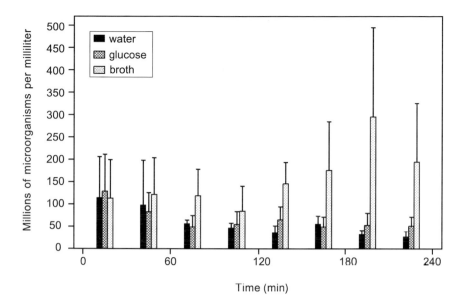

Figure 1. Feeding of the microorganisms in the mouth. Water, a 1-% glucose solution, or a peptone-yeast-extract-glucose solution was pumped into the mouths of six individuals at a rate of 25 ml·h^{-1}. The accumulated fluid in the mouth was continuously collected and the number of viable microorganisms was determined. During the last three hours of the water period the number of microorganisms in the fluid was 44 (±15; SD) million·ml^{-1} and the amount of collected fluid was 87 (±13; SD) ml·h^{-1}. The corresponding figures for the glucose period was 49 (±21; SD) million of microorganisms ml^{-1} and the amount of fluid was 92 (±7; SD) ml·h^{-1}. During the broth period the number of microorganisms significantly increased in five out of six subjects. The mean number of microorganisms was 169 (±125; SD) million·ml^{-1} and the amount of fluid was 90 (±10; SD) ml·h^{-1} (The data originate from Carlsson and Johansson, 1973).

Conditions for Growth in Saliva

In the oral cavity one would envisage that conditions close to a theoretical growth kinetic model would prevail in saliva. The microorganisms are in suspension and nutrients may be freely accessible. On a closer look, however, it is obvious that the conditions in saliva are not compatible with microbial growth because of the rheology of saliva. In the oral clearance model of Dawes (1983) the most important parameters affecting clearance are salivary flow rate, and the volumes of saliva in the mouth before swallowing (VMAX) and immediately after swallowing (RESID). In a large population the mean unstimulated salivary flow rate was 0.29 ml·min^{-1} (±0.24; SD) and mean stimulated salivary flow rate 2.25 ml·min^{-1} (±1.02; SD) (Bergdahl, 2000). In a group of 40 adult subjects Lagerlöf and Dawes (1984) found a VMAX of 1.07 ml (±0.39; SD) and RESID of 0.77 ml (±0.20; SD). When applying these values to the clearance model, one finds that the concentration of any substance in saliva is halved within 2 min by the unstimulated salivary flow and within seconds by stimulated salivary flow (Figure 2). Microbial cells, which are dispersed in saliva, will thus be eliminated from the mouth long before they have a chance to replicate.

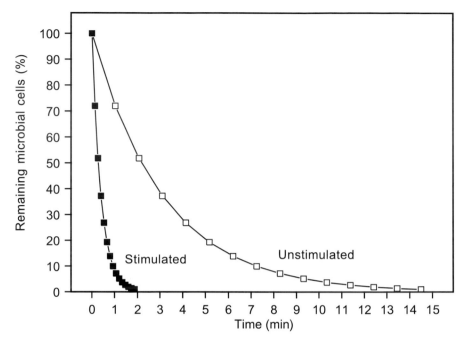

Figure 2. Clearance from the mouth of microbial cells dispersed in saliva. Details are provided in the text.

The only microorganisms that may be seen to replicate when present in saliva may be those that shuttle between a planktonic state in saliva and a sessile state on some surface in the mouth.

Initial Growth and Replication of Microorganisms on Tooth Surfaces

Microorganisms of clinical significance in caries and periodontal disease are most often firmly attached to the teeth. Determination of the growth and replication of these microorganisms is an intricate task. In the colonization of a cleaned tooth surface there is an initial phase of deposition of microorganisms from saliva, followed by growth and replication of the deposited microorganisms, and abrasion of accumulated microbial mass by external shear forces. One of the first well-controlled studies on microbial growth on teeth was done in germ-free rats (Beckers and van der Hoeven, 1982). After inoculation of the oral cavity of the rats with *Streptococcus mutans* there was an initial 2-h period during which there was a decrease in the number of microorganisms on the tooth surfaces. After that, the number of microorganisms increased corresponding to a doubling time of 1.1 h (period 2 to 12 h). In the following 24 h, the doubling time was 7.5 h.

In a study of the accumulation of various streptococcal species in palatal grooves on the molar teeth of conventional monkeys (*Macaca fascicularis*), Beighton and Hayday (1986) found median doubling times of *S. mutans,* and sanguis streptococci *(S. sanguinis, S. gordonii* and *S. oralis)* to be around 4 h during the first 18-h period. Fasting of the animals during that period did not influence the doubling time.

In a study of the colonization of the buccal surface of the upper first molar of one individual, Socransky and co-workers (1977) found the doubling time of sanguis streptococci to be 3 to 4 h during the second day after cleaning of the tooth surface. In a more recent study (Weiger *et al.*, 1995) of six teeth in each of 13 individuals the median doubling time of microorganisms on tooth surfaces was 0.8 h for the interval (1 - 4 h), 2.2 h for the second interval (4 – 8 h), 7.5 h for the third interval (8 – 24 h), and 14.8 h for the fourth interval (24 – 48 h). The main problem in this type of study is to distinguish between the various phases of tooth surface colonization, *i.e.,* deposition of microorganisms on the tooth surface, actual replication of the deposited microorganisms, and abrasion of microorganisms from the tooth surface.

A study to approach this problem was designed by Bloomquist and co-workers (1996). Enamel pieces were placed on the buccal surfaces of teeth in the upper jaw. After various time intervals the enamel pieces were removed from the oral cavity and were exposed to a radiolabeled nucleoside for 30 min. The incorporation of the nucleoside was taken as a measure of DNA synthesis and with concurrent determination of viable cell count the growth of the microorganisms was estimated. The results showed that *in-vivo* biofilm formation began with a rapid attachment of microorganisms until 12 to 32% of the enamel surface was covered (2.5 to 6.3 x 10^5 cells per mm^2). The predominant microorganisms were sanguis streptococci. At these densities the microorganisms incorporated low levels of radiolabeled nucleoside per viable cell. As microbial numbers reached densities between 8.0 x 10^5 and 2.0 x 10^6 cells per mm^2 of enamel surface, there was a small increase in the incorporation of radiolabeled nucleoside. At 3.5 x 10^6 to 4.5 x 10^6 cells per mm^2 of enamel surface, there was a marked increase in the incorporation of radiolabeled nucleoside (Figure 3). At higher cell densities the incorporation of radiolabeled nucleoside again decreased.

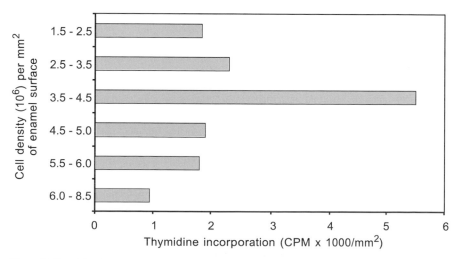

Figure 3. Thymidine incorporation of microorganisms on enamel pieces, which had been attached to the buccal surface of premolars and first molars in the upper jaw of 18 adult individuals (Adapted from Bloomquist *et al.*, 1996). Details are provided in the text.

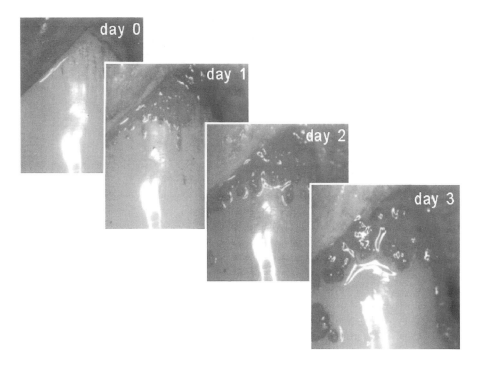

Figure 4. Biofilm formation on tooth surfaces. The subject refrained from tooth brushing and the biofilm was photographed each day after staining with a 0.1% aqueous solution of basic fuchsin. The diet was supplemented with one lump of sugar (sucrose) every hour during the daytime (Adapted from Carlsson and Egelberg, 1965).

These findings may indicate that microorganisms from saliva attached to the enamel and as the microbial densities approached two to six million microorganisms per mm^2, there was a short period of growth and replication. This growth seemed to be dependent only on cell density, since there was no correlation between the incorporation of radiolabeled nucleoside and the time the enamel pieces were present in the oral cavity (2, 4, 8, or 24 h) or on which tooth the enamel pieces were placed (Liljemark *et al.*, 1997).

Such a density-dependent growth may have various explanations. The most probable one is that the attached microorganisms at a certain density efficiently degraded the salivary components accumulated on the enamel pieces and in that way obtained all of the nutrients necessary for growth and replication. By this growth the nutrients on the enamel surface were used up and the microorganisms thereafter became dependent on nutrients from other sources. It is possible that the initiation of this density-dependent growth may be regulated by some quorum-sensing mechanism (see "Stress Handling by Microbial Communities" on page 93).

The colonization of a cleaned tooth surface, as observed in a stereomicroscope, follows a distinct morphological pattern. First a conditioned substratum comparatively free of microorganisms is deposited on the surface and then microcolonies are formed. Together with microbial deposits associated with cracks

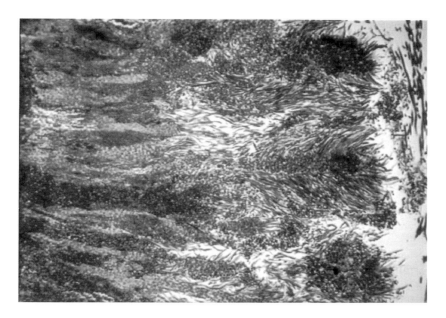

Figure 5. One-week biofilm formed on an epoxy crown in the mouth, tooth surface to the left (Listgarten *et al.,* 1975).

in the enamel surface these colonies increase in size until the entire surface is covered by a biofilm (Björn and Carlsson, 1964, Carlsson and Egelberg, 1965).

During the initial colonization of the tooth surface coccal forms predominate, and then substantial proportions of rods and filaments appear (Listgarten *et al.,* 1975; Nyvad, 1993). Within a week the microbial masses in the biofilm, as observed in the microscopic sections, have a definite structure with loosely and densely packed column of microorganisms arranged perpendicular to the tooth surface (Figure 5). Structures exposed in sections of dental biofilms like those in Figure 5 are the result of microbial adherence mechanisms, adaptation of the microorganisms to their environment, as well as microbial growth and replication. The microbial cells of such structures have the capacity to adapt not only to the actual supply of nutrients, but also to the presence of other microorganisms and to physical-chemical factors such as temperature, pH, oxidation-reduction potential, osmotic pressure, and oxygen tension. A most fascinating thing is that the microorganisms seem to coordinate the information from these various effectors into expedient responses. The various microbial populations living on the tooth surface can be considered as a community of individuals with mutual interests forming a multicellular organism (Palmer Jr, and White, 1997; Shapiro, 1998; Ben-Jacob *et al.,* 1998; Costerton *et al.,* 1999). The section of a biofilm illustrated in Figure 5 gives the appearance of a tissue structure.

However, there is a big difference between a tissue-like biofilm and a tissue of eukaryotic cells. The microorganisms in a biofilm have the ability to replicate in direct response to nutrient availability, while eukaryotic cells in a tissue only respond to nutritional or endocrinological stimuli during certain periods of development (Costerton *et al.,* 1995).

73

Multi-Nutrient Limitation of Growth

The nutritional supply of the microorganisms should meet the requirements for energy, carbon, and nitrogen, as well as essential growth factors such as minerals and vitamins. The form and quantity in which the nutrients have to be supplied is dependent on the armament of housekeeping enzymes of the individual microbial cell. In their natural habitat microbial communities are made up of many different populations that mostly live under famine conditions. The reason for this is that such communities contain many physiological types of microorganisms, which efficiently utilized the available nutrients. The experiment described in Figure 1 illustrates this multi-nutrient limitation of growth. A supply of glucose to the mouth over a four-hour period did not increase the yield of microorganisms compared with a period with only water. There may have been a shortage of sugar for the growth of many of the microorganisms, but they were also missing other nutrients and were not able to replicate when sugar was supplied. In the period with a supply of peptone-yeast-extract-glucose solution all of these nutrients become available and the microorganisms started to replicate.

Microbial Recognition of the Environmental Situation

Sensing Mechanisms

The microbial sensing of the environment is primarily dependent on the molecular diffusion of solutes to the cell and the rate of collisions between the solute molecules and the cell. According to collision frequency theory this rate is set primarily by the radius of the cell, r_x, together with the solute concentration, A, and the molecular diffusion constant of the solute, D (Koch, 1997; Button, 1998). This constant has the dimension $cm^2 \cdot s^{-1}$. When the collisions are expressed as moles of solute, the rate of collisions, v, with a single microbial cell will be

$$v = (4 \pi r_x D/1000) A = moles \cdot cell^{-1} \cdot s^{-1}$$

This would be the uptake rate of a solute, which freely passes the lipid bilayer of the cytoplasmic membrane. However, most solutes only enter the cell when they collide with cognate membrane transport proteins, and the active sites of these transport proteins only cover a part of the cell surface. The uptake rate will be determined by $N \pi r_s^2 /4 \pi r_x^2$, where N is the number of transport proteins, r_s is the radius of the effective solute collecting area of the transport protein, and r_x is the radius of the cell (Button, 1998). However, before the solutes reach the transport proteins of the cytoplasmic membrane, they have to pass cell wall structures. In Gram-positive microorganisms there is a 30- to 40-nm murein layer outside the cytoplasmic membrane. Gram-negative microorganisms are surrounded by a 2- to 2.5-nm murein layer, a periplasmic space and an outer membrane (Dmitriev *et al.,* 1999). The outer membrane lipid bilayer has lipopolysaccharides on its outer face. This shields the Gram-negative microorganisms from assault by environmental hazards like degradative enzymes while allowing solutes (nutrients) <600 Da to passively diffuse through numerous protein pores (different sizes from 10 to 20 Å in diameter) in the outer membrane. Osmotic conditions as well as nutritional limitation and excess

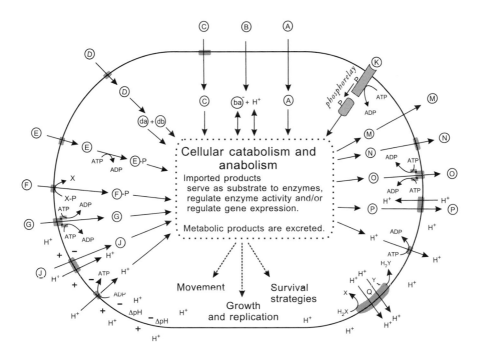

Figure 6. Microbial uptake and excretion of solutes. These solutes may serve both as nutrients and as information carriers responding to environmental changes. Details are provided in the text.

regulate the proportion of pores with larger and smaller size (Death *et al.*, 1993, Liu and Ferenci, 1998). For larger molecules like vitamin B_{12}, bacteriocins and some siderophores there is active transport through the outer membrane. In this transport, specific proteins transduce the energy of the inner membrane across the periplasmic space to a high-affinity outer membrane receptor to energize the transport (Postle, 1999).

Small uncharged molecules like O_2, CO_2, H_2O_2, ethanol and urea passively diffuse down their concentration gradient across the lipid bilayer of both the outer and cytoplasmic membranes (Figure 6; solute A). There is also a group of metabolic products of the cell, which in a similar way passively diffuse across the lipid bilayer of the cytoplasmic membrane (solute M). The best-known example is the *N*-acyl-homoserine lactones (see "*N*-Acyl-Homoserine Lactones as Pheromones" on page 99). They have the capacity to modulate discrete and diverse metabolic processes, when the cell density and consequently the concentrations of these metabolites exceed certain levels (Fuqua *et al.*, 1996). 'Quorum sensing' has been coined for this mode of cell density-dependent regulatory mechanisms. Small-undissociated carboxylic acids and hydrogen fluoride may also enter the cell by passive diffusion through the lipid bilayer of the cytoplasmic membrane (Solute B). If the pH is higher inside than outside the cell, the molecules will dissociate into charged ions and become 'trapped' within the cell. However, most other solutes have to traverse the cytoplasmic membrane through transport proteins. The driving force may be the concentration gradient of the solute across the cytoplasmic membrane (Figure 6; solute C). This type of transport serves both for uptake of nutrients and excretion of metabolic waste products (solute N). Solutes D and E enter the cell in the same way as solutes C, but they may be 'trapped' in the cell because they are cleaved or phosphorylated after entering the cell.

The driving force for uptake of solutes A to E is thus the concentration gradient of the actual solute across the cytoplasmic membrane. To create intracellular conditions conducive with growth, or as a part of a survival strategy, the cell must have the capacity to take up or excrete solutes against a concentration gradient of the solute. Such active transport requires metabolic energy in the form of energy-rich phosphate bond intermediates like ATP, or in the form of electrochemical energy stored in ion gradients across the cytoplasmic membrane. An example is the uptake of solute F. It is a sugar and its transport is energized by energy-rich phosphate bond intermediates. The sugar enters the cell phosphorylated. Many sugars are taken up by such phosphoenolpyruvate:sugar phosphotransferase systems (PTS) and these multicomponent phosphorelay systems are also involved in the control of various phases of metabolism and growth (see "Sugars" on page 98) (Postma *et al.,* 1993; Vadeboncoeur and Pelletier, 1997; Reizer *et al.,* 1998).

Many types of solutes *e.g.*, peptides, amino acids, sugars, and metal ions (Figure 6; solute G) are taken up against their concentration gradients by multicomponent transporters with ATP-binding cassettes (ABC). The driving force comes from ATP hydrolysis (Linton and Higgins, 1998). Such ABC transporters (solute O) also excrete metabolic waste products and products that may be of decisive importance in biofilm development. Such products include bacteriocins (see "Killing of Neighbors" on page 95) like mutacins of *S. mutans* (Qi *et al.,* 1999), and peptides involved in quorum sensing and development of natural competence (see "Peptides as Pheromones" on page 93) in Gram-positive microorganisms, *e.g.*, members of the mitis and anginosus groups of oral streptococci (Håverstein *et al.,* 1997). From genomics we have learnt that there are as many as 80 ABC transporters within the genome of *Escherichia coli* (Blattner *et al.,* 1997) and 77 in the genome of *Bacillus subtilis* (Kunst *et al.,* 1997). Sometimes genomics yields real surprises and new angles of approach. Thus, a gene encoding an oral streptococcal adhesin was found to be a part of an ABC transporter (Kolenbrander *et al.,* 1994; Fenno *et al.,* 1995). This transporter proved to be an inducible high-affinity uptake system for Mn^{2+} (Kolenbrander *et al.,* 1998).

When electrochemical energy is used for transport of a solute against its concentration gradient (Figure 6; solute J), the driving force can be the voltage across the membrane (the membrane potential) and/or a proton gradient or a sodium gradient (Dimroth, 1997, Kakinuma, 1998). Proton-translocating ATPases (Figure 6) can create a proton gradient across the cytoplasmic membrane (ΔpH) with the highest proton concentration outside the membrane. ATP hydrolysis is then the driving force (Konings *et al.,* 1997). When protons pass through the ATPase from outside into the cell, ATP will be synthesized (Figure 6). In the caries-inducing lactobacilli and mutans streptococci ATPases transporting protons out from the cell work more efficiently at lower pH than in any other oral microorganisms (Bender *et al.,* 1986; Sturr and Marquis, 1992; Smith *et al.,* 1996).

Solute J can represent inorganic ions, sugars, amino acids or peptides. In transport of solute J, protons may be translocated in one direction and the solute J in the other direction (antiport), or both protons and solute J are translocated in the same direction (symport). Such ion gradient-dependent transport systems are also used for excretion of solutes against a concentration gradient of the solute (solute P). Antiport systems can also be used for uptake of one solute (precursor) and the excretion of another solute (product). Examples of such precursor/product antiport systems are sugar-phosphate/phosphate, and arginine/ornithine (Konings *et al.,* 1994).

A solute can thus enter the cell with the concentration gradient as the driving force. However, it has to be noted that this is true for solutes A to E, but also for solutes F, G, and J. The integral membrane proteins of the transport systems for solutes F, G, and J can serve as facilitators for diffusion of the solutes into the cell, when the concentrations of these solutes are higher outside than inside the cell. This is probably the way most dietary sugars in high concentrations enter into microorganisms (Chaillou *et al.,* 1999), an essential step in the caries process.

By collisions with molecules from the outside and uptake of these molecules the microbial cell will obtain nutrients as well as information about its environment. On the other hand, solute K is solely dedicated to provide information about the environment (Figure 6; solute K). The solute does not enter the cell. It interacts with the exterior of a transmembrane sensor and the signal is transduced via an intracellular histidyl-aspartyl phosphorelay to a response regulator. The signal acquisition involves autophosphorylation of a histidine protein kinase of the sensor. Signal transduction takes place when the histidine protein kinase serves as a phospho-donor for the cognate response regulator. This may lead to specific alteration of gene expression or cellular activity (Egger *et al.,* 1997). This is the 'classical' two-component system, but there are also histidyl-aspartyl phosphorelay systems, in which additional signaling domains are employed, leading to histidyl-aspartyl-histidyl-aspartyl phosphorelays between the sensor and the response regulator (Fabret *et al.,* 1999; Perraud *et al.,* 1999).

Analysis of complete microbial genome sequences indicates that the number of histidyl-aspartyl phosphorelay systems in a microorganism is related to genome complexity. In *Haemophilus influenzae* and *Helicobacter pylori* there are four sensor/response regulator pairs and in *E. coli* and in *B. subtilis* there are at least 30 such pairs (Kunst *et al.,* 1997; Msadek, 1999). In *B. subtilis* these systems have overlapping roles forming signal transduction networks. This overlap between different regulons provides opportunities for signal integration and/or amplification, fine-tuning of specific responses and filtering of sub-threshold 'noise'. In such signal transduction networks there are at least three levels of regulation: (1) transmembrane sensors, allowing detection of chemicals or ligands in the environment, (2) intracellular protein-protein interactions, involving phosphorylation and dephosphorylation of regulatory proteins as well as inhibition and degradation of enzymes, and (3) protein-DNA interactions, which modulate genetic expression (Msadek, 1999).

Such signal transduction networks are successfully used by *B. subtilis* for its survival in harsh environments. The adaptive responses include synthesis of extracellular macromolecule-degrading enzymes to scavenge alternative sources of essential nutrients, secretion of antibiotics to eliminate competitors within the actual ecosystem, and induction of mobility and chemotaxis to swim towards nutrients or away from hazardous conditions. *B. subtilis* can also respond to environmental conditions (1) by development of natural competence for genetic transformation, in which specific proteins for DNA uptake, repair and recombination are synthesized, or (2) by sporulation, a complex genetic program during which the cell differentiates into a dormant, heat- and stress-resistant form (the spore), providing its genome the ultimate shelter from a hostile environment (Shapiro, 1998; Msadek, 1999). Spore-forming microorganisms are not known to colonize the healthy human mouth. A capacity to exchange genetic material by natural competence is best known among oral microorganisms in the mitis and anginosus groups of streptococci (see "Peptides as Pheromones") (Håverstein *et al.,* 1997).

Of all phosphorelay systems in microorganisms the one used for induction of mobility and chemotaxis in *E. coli/Salmonella typhimurium* has emerged as a prototype for elucidating the mechanisms by which microorganisms sense and process information about their surroundings (Silversmith and Bourret, 1999; Stock, 1999). The molecular mechanism of chemotaxis has actually been worked out to such a sophistication today that it is possible to simulate in a computer the migration of a 'virtual microbial cell' in a 'virtual gradient' and obtain rates very similar to those observed in real life (Stock, 1999). The proteins involved in these systems are highly conserved among microorganisms (Greene and Stamm, 1999; Li *et al.,* 1999). One of the proteins (CheA) of the oral spirochete *Treponema denticola* has an amino acid sequence so similar to CheA of *E. coli* that it suggests that these proteins function in an analogous manner (Greene and Stamm, 1999). Histidyl-aspartyl phosphorelay systems also have important roles in the virulence of many microorganisms, *e.g.,* the ability of microorganisms to survive and grow in the hostile site of infection. Since the histidyl-aspartyl phosphorelay systems are not found in humans and have high degrees of homology across Gram-positive and Gram-negative genera, they are obvious targets for new antimicrobial agents (Barrett and Hoch, 1998).

Competition for Nutrients and Kinetics of Nutrient Uptake

Of all ecological determinants of a microbial community the ability to compete for nutrients may be the most important. The outcome of the competition for nutrients is determined by the capacities of the uptake systems of the various populations. For nutrients crossing the lipid bilayer of the cytoplasmic membrane by passive diffusion, the rate of uptake will be the same in all cells and directly proportional to the concentration of the solute outside the cells (Figure 6; solute A). When nutrients enter the cell via transport proteins and are driven by their own concentration gradients (Figure 6; solutes C to E) or by active transport systems, the process is saturable. The uptake would then follow the Michaelis-Menten formula for enzyme kinetics,

$$v = V_{max} \frac{A}{K_m + A}$$

where v is the rate of the reaction, V_{max} is the rate when the enzyme is fully saturated, and A is the concentration of the solute. An example of such kinetics is the uptake of the tripeptide L-γ-glutamyl-L-cysteinylglycine (glutathione) by *Peptostreptococcus micros* with a K_m of 7.4 (±0.8; SD) µmol glutathione (Carlsson *et al.*, 1993) (Figure 7). The kinetics of L-cysteine uptake by *P. micros* was quite different and no maximal uptake rate could be established (Carlsson *et al.*, 1993). This is a common finding in studies of microbial nutrient uptake (Button, 1998). However, it is important to characterize the kinetics of solute uptake under such conditions in order to understand the ecological role for various nutrients. As stated above the rate of nutrient uptake (*v*) is dependent on the frequency of collision of the nutrient molecules with the transport proteins on the cell surface, and then

$$v = -\frac{dA}{dt} = a^o{}_A A X$$

where *A* is the concentration of the nutrient and *X* is the microbial mass (Button, 1985).

If one assumes that saturation of the uptake system is negligible when the nutrient concentration *A* approaches zero, the base value for *specific affinity* for the nutrient is defined as $a^o{}_A$. The dimensions of this parameter could be liters per mg of cells per hour (Button, 1985). The *affinity constant* K_A is defined as the nutrient concentration, where the base value for specific affinity is reduced by half, $a^o{}_A/2$ (Figure 7). With this model it is possible to evaluate the ability of one population to compete for nutrients with another population in a microbial community in an ecological niche even in situations where the nutrient uptake does not follow Michaelis-Menten kinetics. When comparing the uptake of L-cysteine and L-γ-glutamyl-L-cysteinylglycine by *P. micros*, the affinity constant K_A was 1800 µM for L-cysteine and 7 µM for the tripeptide L-γ-glutamyl-L-cysteinylglycine (Carlsson *et al.*, 1993). The latter value follows from the fact that the uptake of L-γ-glutamyl-L-cysteinylglycine followed Michaelis-Menten kinetics and then K_A equals K_m (Button, 1985). It is thus very likely that L-cysteine-containing peptides and not L-

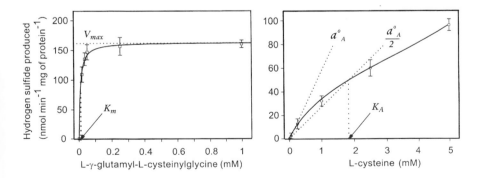

Figure 7. Uptake of L-γ-glutamyl-L-cysteinylglycine (glutathione) and L-cysteine as determined by the production of hydrogen sulfide by washed cells of *P. micros* harvested from the surface of blood agar after growth for 2 days. The microbial mass was expressed as mg of protein (The data originate from Carlsson *et al.*, 1993).

cysteine will be the L-cysteine source of *P. micros* in the periodontal pockets. It may be noted that *S. mutans* has a similar high affinity glutathione uptake system as *P. micros* (Sherrill and Fahey, 1998).

Stress Handling by Individual Microbial Cells

Coordination of the Responses in Individual Cells (Global Regulation)

The most intriguing feature about biofilms on teeth is how their structural and functional integrity can be maintained over long periods of time. Not only do all necessities for the survival of microorganisms have to be supplied, the microorganisms also have to withstand the stresses of the environment, *i.e.,* all conditions not optimal for growth. When we try to understand how this is possible, the main problem is that our knowledge about growth and nutrition of oral microorganisms is sketchy. For most of the microorganisms we do not even have a keyhole view of very fundamental processes. However, in the entire microbial world there are actually not more than a handful of microorganisms whose physiology is thoroughly understood. Moreover, this knowledge mostly refers to conditions of exponentially growing populations. Very little is known about conditions prevailing in multi-nutrient limited polymicrobial consortia such as those on the teeth.

To understand how a single microbial cell is able to handle its situation on a tooth surface is a daunting challenge; yet, some predictions can be made by using the entire microbial world as a reference. The rational behind this exercise is that recent studies on the whole genomes of a number of microorganisms have shown that there are more similarities than differences in the way various microorganisms handle their life situations. The compilation of the sequences of whole genomes of some oral microorganisms is currently in progress. The day this is accomplished the code may be interpreted and the present knowledge of microbial physiology applied to oral microorganisms in a way similar to that recently done with the stomach pathogen *H. pylori* (Marais *et al.*, 1999; Doig *et al.,* 1999). An example of what can be achieved when the whole genome is deciphered (Goffeau *et al.*, 1996) is the study of the yeast *Saccharomyces cerevisiae* (DeRisi *et al.*, 1997). In this organism it was possible to explore the changes in expression of virtually every gene (about 6000) during a metabolic shift from fermentation to respiration. For these experiments DNA microarrays ('gene chips') were used. In this technique, an area of 18 mm x 18 mm on a microscopic glass slide was covered with 6400 distinct DNA sequences from the *S. cerevisiae* genome. Messenger RNA was isolated from the growing organism and fluorescently labeled cDNA was prepared for hybridization with the DNA sequences on the slide. During exponential growth of *S. cerevisiae* in a glucose-rich medium the global pattern of gene expression was remarkably stable. Messenger RNA levels differed by a factor of two or more for only 19 genes (0.3%). When glucose was progressively depleted from the growth medium, a marked change was seen in the global pattern of gene expression. Messenger RNA levels for 710 genes were induced by a factor of at least two, and the mRNA levels for 1030 genes declined by a factor of at least two. Messenger RNA levels for 183 genes increased by a factor of at least four, and mRNA levels for 203 genes diminished by a factor of at least four. One can easily envisage how powerful DNA microarray techniques will be for the evaluation of changes in gene expression induced in an organism by environmental signals (Gerhold *et al.*, 1999). A valuable complement is a technique where the level of specific mRNA in individual microbial cells can be studied (Tolker-Nielsen *et al.*, 1997; Holmstrøm *et al.*, 1999).

An international consortium has been organized to generate deletion strains for each of the more than 6000 genes of the *S. cerevisiae* genome. In a first report a collection of isogenic strains with a precise deletion of one of 2026 genes was studied (Winzeler *et al.*, 1999). Of the deleted genes 17 per cent were essential for viability in rich media. In these studies all the strains were provided with two 'molecular barcodes'. These barcodes are unique 20-base oligomer sequences and serve as strain identifiers. This allows studies where the deletion strains can be pooled and analyzed in parallel in competitive growth assays. In one study 558 strains were pooled and grown in rich and minimal media for about 60 generations. Almost 40 per cent of the strains then showed some sort of growth defect (Winzeler *et al.*, 1999). It is obvious that genes encoding proteins essential for viability, and lacking human homologs are very interesting as targets for antifungal drugs. A set of isogenic mutants involving each of the 6000 genes of *S. cerevisiae* will provide a solid foundation for future studies of yeast physiology, pathology and ecology. The barcode technology as such can be a very useful tool in many types of microbial ecological studies, including those in oral microbiology.

A metabolic shift from fermentation to respiration studied by DeRisi and co-workers (1997) is only one of numerous responses an organism has to adapt to in order to survive in its natural environment. The cellular machinery also has to adjust to various types of feast and famine conditions, shifts from aerobic to anaerobic conditions, and rapid fluctuations of pH, temperature, and osmotic conditions. An immediate response to environmental signals can often be achieved by regulation of the activities of preformed enzymes. This may involve protein-protein interactions, allosteric changes generated by ligand binding, and chemical modifications such as phosphorylation. However, continual changes in the environment have to be accommodated by alterations in the pattern of gene expression. The key unit is then a gene with its promoter site. However, an environmental signal usually requires changes in expression of multiple genes. This can be accomplished in various ways. It can be via an operon, where all related genes are located adjacent to each other in the chromosome and transcribed as a single transcript controlled by a single promoter site (Figure 8; 1). A regulatory unit can also be genes situated in different locations on the chromosome or on extrachromosomal elements as single genes or as operons. If they all have promoter sites that respond to a single regulatory protein, it is called a regulon. A number of genes may respond to the same signal (*e.g.*, nutrient limitation as was the case with *S. cerevisiae* in the example given above), but some genes are transcribed at an elevated level, whereas other genes are transcribed at a reduced level. Such a set of genes is called a stimulon (Moat and Foster, 1995).

One of the fundamental features of the regulation of gene expression in a microbial cell is the fact that one RNA polymerase transcribes all genes (Figure 8; RNAP). This enzyme is made up of four essential subunits, two copies of α, and one copy each of β and β', and one copy of σ (sigma). The enzyme without σ is called the core RNA polymerase (E). The σ subunit directs the enzyme to the promoters. Most microorganisms contain multiple σ factors including a number of σ factors activated by specific stress conditions and a housekeeping σ factor responsible for transcribing genes of the essential biosynthetic pathways, such as genes required for biosynthesis of amino acids, nucleotides, enzyme cofactors, cell wall and membrane components, as well as those required for carbon source utilization (Hughes and Mathee, 1998). In *B. subtilis* there are at least 14 different σ factors. In *E. coli* the enzyme bearing the σ^{70} factor is the housekeeping enzyme. The alternative sigma factors conferring differential promoter specificity to RNA polymerase allow sets of genes (regulon) to be coordinately regulated in response to specific environmental conditions. A response to ammonia limitation may involve sigma factor σ^{54}, tolerance to abrupt temperature changes sigma factors σ^{32} and σ^E, and a switch to maintenance metabolism in stationary phase of growth as well as stress protection, sigma factor σ^s (Wösten, 1998).

In the transcription of a gene, RNA polymerase first recognizes and binds to the double-stranded promoter DNA, forming a complex that is referred to as the 'closed' complex. This complex isomerizes to form a transcriptionally active open complex in which the DNA strands are melted. The next steps are nucleoside triphosphate binding, release of the sigma factor, and when the transcript is 8 to 9 nucleotides in length the enzyme complex escapes from the promoter (DeHaseth *et al.*, 1998; Mooney *et al.*, 1998). The recruitment of the transcriptional machinery to a gene is thus primarily determined by the interaction between the sigma factor of the RNA polymerase and the cognate promoter. It is also influenced by the interaction of DNA-binding proteins, transcriptional activators and repressors (Figure 8; A and O) adjacent to the promoter site (Figure 8; P). The RNA polymerase activity is also influenced by anti-σ factors and anti-anti-σ factors with the ability to modulate their cognate σ factors to compete for core RNA polymerase (Figure 8; anti-sigma). Some of these anti-σ factors are located in the cytoplasmic membrane, and are presumed to serve as sensors and signaling molecules for extracellular changes (Hughes and Mathee, 1998; Missiakas and Raina, 1998; Helmann, 1999).

The transcriptional activator usually binds to a site on the DNA upstream of the promoter site, contacts RNA polymerase, and assists in transcriptional initiation by facilitating closed complex formation, open complex formation, or the escape of RNA polymerase from the promoter (Hochschild and Dove, 1998; Struhl, 1999). The rate of transcription is further influenced by regulatory elements that target the transcription elongation complex or alter termination efficiency of the transcript (Weisberg and Gottesman, 1999). The repressors bind to the promoter site or regions downstream of the promoter site (Figure 8; O) and in this way inhibit binding of RNA polymerase or prevent bound RNA polymerase from starting transcription (Rojo, 1999; Struhl, 1999).

Starvation

Microorganisms in biofilms on teeth are likely starved for various combinations of nutrients. Intracellular accumulation of polysaccharides is a common finding in biofilms on teeth (Gibbons and Socransky, 1962) and is a sign of limited supply of some nutrients in the presence of an excess of sugar (Holme and Palmstierna, 1956). Another indication of starving conditions is the thickening of the cell wall of Gram-positive organisms (Shockman *et al.*, 1976). In the inner layers of the biofilms on teeth there are many thick-walled bacteria with intracellular polysaccharide storage granules (van Houte and Saxton, 1971). It is known, however, that cell wall thickening does not occur in an organism at the same time as accumulation of intracellular polysaccharides (Shockman *et al.*, 1976). The question is whether the microorganisms in biofilms with intracellular polysaccharides and thick walls are normally thick walled, or if they have gone through two sequential processes during the organization of the biofilm; one resulting in thick cell walls and the other favoring formation of intracellular polysaccharides. At present, these two alternatives cannot be discriminated. Before such a question can be approached we need more basic information on stress handling in these microorganisms. In *S. mutans* there is an intracellular α-amylase involved in turnover of intracellular polysaccharide (Simpson and Russell, 1998b). When the gene (*amy*) coding for this enzyme was knocked out, there was a three-fold increase in the amount of intracellular polysaccharide, and the mutant was still able to digest and produce acid from intracellular polysaccharides. In *S. mutans* there is also a locus *dlt* coding for products involved in synthesis, storage, and/or breakdown of intracellular polysaccharide, as well as in synthesis of D-alanyl-lipoteichoic acid (Spatafora *et al.*, 1999). A similar locus has been found in *Lactobacillus casei* (Debabov *et al.*, 1996) and in *B. subtilis* (Perego *et al.*, 1995). The expression of the genes of this locus is dependent on cell density and is subjected to regulatory control by sugars transported by the phosphotransferase system (PTS) but not regulated by sugars transported by other systems (Spatafora *et al.*, 1999). It is suggested that membrane-associated D-alanyl-lipoteichoic acids have roles in sensing cell density by being a part of a histidyl-aspartyl phosphorelay system regulating the accumulation of intracellular polysaccharides. The pattern of expression of the *dlt* locus in *S. mutans* is similar to that of the *dlt* operon of *B. subtilis,* where the modulation of transcription is a stress response (Perego *et al.*, 1995). It may be noted that glycogen synthesis and its catabolism in enteric bacteria are under the control of the *Csr* (carbon storage regulator) system. This global regulation system is unique and controls posttranscriptionally the expression of the genes by modulating the stability of mRNA. The system includes CsrA and *CsrB*. CsrA binds to specific mRNA transcripts and facilitates their decay. *CsrB* is a non-coding RNA molecule, which forms a globular complex with 18 CsrA polypeptides and antagonizes CsrA activity (Romeo, 1998).

Storage of intracellular polysaccharides can thus be a part of a stress response. Stress responses have been thoroughly studied in enteric bacteria. The sigma factor σ^S is a key factor in many of these responses. It is a rapidly inducible emergency co-ordinator and a master regulator of a long-term adaptation process with complex physiological consequences (Hengge-Aronis, 1999). In rapidly growing *E. coli* there is very little, if any, σ^S and most of the core RNA polymerase is complexed to the housekeeping σ^{70}, $E\sigma^{70}$. Under starving conditions the level of σ^S increases. This allows the formation of $E\sigma^S$ and a large number of σ^S-dependent genes are transcribed (Figure 8; 1 the σ^S regulon). The products of these genes are crucial for the survival of the organisms. However, $E\sigma^{70}$ will still be responsible for the maintenance of a certain number of housekeeping functions and even control of some stress-induced

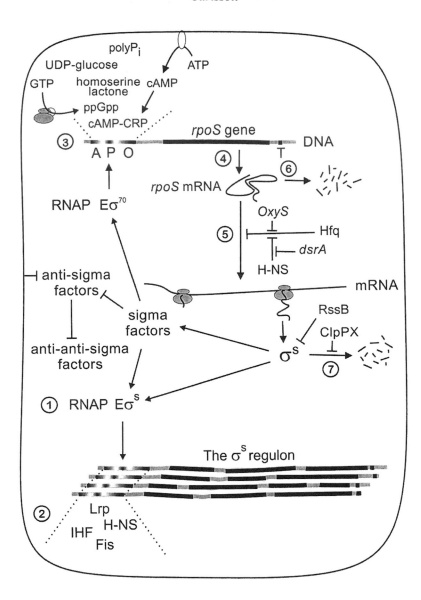

Figure 8. Interplay of some global regulators in a stress response, which is dependent on starvation sigma factor, σ^S, in *E. coli*. Details are provided in the text.

genes. There is thus a competition between σ^{70} and σ^S for the core enzyme of RNA polymerase. It may also be noted that the promoter sequences of genes controlled by $E\sigma^S$ are not clearly different from those transcribed by $E\sigma^{70}$. Various combinations of a number of global regulators (Figure 8; 2) such as the histone-like nucleoid structuring protein (H-NS), leucine-responsive regulatory protein (Lrp), cAMP receptor protein (CRP), integration host factor (IHF) and the factor for inversion stimulation (Fis) contribute to the recognition of specific promoters by Eσ^S and Eσ^{70}. The sophistication of this regulation can really be anticipated, when one recognizes that the cellular levels of each of these global regulators are also subjected to intricate metabolic control (Calvo and Matthews, 1994; Nyström, 1994; Aviv *et al.*, 1994; Williams and Rimsky, 1997; Loewen *et al.*, 1998; Hengge-Aronis, 1999; Dorman *et al.*, 1999).

Modulation of the cellular level of σ^S is a primary event in the adaptation to starvation in enteric bacteria. The level is determined by series of mechanisms that affect transcription of the gene for the σ^S protein *rpoS*, translation of *rpoS* mRNA, and stability of the σ^S protein (Lange and Hengge-Aronis, 1994; Loewen *et al.*, 1998). The transcription of the *rpoS* gene seems to be influenced by several factors (Figure 8; 3), where guanosine 3',5'-bispyrophosphate (ppGpp) and possibly inorganic polyphosphate, homoserine lactone (non-acylated), and UDP-glucose act as activators, and cAMP-CRP as inhibitor (Loewen *et al.*, 1998).

The formation of ppGpp and pppGpp in the cell can be a response to starvation for amino acids, and is catalyzed by the RelA protein when associated with ribosomes in the presence of non-aminoacylated tRNA. The formation of (p)ppGpp may also be catalyzed by SpoT protein as a response to fluctuations in the intracellular energy pool. ppGpp may modulate transcription of some operons by binding directly to RNA polymerase resulting in decreased synthesis of stable RNA, ribosomal proteins, peptidoglycan, phospholipids, and lipids. It may also increase proteolysis and synthesis of some amino acids (Cassels *et al.*, 1995; Condon *et al.*, 1995; Martínez-Costa *et al.*, 1998; Sze and Shingler, 1999). How ppGpp works in activating transcription of *rpoS* is not yet established.

Translation of *rpoS* mRNA is influenced by the fact that this mRNA has a self-complementary sequence, which allows an extensive branched stem and loop structure (Figure 8; 4). This means that the ribosome-binding site and the initiation codon will be inaccessible for ribosome binding. Under conditions of stress an RNA-binding protein (Hfq) may disrupt the secondary structure of *rpoS* mRNA and this allows translation (Figure 8; 5). The histone-like nucleoid structuring protein H-NS prevents this effect of Hfq. In addition, the small regulatory RNA, *dsrA* RNA, interacts with H-NS and antagonizes the role of H-NS in repressing translation of *rpoS* mRNA (Loewen *et al.*, 1998; Hengge-Aronis, 1999). In addition, the small *oxyS*, induced in a response to H_2O_2-stress (see "Reactive Oxygen Species" on page 87), binds to Hfq and prevents its effect (Zhang *et al.*, 1998; Wassarman *et al.*, 1999). The cellular level of *rpoS* mRNA and consequently its translation is also influenced by its degradation. Compared to eukaryotic mRNA, the mRNAs of prokaryotes show considerable instability with average half-lives in the order of only a few minutes (Figure 8; 6). The rate of degradation is influenced by structural features of the RNA and by environmental parameters. The mRNA degradation machinery in *E. coli* is organized in complex structures called degradosomes, which contain the enzymes ribonuclease, helicase, enolase, polyphosphate kinase, and polynucleotide phosphorylase (Rauhut and Klug, 1999). How the degradosomes handle *rpoS* mRNA awaits elucidation.

The cellular level of σ^S will finally be dependent on a controlled degradation of the σ^S protein (Figure 8; 7). In an ATP-dependent proteolysis ClpPX protease is assisted by RssB, a protein with homology to the response regulator of histidyl-aspartyl phosphorelay systems. The role of RssB is not fully established but it appears that RssB attaches to the σ^S protein and serves as a substrate recognition factor for the ClpPX protease. When RssB is phosphorylated, σ^S protein is degraded. Under certain stress conditions RssB will be dephosphorylated and the σ^S protein becomes less sensitive to degradation. Among all the control mechanisms of σ^S levels this 'final' step of controlled proteolysis seems to be the most important for the actual cellular level of σ^S (Bouché *et al.*, 1998; Loewen *et al.*, 1998; Hengge-Aronis, 1999).

The σ^S factor realm of stress response in enteric bacteria includes many complex physiological processes, which are essential for survival under various environmental conditions. This response seems to be induced whenever the conditions are not optimal for growth. It includes protein-protecting systems, membrane-protecting mechanisms and DNA repair systems (Loewen and Hengge-Aronis, 1994). For survival these cellular mechanisms have to be supplemented with protection mechanisms against extreme environmental changes in nutrient supply, temperature, osmotic pressure, pH, and oxygen level. Some of these responses are more or less universal among microorganisms, while others are only found in few of them. It will be interesting in the future to learn how oral microorganisms on a molecular level handle these situations.

The Anaerobic-Aerobic Interface
It is rare that oral microorganisms are freely exposed to molecular oxygen. Most of them live in biofilms, such as in the periodontal pockets, where anaerobic conditions prevail most of the time with only short intervals with significant levels of molecular oxygen (Loesche *et al.*, 1983). Most microorganisms colonizing the body have mechanisms to cope with the conditions at aerobic/anaerobic interfaces. The redox-status of the cell is continuously monitored and the metabolic pathways are modulated in response to fluctuations in oxygen concentration (Sawers, 1999). In *E. coli* FNR (fumarate and nitrate reductase regulation) and ArcA (aerobic respiratory control) proteins are transcriptional modulators. FNR becomes active at an intracellular molecular oxygen concentration of 1-5 µM and the Arc system at 5-10 µM (Sawers, 1999). The half-maximal synthesis of the fermentation products ethanol, acetate, and succinate is reached when the oxygen concentration in the medium is 0.2-0.4 µM, and 1 µM for formate (Becker *et al.*, 1997).

In its active form FNR has two cubic ($[4Fe\text{-}4S]^{2+}$) clusters in a dimeric protein that binds to DNA, and genes involved in fermentation or in anaerobic respiration with nitrate, nitrite, and fumarate as electron acceptors are transcribed. Upon exposure to stoichiometric levels of molecular oxygen the cubic ($[4Fe\text{-}4S]^{2+}$) clusters are converted to planar ($[2Fe\text{-}2S]^{2+}$) clusters and FNR loses its DNA-binding activity (Spiro and Guest, 1990; Unden and Schirawski, 1997; Kiley and Beinert, 1999). FNR-like proteins have been found in many Gram-negative microorganisms and also in the Gram-positive organisms, *L. casei* and *B. subtilis* (Spiro, 1994; Gostick *et al.*, 1998; Nakano and Zuber, 1998).

The Arc system is a histidyl-aspartyl phosphorelay system (see "Sensing Mechanisms"), where ArcB is a membrane-bound sensor protein that autophosphorylates upon changes in redox conditions and/or changes in intracellular levels of metabolites like pyruvate, NADH, lactate, and acetate. The sensor module in ArcB is a PAS domain and this is located in the cytosolic part of the molecule. PAS is an acronym formed from the names of the proteins in which it was first recognized: the *Drosophila* period clock protein (PER), vertebrate aryl hydrocarbon receptor nuclear translocator (ARNT), and *Drosophila* single-minded protein (SIM). PAS domains in various sensor proteins may monitor energy changes in the cells by sensing oxygen, light, redox potential, or proton motive force (Taylor and Zhulin, 1999). After the ArcB protein is autophosphorylated, the phosphoryl residue is transferred to the response regulator ArcA. ArcA then serves as a repressor of a wide variety of aerobic enzymes such as the dehydrogenases of the flavoprotein class, several enzymes of the TCA cycle and the cytochrome *o* oxidase complex (Iuchi and Lin, 1991). It may be noted that this repressor is also activated under carbon starvation (Nyström *et al.*, 1996). ArcA is also required together with FNR and IHF (integration host factor) proteins for optimal anaerobic induction of the pyruvate formate-lyase (*pfl*) operon (Knappe and Sawers, 1990; Drapal and Sawers, 1995).

In the regulation of many genes by molecular oxygen, regulation due to growth rate or growth phase is also superimposed. Regulation is thus also effected by IHF, H-NS, Fis (factor of inversion stimulation), and StpA (suppressor of td phenotype) (Unden and Bongaerts, 1997). The FNR and Arc systems and the global regulators thus provides *E. coli* with a coordinated network of control that allows it to respond rapidly and efficiently to a wide range of concentrations of molecular oxygen in its environment (Unden *et al.*, 1994; Iuchi and Lin, 1991; Sawers, 1999).

Reactive Oxygen Species

Of all the solutes oral microorganisms have to handle, molecular oxygen may be the one that requires the most incessant monitoring. The reason for this is that molecular oxygen freely traverses the cell envelope into the cell. In this molecule one electron pair is shared, while there are two orbitals with only one electron. Oxygen is thus a diradical, but as such it is quite sluggish in its reaction, because the two unpaired electrons have the same spin quantum number, 'parallel spin' (Figure 9). This imposes a restriction on the reactions of the molecule since it will usually accept only one electron at a time. However, the reactivity of oxygen is significantly increased, if one of the unpaired electrons moves to the other orbital with one electron forming singlet oxygen (Figure 9) (Naqui, *et al.*, 1986). When the first form of life appeared on our planet four billion years ago, the atmosphere contained hydrogen, helium, ammonia, water, and methane, but no oxygen. Not until two billion years ago did significant quantities of oxygen appear in the atmosphere (Tappan, 1974). Some organisms then took advantage of oxygen as an electron sink in their energy metabolism, but all organisms had to cope with the toxicity of the products formed from molecular oxygen.

The intracellular milieu of all organisms has a surplus of electrons. Electrons suitable for reduction of molecular oxygen are present in various cell constituents. When one electron is taken up by molecular oxygen, the superoxide anion radical ($O_2\cdot^-$) is formed. $O_2\cdot^-$ has one unpaired electron and is much more reactive than oxygen. When another electron is taken up by $O_2\cdot^-$, hydrogen peroxide (H_2O_2) is

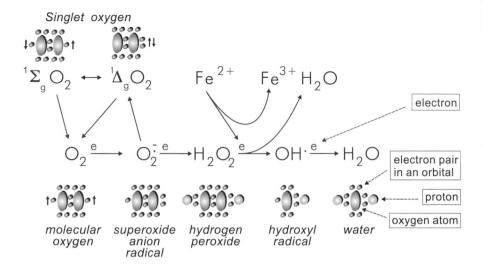

Figure 9. Conversion of molecular oxygen into reactive oxygen species when oxygen enters the cytoplasm of a living cell with a surplus of electrons.

formed. Single electrons in transitional metals are usually required for reduction of H_2O_2. In living cells ferrous iron usually reduces H_2O_2 in the so-called Fenton reaction to form ferric iron and the extremely toxic hydroxyl radical ($\cdot OH$). The reactivity of $\cdot OH$ is so great that it never diffuses away from its site of formation. It immediately reacts with molecules in its vicinity. If it targets vital parts of DNA, membrane lipids, or proteins, the consequences could be devastating.

Very few oral microorganisms, if any, make use of molecular oxygen as the normal electron sink of their energy metabolism. However, all oral microorganisms have to be prepared to encounter oxygen in their habitat. To be able to survive and grow in such an environment they have to (1) prevent intracellular formation of toxic intermediates of oxygen reduction, (2) scavenge these intermediates, or (3) repair the sites damaged by the intermediates. An example of the first level of defense is cytochrome oxidase of eukaryotic cells. This enzyme has the unique capacity of reducing O_2 to water without releasing any toxic intermediates (Naqui *et al.*, 1986). NADH oxidase/peroxidase complexes may play a similar role in both anaerobic and facultatively anaerobic microorganisms (Hoshino *et al.*, 1978, Abbe *et al.*, 1991; Matsumoto *et al.*, 1996; Caldwell and Marquis, 1999). Another example of a first level of defense is the sequestering of transitional metal ions into structures that do not reduce O_2. A recently detected microbial iron-scavenger is the DNA-binding protein Dps, which is a product both of the OxyR and the σ^s regulons of *E. coli* (Grant *et al.*, 1998; Michán *et al.*, 1999). The second level of defense is reactive oxygen species scavengers like Mn(II) and carotenoides (Archibald, 1986; Bridges and Timms, 1998). However, the most important part of the intracellular defense against oxygen toxicity would be the concerted action of the enzymes catalase, glutathione peroxidase, and superoxide dismutase (SOD) or superoxide reductase (SOR; Figure 10). NADH peroxidase may be an alternative or a supplement to

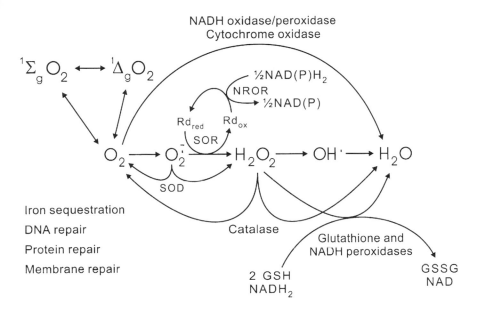

Figure 10. Cellular defense mechanisms against reactive oxygen species. SOD, superoxide dismutase; SOR, superoxide reductase; Rd_{red}, reduced rubredoxine; Rd_{ox}, oxidized rubredoxine; NROR, reduced nicotinamide adenine dinucleotide phosphate (NAD(P)H)-rubredoxine oxidoreductase; GSH, reduced glutathione; GSSG, oxidized glutathione.

glutathione peroxidase in many microorganisms (Ross and Claiborne, 1997). However, it may be noted that DNA repair mechanisms sometimes can be more important than superoxide dismutase and catalase in protecting microorganisms from being killed by oxygen reactive species (Carlsson and Carpenter, 1980).

When these enzymes and the intracellular iron-sequestering compounds work efficiently, the hydrogen peroxide level will be kept low and no hydroxyl radicals will be formed. Most aerobic and many anaerobic microorganisms colonizing the human body express superoxide dismutase activity (Gregory *et al.*, 1978; Amano *et al.*, 1986). However, in the complete genome sequences for anaerobic organisms now available, *i.e.*, *Methanococcus jannaschii*, *Archaeoglobus fugidus*, *Pyrococcus horikoshi*, *P. abysii*, and *Thermotoga maritime*, there are no genes for superoxide dismutase or catalase (Jenney Jr. *et al.*, 1999). What appeared to be superoxide dismutase activity in these anaerobic organisms was instead superoxide reductase (SOR) activity (Jenney Jr. *et al.*, 1999). It will be interesting to learn whether the anaerobic organisms colonizing the human body express superoxide reductase or superoxide dismutase. Catalase activity is a trait often used in taxonomy, but it is not very common among oral microorganisms.

Not only oxygen, but also the uncharged H_2O_2 molecule diffuses freely through cytoplasmic membranes. This fact is a potential threat against oral microorganisms. Among those that colonize oral surfaces the mitis group of oral streptococci, *i.e.*, *S.*

sanguinis, S. gordonii, S. oralis, S. parasanguinis, and *S. mitis,* have a quite unique mechanisms to produce hydrogen peroxide (Carlsson and Edlund, 1987). They express a cytoplasmic pyruvate oxidase that catalyzes the reaction:

$$pyruvate + O_2 + inorganic\ phosphate \rightarrow H_2O_2 + CO_2 + acetyl\ phosphate$$

and they get energy from acetyl phosphate in the reaction

$$acetyl\ phosphate + ADP \rightarrow acetate + ATP$$

These streptococci will thus excrete acetate and H_2O_2, when they are exposed to molecular oxygen in the oral cavity in the presence of sugars. A cytoplasmic pyruvate oxidase is, in addition to the mitis group of oral streptococci, only found in some lactobacilli (Sedewitz *et al.*, 1984) and in *Treponema pallidum* (Barbieri and Cox, 1979). Interestingly, the enzyme was not recognized in the complete genome sequence of *T. pallidum* (Fraser *et al.*, 1998). Any microorganisms and mucosal cells in the vicinity of these oral streptococci thus run the risk of being exposed to high levels of H_2O_2. However, it is very likely that any such H_2O_2 exposure is prevented by salivary peroxidase. This enzyme is secreted by the salivary glands together with thiocyanate ions (SCN^-) and catalyzes the reaction

$$SCN^- + H_2O_2 \rightarrow OSCN^- + H_2O \text{ (Aune and Thomas, 1977)}$$

The hypothiocyanite ion ($OSCN^-$) is less toxic than H_2O_2 since unlike H_2O_2 it is not a source of toxic hydroxyl radicals. The reaction of hypothiocyanite ion in the cell is usually restricted to oxidation of sulfhydryl groups to sulfenyl-thiocyanate derivatives, and these reactions are readily reversible (Thomas and Aune, 1978).

Each time a cell is confronted with molecular oxygen is a potentially risky situation. Oxygen will be reduced into toxic intermediates and the organism has to organize its defense against this stress situation by expressing enzymes, which scavenge the intermediates, and repair damaged sites. This defense has been best characterized in *E. coli,* where one transcriptional factor OxyR activates defense genes in a response to hydrogen peroxide, and another transcriptional factor SoxR activates genes in response to superoxide-generating compounds.

Upon exposure to H_2O_2, disulfide bonds are formed in the tetramer OxyR and it activates the *oxyR* regulon with expression of *e.g.*, a catalase, an alkyl hydroperoxide reductase, a glutathione reductase, glutaredoxin 1, and a small nontranslated regulatory RNA *oxyS* (Wassarman *et al.*, 1999). The entire response is autoregulated, since OxyR is deactivated by reduction of its disulfide bonds by the glutaredoxin system (glutaredoxin 1 and glutathione reductase) expressed in the presence of active OxyR (Zheng *et al.*, 1998).

The transcriptional factor SoxR contains two [2Fe-2S] centers per dimer. It appears to be activated by oxidation of $[2Fe\text{-}2S]^{1+}$ to $[2Fe\text{-}2S]^{2+}$. The SoxR is involved in a two-stage process, where the oxidized form enhances the expression of SoxS, which in turn activates the *SoxRS* regulon with expression of *e.g.*, manganese superoxide dismutase, the DNA-repair enzyme endonuclease IV, glucose-6-

phosphate dehydrogenase, and the Fur repressor. The latter decreases the uptake of iron and in that way diminishes the formation of hydroxyl radicals (see Figure 9). Although the defense against oxidative stress is the immediate concern of the *OxyR* and *SoxRS* regulons, important antioxidant activities are also provided by the *FNR*, *ArcAB*, and *rpoS* regulons in *E. coli* (Storz and Imlay, 1999).

Molecular oxygen can be expected to be an important determinant of oral microbial ecology. Sessile organisms living in biofilms with an interface to saliva are probably able to convert most molecular oxygen into water without forming reactive oxygen species. Cytochrome oxidase systems with this capacity are not known in oral microorganisms, but NADH oxidases/peroxidases may fulfill this role (Figure 10). Due to the low solubility of molecular oxygen in saliva it is also likely that only a few layers of cells living next to the salivary interface of the biofilm are required to convert molecular oxygen into water thereby creating anaerobic conditions in deeper layers of the biofilm. The same may be true for biofilms with an interface to oxygenated tissues. H_2O_2 produced by the mitis group of streptococci colonizing oral surfaces exposed to saliva may not do any harm or influence ecology because of the efficient conversion of H_2O_2 into hypothiocyanite (Adamson and Carlsson, 1982; Carlsson *et al.*, 1984). Hypothiocyanite as such may influence ecology. It perturbs glycolysis of microorganisms by inhibiting glyceraldehyde 3-phosphate dehydrogenase (Carlsson *et al.*, 1983). Among oral streptococci there are big differences in sensitivity to hypothiocyanite. The H_2O_2-forming mitis group of streptococci is very resistant because they express a NADH-$OSCN^-$ oxidoreductase, which reduces any hypothiocyanite entering the cell into the non-toxic thiocyanate. Mutans streptococci and *S. salivarius* do not have this enzyme and are very sensitive to hypothiocyanite (Carlsson *et al.*, 1983). Mutans streptococci are not able to compete successfully with the mitis group of streptococci in the interface between biofilm and saliva (van der Hoeven and Camp, 1993) due to the salivary levels of hypothiocyanite (Tenovuo *et al.*, 1982; Pruitt *et al.*, 1983). In deeper layers of the biofilm hypothiocyanite will be converted by the reducing environment into thiocyanate and will then not influence the microbial ecology.

Damaged Macromolecules

Many stress situations such as exposure to heat, oxidative agents, or changes in pH or osmotic pressure result in damage to cellular macromolecules including nucleic acids, proteins and lipids of the cytoplasmic membrane. The damage to nucleic acids and their repair are thoroughly discussed in standard textbooks and will not be elaborated upon here. Damage to the cytoplasmic membrane and its repair is a central topic in eukaryotic cell biology. However, alteration of microbial membranes and their repair are mostly overlooked in the literature. The present section will be devoted to protein damage and repair. A protein may be conformationally changed without any change in its chemical structure. There may also be a covalent damage, where the primary structure of the protein is modified (Visick and Clarke, 1995). In exponentially growing cultures such damaged proteins will be degraded by proteolysis and replaced by newly synthesized proteins. However, under conditions of starvation and growth arrest, protein repair systems permit the microorganisms to survive and make the best possible use of their existing proteins. Conformationally changed proteins can be properly refolded by the assistance of molecular chaperons (Sigler *et al.*, 1998). In almost all known stress responses and in nearly all studied microorganisms there is an increased production of such chaperones upon stress, typified by the stress (heat shock) proteins DnaK, and GroEL (Narberhaus, 1999; Derré *et al.*, 1999). In *E. coli* transcription of the heat shock regulon, consisting of over 20 genes, requires the sigma factor σ^{32}. It may be noted that not only an increased temperature but also proteins damaged by oxidation can serve as signals for induction of the heat shock regulon under growth arrest (Dukan and Nyström, 1998; Nyström, 1998; 1999). Another heat shock regulon in *E. coli* responds to unfolded proteins in the periplasm and is controlled by σ^E. The heat shock response is also dependent on alternative sigma factors in Gram-positive microorganisms (Hecker and Völker, 1998), but in many cases it is also controlled by highly conserved repressor mechanisms. One such repressor is CtsR (<u>c</u>lass <u>t</u>hree <u>s</u>tress gene <u>r</u>epressor). This protein may act as a direct heat sensor. A conformational change at elevated temperature leads to an inactive form of the repressor and derepression of the target genes (Derré *et al.*, 1999). Among oral microorganisms repressor mechanisms for stress responses have been described in *S. salivarius* (Narberhaus, 1999; Derré *et al.*, 1999). Various stress proteins have been demonstrated also in *Porphyromonas gingivalis, Bacteroides forsythus, and Actinobacillus actinomycetemcomitans* (Koga *et al.*, 1993; Lu and McBride, 1994; Hinode *et al.*, 1998).

Some proteins in the cell envelope require disulfide linkages to be functional. Such linkages are formed in *E. coli* by the enzyme thiol-disulfide oxidoreductase, DsbA. In the cytoplasm there are few functional proteins with disulfide linkages. However, oxygen-reactive species may accidentally create such linkages, and the function of the protein is inhibited. To restore the function these linkages are cleaved in a reaction catalyzed by thioredoxin reductase. Genetic and biochemical studies coupled with sequence analysis have revealed that there are ten genes in *E. coli* coding for thioredoxin-like proteins. Some of these proteins may be redundant and part of a 'backup system', but there are reasons to believe that others have specific roles to play under some, so far unknown, environmental conditions (Åslund and Beckwith, 1999a, 1999b). This is an example of the goal of functional genomics: to deduce the functions of thousands of individual open reading frames (ORFs) in organisms where the entire or essential parts of the genome have been sequenced.

Methionine residues in proteins may be oxidized to methionine sulfoxide by a large number of different oxygen reactive species, including H_2O_2, hypochlorous acid and the hydroxyl radical. This damage may be repaired in a reaction catalyzed by the enzyme peptide methionine sulfoxide reductase (Moskovitz *et al.*, 1999). Oxidative modifications of cysteine and methionine residues in proteins are the only oxidative damages that can be repaired. However, all amino acids are susceptible to oxidative damage. Aromatic amino acid residues are among the preferred targets for attack by reactive oxygen species. Tryptophan, phenylalanine, tyrosine, and histidine residues yield a number of hydroxy derivatives. Oxidation of lysine, arginine, proline, and threonine yield carbonyl derivatives. These reactions are often initiated by hydroxy radicals formed by a reaction between H_2O_2 and transitional metal ions in the vicinity of amino acids (Stadtman, 1992; Berlett and Stadtman, 1997). This lack of repair mechanisms against oxidatively damaged proteins is one

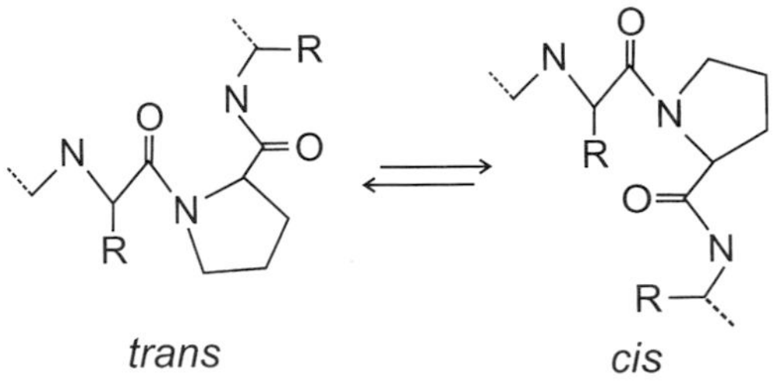

Figure 11. Isomerization of a prolyl residue in a protein.

trans   *cis*

of the reasons why the microorganisms have to have such elaborate defense systems against oxidative stress.

Another potential damage to protein is an isomerization of proline peptide bonds between *trans*- and *cis*-configurations (Figure 11). This affects both the folding of the protein and its protection against proteolysis. The enzyme peptidyl-propyl *cis-trans* isomerase catalyzes the proper proline isomerization in the initial folding of nascent proteins and in the repair of damaged proteins (Visick and Clarke, 1995). Other changes in proteins that can be detrimental to function are covalent alteration of asparaginyl and aspartyl residues with formation of succinimide derivatives and the ultimate products D- or L-aspartyl or D- or L-isoaspartyl residues (Figure 12). Of these derivatives the L-isoaspartyl residues can be repaired by an L-isoaspartyl protein methyltransferase. This enzyme has been shown to be important in the survival of *E. coli* in various stress situations (Visick *et al.*, 1998).

There are thus situations, when the organism has no way of repairing damaged proteins. The only choice is then degradation of the protein. There are various proteases induced as a response to stress situations. Some of these proteases are ATP-dependent and have also important regulatory functions (Spiess *et al.*, 1999; Gottesman, 1999).

Stress Handling by Microbial Communities

Peptides as Pheromones

To survive in its natural environment, a microorganism has to accurately monitor and rapidly respond to environmental signals. As was discussed in "Sensing Mechanisms" solutes enter the cell in various ways and provide information about the environment. The most sophisticated form of signal transduction is the histidyl-aspartyl phosphorelay system (Figure 6, K). This consists of a membrane-bound histidine protein kinase sensor and a response regulator, which modulate gene

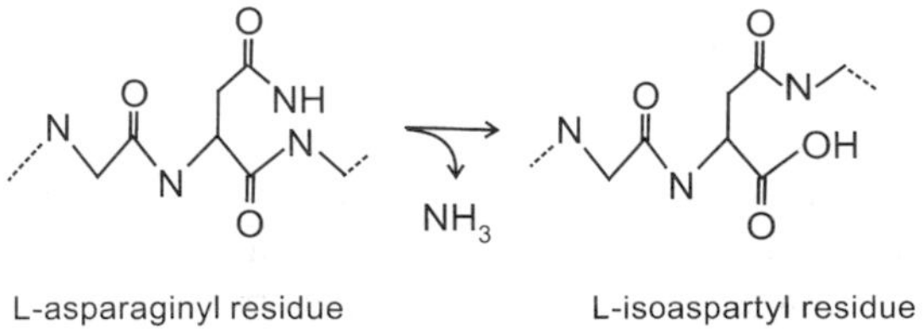

Figure 12. Deamination of asparaginyl residue followed by hydrolysis and formation of L-isoaspartyl residue in a protein.

L-asparaginyl residue   L-isoaspartyl residue

expression patterns and cell activity. Some Gram-positive microorganisms also use such histidyl-aspartyl phosphorelay systems in their monitoring of population density. They secret peptides (pheromones) and some of these peptides modulate gene expression in other cells by acting via histidyl-aspartyl phosphorelay systems (Lazazzera and Grossman, 1998). Other secreted peptides are taken up by ATP-dependent oligopeptide permeases and interact with intracellular receptors to modulate gene expression. The importance of an accurate assessment of cell population density may be illustrated by *B. subtilis* in its decision to sporulate or develop natural competence. When *B. subtilis* is crowded and starved for essential nutrients, sporulation is a reasonable response. However, under low-density conditions it may be beneficial to wait for new food sources instead of sporulating. For development of competence, high cell density implies that homologous DNA might be present as a result of cell lysis. Indeed, the first step in the development of competence is the production of surfactin synthetase, leading to synthesis of the antibiotic surfactin (see "Killing of Neighbors"), which lyses surrounding microorganisms (Dunny and Leonard, 1997; Perego, 1998; Msadek, 1999; Kroos *et al.*, 1999).

In the development of dental biofilms the ability to monitor cell density would be expected to be important. However, we are completely ignorant as to whether such a mechanism actually exists in this habitat. However, among the members of the mitis group of streptococci, *i.e.*, *S. sanguinis, S. gordonii, S. oralis, S. parasanguinis, S. pneumoniae* and *S. mitis,* and the anginosus group of streptococci, *i.e.*, *S. anginosus, S. constellatus*, and *S. intermedius* (Kawamura *et al.*, 1995) it is known that the development of natural competence is cell-density dependent. This characteristic is dependent on a regulon, which includes a strain-specific peptide pheromone, a secretion apparatus for the pheromone, and a histidyl-aspartyl phosphorelay recognition and signal transduction system for the pheromone (Håverstein *et al.*, 1996, Håverstein *et al.*, 1997; Håverstein, 1998). Although natural competence has also been demonstrated in *S. mutans* among oral streptococci (Perry and Kuramitsu, 1981), its regulation has not been examined.

N-Acyl-Homoserine Lactones as Pheromones

The signal molecules of the quorum sensing systems in Gram-negative microorganisms could be a variety of *N*-acyl-homoserine lactones (Fuqua *et al.*, 1996; Gray, 1997).

This system was first described for luminescence in *Vibrio fischeri.* The genes are organized as a single operon controlled by the transcriptional activator protein LuxR. The *luxR* gene is linked to the *lux* operon (*luxICDABEG*) but it is transcribed as a separate unit. LuxR requires an inducer *N*- (3-oxo) hexanoyl homoserine lactone to activate transcription of the *lux* operon. This homoserine lactone is synthesized from *S*-adenosylmethionine and 3-oxohexanoyl-acyl carrier protein (ACP) by the product of *luxI*, the first gene within the *lux* operon (Figure 13). *N*- (3-oxo) hexanoyl homoserine lactone can diffuse freely across the cell envelope and the concentration will be the same inside and outside the cell. When its concentration reaches a certain level the genes of the *lux* operon will be expressed. This expression is also modulated by various growth conditions and it is under the direct control of cAMP-CRP. *N*-acyl-homoserine lactones have not yet been described among oral microorganisms.

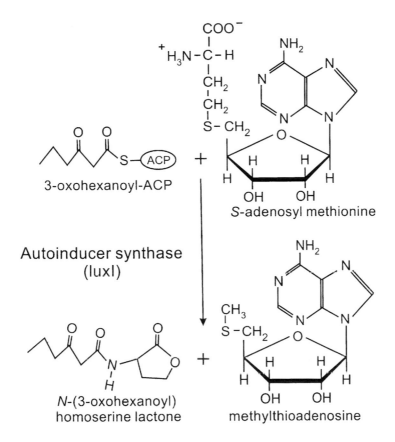

3-oxohexanoyl-ACP

S-adenosyl methionine

Autoinducer synthase
(luxI)

N-(3-oxohexanoyl)
homoserine lactone

methylthioadenosine

Figure 13. Synthesis of *N*- (3-oxo) hexanoyl homoserine lactone, a pheromone conveying quorum sensing in *Vibrio fischeri*.

Killing of Neighbors

In the search for nutrients and space an organism may have to kill some of its neighbors in the microbial community. Therefore, various antimicrobial peptides are secreted. There are principally two types of such peptides: ribosomally and non-ribosomally synthesized bactericidal agents. An example of the latter type of anti-microbial peptides is surfactin. As described in "Peptides as Pheromones" (p. 93), it is produced by *B. subtilis* as a response to high cell density. It is a lipopeptide with bactericidal and antiviral activities and exerts this effect by altering membrane integrity. However, the most conspicuous property of this molecule is its exceptional surfactant power. It lowers the surface tension of water from 72 mN·m^{-1} to 27 mN·m^{-1}. Surfactin is a heptapeptide interlinked with a β-hydroxy fatty acid to form a cyclic lactone ring structure (Figure 14). It is synthesized by the multi-enzyme complex surfactin synthetase (Desai and Banat, 1997; Peypoux *et al.*, 1999). It is open to speculation as to whether or not such lipopeptides are produced in dental biofilm.

Ribosomally synthesized antimicrobial peptides and proteins are called bacteriocins. They are usually subjected to extensive post-translational processing

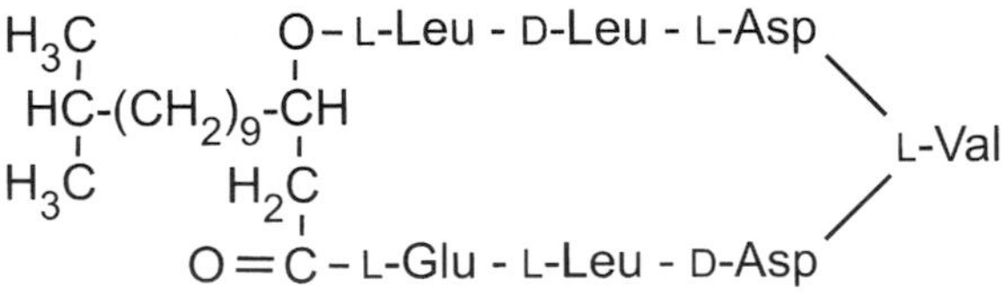

$$H_3C$$
$$HC\text{-}(CH_2)_9\text{-}CH$$
$$H_3C \quad H_2C$$
$$O\text{-}L\text{-}Leu\text{-}D\text{-}Leu\text{-}L\text{-}Asp$$
$$L\text{-}Val$$
$$O=C\text{-}L\text{-}Glu\text{-}L\text{-}Leu\text{-}D\text{-}Asp$$

Figure 14. The antimicrobial agent surfactin produced by *Bacillus subtilis*.

giving them their unique structural and toxic properties. They cause a highly selective killing of competing microorganisms but little or no harm to the host cells. In Gram-negative microorganisms these are represented by colicins and microcins, and in Gram-positive microorganisms lantibiotics and nonlantibiotic bacteriocins (Baba and Schneewind, 1998). Bacteriocins containing the thioether amino acid lanthionine are called lantibiotics (Figure 15).

The thioether bond is formed post-translationally in the polypeptide chain between serine and cysteine residues after enzymatic dehydration of the serine residue. Methyl-lanthionine is formed in the same way as lanthionine except that it is derived from a threonine residue instead of a serine residue. These thioether bonds stabilize the three-dimensional structure of the lantibiotics. Most lantibiotics are toxic because they form pores in the target cytoplasmic membrane and then cause a rapid efflux of ions, solutes and small metabolites, which leads to an inhibition of all biosynthetic processes (Nissen-Meyer and Nes, 1997; Sahl and Bierbaum, 1998).

The lantibiotics are synthesized from prepeptides with characteristic N-terminal leader peptides. The gene cluster for lantibiotic synthesis includes: (1) the structural gene encoding the preform of the bacteriocin, (2) one or more proteins involved in self-protection, 'immunity', (3) an ABC transporter for export of the bacteriocin, (4) a serine protease for cleaving off the leader peptide, (5) one or two proteins catalyzing the dehydration for lanthionine formation, and (6) a histidyl-aspartyl phosphorelay system for transmission of extracellular signals and thereby inductions of bacteriocin expression (Nissen-Meyer and Nes, 1997; Sahl and Bierbaum, 1998). As with most bacteriocins, lantibiotics are active against strains that are closely related to the producer strain. The 'immunity' peptides or proteins provide protection by attaching to the outside of the cytoplasmic membrane of the producer strain. Lantibiotics may also play a role as signal molecules between microorganisms in a community. Nisin from *Lactococcus lactis* thus serves as a signal molecule for cell density and the synthesis of nisin is an autoregulated process. When the concentration of nisin reaches a threshold level due to a certain cell density, nisin serves as an input signal to a membrane bound sensor histidine kinase and the response regulator serves as a transcriptional activator. Other lantibiotics and bacteriocins that contain phosphorelay systems in their respective gene cluster probably have a similar

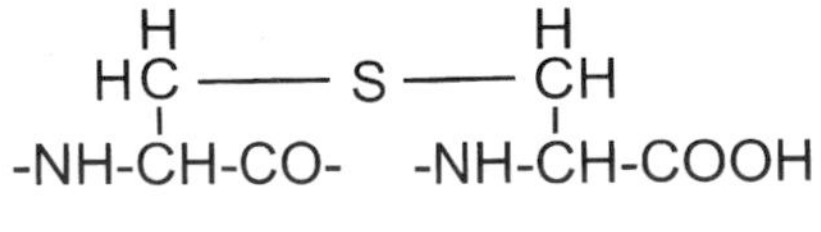

$$HC\text{---}S\text{---}CH$$
$$\text{-NH-CH-CO-} \quad \text{-NH-CH-COOH}$$

Figure 15. *meso* – Lanthionine, a constituent of lantibiotics.

pheromone-like function with cell-density-dependent gene expression (Nissen-Meyer and Nes, 1997; Sahl and Bierbaum, 1998).

Most of the bacteriocins of the Gram-positive microorganisms are bactericidal as a result of producing membrane damage. Bacteriocins of Gram-negative microorganisms target much more diverse sites: colicins E1, Ia, A, and N are pore-forming; colicins E2, E7, E8, and E9 are nonspecific DNases; microcin B17 functions as a DNA gyrase inhibitor; colicin E3 cleaves 16S ribosomal RNA at the 49th phosphodiester bond from the 3' end; and colicin E5 cleaves transfer RNAs for the amino acids Tyr, His, Asn, and Asp (Baba and Schneewind, 1998; Ogawa *et al.*, 1999). Colicins of 50 to 70 kDa gain entrance into the cell by parasitizing uptake systems for nutrients such as vitamin B_{12}, siderophores or nucleosides. Each of these colicins has three distinct domains: (1) one involved in the recognition of a specific receptor in the outer membrane of the cell envelope, (2) one involved in translocation through the cell envelope or to the inner membrane, and (3) another responsible for its lethal action (Lazdunski *et al.*, 1998).

When microorganisms secrete these toxins in the competition for survival in a microbial community, they have to protect themselves against their own toxic products. In the case of pore-forming bacteriocins they produce 'immunity proteins' which protect by integrating into the cytoplasmic membrane. Protection against enzymatic bacteriocins is achieved by co-synthesis of immunity proteins, which bind to the toxin in the producing cell and are exported as a heterodimeric complex into the environment of the cell. The affinity of these immunity proteins for their cognate toxins is very high (equilibrium dissociation constants [K_d] up to 9.3×10^{-17}). How and where the dissociation of immunity protein from the toxin occurs and when the toxin attacks its target is not known (Kleanthous *et al.*, 1998). The colicins have a very narrow target range and only kill *E. coli* and its close enteric relatives. These microorganisms can develop resistance against colicins by producing proteins that block the cell surface receptors for the colicins or through changes in the transport mechanisms that colicins use to cross the cell envelope. In natural populations there is a continual flux in the abundance of sensitive, producer, and resistant cell types. The ecological role of the colicins is not obvious, but thorough analysis indicates that colicins provide a competitive edge in nutrient-poor environments and that there might be a trade-off between the cost and the benefits of colicin production (Riley and Gordon, 1999).

The interest in bacteriocins in the oral cavity began in 1969 when Kelstrup and Gibbons demonstrated bacteriocin-like activity in oral streptococci (Kelstrup and Gibbons, 1969). The potential of bacteriocins in providing space and nutrients to an organism is a complex ecological question. It has been shown both theoretically and experimentally that a bacteriocin-producing strain is unable to invade a sensitive population in a homogeneous well-mixed environment such as liquid culture unless the initial density of the bacteriocin-producing strain is very high. Similarly a sensitive strain will normally not be able to invade a culture of a bacteriocin-producing population (Durrett and Levin, 1997; Riley, 1998). This has also been shown to be true for gnotobiotic rats infected by a bacteriocin-producing strain of *S. mutans* and/ or a bacteriocin-sensitive mutant of this strain (Rogers *et al.*, 1979). Theoretically it is possible and in nature it is evident that bacteriocin producing strains and sensitive strains can co-exist in environments where there are solid surfaces and a spatial heterogeneity in the abundance of nutrients (Durrett and Levin, 1997). This is also true for the human oral cavity. Bacteriocin-producing strains of *S. mutans* can be implanted and established in humans (Hillman *et al.*, 1987) and they are spontaneously more easily transferred from mother to child than non-producing strains (Grönroos *et al.*, 1998). In recent years lantibiotics of oral streptococci have been genetically characterized, *e.g.*, salvaricin A in *S. salivarius* (Ross *et al.*, 1993; Nes and Tagg, 1996), and mutacins in *S. mutans* (Hillman *et al.*, 1998; Qi *et al.*, 1999).

Altruistic Cell Death

For any microbial community in nature it is not hard to envisage a situation where the cell density reaches a level where nutrients for the survival of the community are not available. Every cell in the community may have mobilized all the survival strategies discussed above, but that does not suffice. What remains would be an 'altruistic lysis' of cells of the various populations in the community and in a manner which would enable the rest of the population to survive (Aizenman *et al.*, 1996; Nyström, 1998; 1999).

There are toxin-antitoxin systems in microorganisms, where the toxin is lethal for the host cell and the antitoxin prevents the effect of the toxin. There are also systems where highly stable mRNAs can be translated into proteins, which are toxic to the producer strain. These systems have so far mostly been identified with 'plasmid stabilization' by killing of plasmid-free cells upon segregation (Franch and Gerdes, 1996; Hochman, 1997; Gerdes *et al.*, 1997; Gotfredsen and Gerdes, 1998). There are indications, however, that such systems may also be activated by (p)ppGpp-dependent mechanisms during starvation (Aizenman *et al.*, 1996).

Recently, it was demonstrated that *H. pylori* harbors an antimicrobial cecropin-like peptide derived from the ribosomal protein L1 (RpL1) (Pütsep *et al.*, 1999). *H. pylori* itself was resistant to this peptide, but it killed both *E. coli* and *B. subtilis*. It was suggested that an altruistic death of a part of a *Helicobacter* population would release antimicrobial peptides that would kill other competing organisms in its natural habitat, the stomach.

Nutrients

Sugars

Sugars serve as the main source of energy and carbon for many oral microorganisms. The level of free glucose in stimulated parotid saliva of healthy individuals may vary from 1 to 70 $\mu mol \cdot l^{-1}$ at a flow rate of around 2 $ml \cdot min^{-1}$. In the gingival exudate the mean level of glucose is around 4.4 $mmol \cdot l^{-1}$ (Borg-Andersson *et al.*, 1998). The amount of free glucose in saliva is too low to account for the growth of the oral microorganisms. However, most proteins in saliva are heavily glycosylated. A glance in any taxonomic scheme of oral microorganisms reveals that these microorganisms are endowed with various exoglycosidase activities. The collective evidence of various studies indicates that the consortia of oral microorganisms have the capacity to degrade and use the sugars of the oligosaccharide side-chains of salivary glycoproteins (Van der Hoeven *et al.*, 1990; Bradshaw *et al.*, 1994; Willcox *et al.*, 1995; Rafay *et al.*, 1996).

Extracellular polysaccharides produced from sucrose made up a significant portion of the biofilm on the teeth illustrated in Figure 4 (Carlsson and Egelberg, 1965; Carlsson and Sundström, 1968). These polysaccharides have a key role in the colonization of the microorganisms on the supragingival tooth surfaces (see Chapter 3). At least 14 of the glycosyltransferases (Gtfs) responsible for polysaccharide production from sucrose have been genetically characterized in mutans and sanguis streptococci as well as in *S. salivarius* (Monchois *et al.*, 1999). The polysaccharides can be dextrans with $\alpha(1\text{-}6)$ glucosidic bonds, mutans with $\alpha(1\text{-}3)$ glucosidic bonds and alternans with alternately $\alpha(1\text{-}6)$ and $\alpha(1\text{-}3)$ glucosidic bonds (Guggenheim, 1970; Monchois *et al.*, 1999). The polymers also differ in the type of branch linkages, resulting from $\alpha(1\text{-}2)$, $\alpha(1\text{-}3)$, $\alpha(1\text{-}4)$, and $\alpha(1\text{-}6)$ glucosidic bonds, the degree of branching, the length of the branch chains, and their spatial arrangement. There are also fructose polymers, inulin-like with $\beta(2\text{-}1)$ linked fructosyl residues in *S. mutans* and levan with $\beta(2\text{-}6)$ linked fructosyl residues in *S. salivarius* and *Actinomyces viscosus* (Birkhed *et al.*, 1979).

The extracellular and to some extent also the intracellular polysaccharides (see above) can be considered a way by which microorganisms in biofilms capture as much of the energy and carbon supplied by dietary sucrose as possible. The polysaccharides can be used when other energy and carbon sources are lacking in the environment of the microorganisms (Parker and Creamer, 1971; Sund *et al.*, 1989). The fructose polymers are more readily utilized than the glucose polymers. It has recently been shown that levan, inulin and sucrose induce the expression of a fructan hydrolase gene of *S. mutans* and that the expression of this gene is repressed by readily fermentable sugars (Burne *et al.*, 1999). Both Gram-positive and Gram-negative microorganisms have glucan hydrolyzing activity (dextranase). This activity is found in streptococci, bifidobacteria and members of genus *Actinomyces* as well as in members of the genera *Capnocytophaga* and *Prevotella* (Igarashi *et al.*, 1998).

Whether the complex structures of the salivary glycoproteins, or the intracellular and extracellular polysaccharides, are used as energy and carbon sources, a variety of different metabolic processes are required for their efficient utilization. The genes coding for the required enzymes are usually expressed only if the respective sugar is present, and actually, only if more easily fermentable sugars are absent in the environment. The regulation of these processes is controlled at the level of enzyme activity and transcription of the carbon catabolic genes and operons. In many microorganisms the phosphoenolpyruvate:sugar phosphotranferase system (PTS) is the hub of this regulation. The PTS is made up of two general proteins, enzyme I (EI) and the heat-stable histidine-containing protein (HPr), as well as several sugar-specific enzymes (EII). EII can have three or four domains in individual or linked polypeptides; two domains (EIIA and EIIB) are involved in phosphorylation of the sugar, and one or two domains (EIIC and, if present, EIID) are active in sugar permeation across the cytoplasmic membrane. In sugar transport the energy-rich phosphate is transferred from the glycolytic intermediate phophoenolpyruvate to the sugar via EI, HPr, EIIA, and EIIB. In this phosphoryl transfer HPr is phosphorylated on the histidine[15] residue (Postma *et al.*, 1993; Reizer *et al.*, 1998).

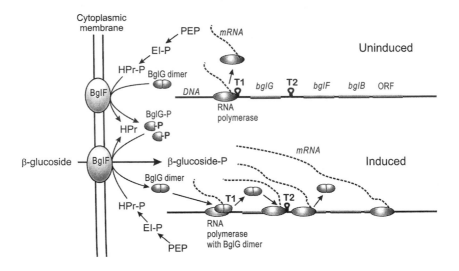

Figure 16. Antiterminator protein BglG involved in the regulation of the *bgl* operon encoding components of PTS specific for an aromatic β-glucoside, *e.g.*, salicin. Details are given in the text. Note that the BglG dimer attaches to nascent mRNA transcribed upstream of terminator T1 and allows the polymerase to bypass the terminator. Upstream of the second terminator, T2, a similar sequence is transcribed and the BglG dimer reassociates with the polymerase and allows it to circumvent the terminator.

There are at least 16 sugar transporters belonging to PTS recognized in the genome of *B. subtilis* (Kunst *et al.*, 1997). PTS of oral streptococci has been extensively studied (Vadeboncoeur and Pelletier, 1997) (see Chapter 5).

In addition to transporting sugar, PTS also works as a signal generator in integrating carbon catabolic functions. The signals lead to regulation of enzyme activities and transcription of carbon catabolic genes and operons. Many microorganisms prefer glucose as the major energy and carbon source. To prevent the uptake or the formation of inducers for the operons of other sugars, the permeases of these sugars are inhibited. In the presence of glucose this is achieved in enteric bacteria by binding of the unphosphorylated form of the glucose specific EIIA (EIIA^{Glc}) to the permeases of these sugars, resulting in inducer exclusion. In the absence of glucose in enteric bacteria the phosphorylated form of EIIA^{Glc} stimulates the activity of adenylate cyclase. cAMP is produced from ATP and cAMP forms a complex with CRP (cAMP receptor protein), which activates the transcription of carbon catabolic genes and operons. In Gram-positive microorganisms glucose is favored over other sugars by a completely different mechanism but the functional result is the same as in enteric bacteria. In the presence of glucose, HPr is phosphorylated on Ser^{46} by an HPr kinase at the expense of ATP. HPr (Ser-P) forms a complex with the catabolic control protein A (CcpA), which binds to carbon catabolic genes via a *cis*-acting catabolite responsive element (*cre*) in their operator sites. This prevents transcription of genes encoding the pathways of sugars other than glucose (Stülke and Hillen, 1998; 1999). CcpA has been found in many Gram-positive microorganisms including lactobacilli and oral streptococci (Monedero *et al.*, 1997; Simpson and Russell, 1998a). Studies of whole microbial genomes have shown that there are organisms, like *T. pallidum*, which have general components of the PTS but do not have PTS-dependent sugar transport. The PTS components may thus be used only for regulatory purposes. This may be an equally important function of PTS as the transport of sugars (Fraser *et al.*, 1998; Stülke and Hillen, 1999). It would be very interesting to learn whether oral treponemas also are equipped with the regulatory component of the PTS system.

By the regulatory functions of PTS the expression of carbon catabolic genes and operons can thus be modulated by the rate of transcriptional initiation. Some carbon catabolic operons in Gram-positive as well as in Gram-negative microorganisms are regulated by antiterminator proteins controlling transcript elongation. An example of how an antiterminator protein works is the regulation of the utilization of aromatic β-glucosides. These sugars are handled by PTS in *E. coli* and the sugar-specific components of PTS are encoded by the *bgl* operon, which contains four genes: *bglG*, encoding an antiterminator protein; *bglF*, encoding the EII^{bgl} of PTS - the transporter of β-glucosides; *bglB* encoding a phospho-β-glucosidase; and an unknown ORF. In this operon the genes are in the sequence *bglG*, *bglF*, and *bglB*. Before and after *bglG* there are transcription terminators (Figure 16). Transcriptions from *bgl* promoters is constitutive, and in the absence of β-glucosides most transcripts terminate at the first terminator within the operon upstream of the first gene, *bglG* (Figure 16; T1), or at the second terminator downstream of *bglG* (Figure 16; T2). In the absence of aromatic β-glucosides there are low levels of BglG and BglF in the cell, and membrane-bound BglF will phosphorylate BglG. The site for phosphorylation and dephosphorylation is a duplicate conserved domain (the PTS regulation domain, PRD). This domain is found in many transcriptional regulators of catabolite repression (activators as well as antiterminator proteins) (Stülke *et al.*, 1998). In the presence of aromatic β-glucoside, BglF dephosphorylates BglG, and BglG forms dimers. The dimeric BglG binds to the nascent mRNA of RNA polymerase complex and the complex can then bypass the terminators allowing the entire *bgl* operon to be transcribed followed by fermentation of the aromatic β-glucoside (Rutberg, 1997; Chen *et al.*, 1997; Weisberg and Gottesman, 1999; Boss *et al.*, 1999). Another example of antiterminator protein regulation of carbon catabolic gene expression is the utilization of sucrose by *B. subtilis* (Rutberg, 1997). This type of regulation has not yet been shown among oral microorganisms, although there is reason to believe that the expression of the urease operon of *S. salivarius* can be regulated in this way (Chen *et al.*, 1998).

Amino Acids and Peptides
All oral microorganisms have to satisfy their requirement for nitrogen with ammonium ions or amino acids. Some of them may also use amino acids as their exclusive carbon and energy sources. To obtain amino acids the microorganisms have to degrade proteins in saliva and gingival fluid by extracellular proteinases. Among oral microorganisms such enzymes have been most extensively studied in *P. gingivalis* (Kuramitsu, 1998; Lamont and Jenkinson, 1998).

101

To appreciate the complexity of a microbial proteolytic system we may consider the example of *L. lactis,* a facultatively anaerobic glycolytic Gram-positive organism. It has a limited capacity to synthesize some amino acids and for optimal growth it must obtain these amino acids from exogenous sources. *L. lactis* has a single cell wall-bound extracellular proteinase (PtrP), which hydrolyzes milk proteins into more than 100 different oligopeptides ranging from 5 to more than 30 amino acid residues in length. For transport of peptides, three translocators have been identified: a proton-motive-force-driven di/tripeptide transporter (DtpT) for relatively hydrophilic peptides; an ATP-driven di/tripeptide permease (DtpP) for more-hydrophobic peptides; and an ATP-driven oligopeptide transport system (Opp) for peptides from 4 up to at least 18 residues. The Opp system consists of five proteins: the integral membrane proteins OppB and OppC, the ATP-binding proteins OppD and OppF, and the substrate-binding protein OppA. It is thus an example of an ABC-transporter (Figure 6; G). In addition, this organism has at least 10 amino acid transport systems, where each system has high specificity for structurally similar amino acids. Some of these systems are driven by hydrolysis of ATP and others by the proton motive force. In *L. lactis* there are a multitude of aminopeptidases present intracellularly but there are no extracellular peptidases. There are two peptidases PepN and PepC, which release N-terminal amino acids from a wide range of di-, tri- and oligopeptides. PepV has a broad specificity for dipeptides, and PepT for tripeptides. The endopeptidases PepO, PepF and PepF2 hydrolyze internal peptide bonds of oligopeptides. In addition, there are peptidases that recognize more-specific sequences (indicated by an arrow): PepA (Glu/Asp$\downarrow$(X)$_n$); PepR (Pro$\downarrow$X); PepI (Pro$\downarrow$(X)$_n$); PepQ (X$\downarrow$Pro); PepP (X$\downarrow$Pro-(X)$_n$); and PepX (X$\downarrow$Pro-(X)$_n$). It is notable that five of these six specific peptidases are devoted to hydrolysis of peptide bonds with proline residues. No carboxypeptidases have been found in *L. lactis.* The genetic regulation of the whole proteolytic system is so far largely unexplored (Mierau *et al.*, 1996; Kunji *et al.*, 1996; Kunji *et al.*, 1998; Detmers *et al.*, 1998). Proteinases, peptide utilization and peptidases have been described among oral streptococci and they may have proteolytic systems similar to the system of *L. lactis.* (Cowman *et al.*, 1975; Andersson *et al.*, 1984; Rogers *et al.*, 1991; Jenkinson *et al.*, 1996; Cowman and Baron, 1997; Juarez and Stinson, 1999).

In some microorganisms the proteolytic systems not only compensate for the auxotrophy of some amino acids but also provide energy and carbon. The members of the genus *Porphyromonas* represent this category. Their extracellular peptidases and proteinases have been thoroughly explored (Kuramitsu, 1998; Lamont and Jenkinson, 1998; Banbula *et al.*, 1999), but the concomitant transport systems and intracellular peptidases are largely unknown. When an organism uses amino acids as energy sources, it may obtain the equivalent of one ATP per molecule of degraded amino acid (Sokatch, 1969; Thauer *et al.*, 1977). Since the import of an amino acid against a concentration gradient requires the energy equivalent of at least one ATP, it is only when amino acids are present in such high concentrations that no energy is required for the uptake that amino acids will be of significance as energy sources. The advantage of peptide transport is then obvious. All the amino acids of a peptide will enter the cell at the same energy expenditure as for a single free amino acid.

To understand the competition for amino acids and peptides among populations in microbial communities information on their affinity for these nutrients is indispensable. However, such information is largely missing. This is mainly due to difficulties in accounting for the concomitant exodus of peptides and amino acids from the cell during uptake studies. There is some information on the affinity for peptides of some oral microorganisms. As shown in Figure 7, the K_m for uptake of the tripeptide γ-Glu-Cys-Gly by *P. micros* was 7 μM (Carlsson *et al.*, 1993). *Fusobacterium nucleatum* has a similar high affinity for the dipeptides Cys-Gly and Met-Met, while the K_m for these peptides in *P. gingivalis* strain W83 were 538 μM and 170 μM, respectively (Tang-Larsen *et al.*, 1995). In *L. lactis* the K_m for the tetrapeptide Lys-Tyr-Gly-Lys of the Opp system was 700 μM, and 8.8 μM for the peptide Arg-Asp-Met-Pro-Ile-Gln-Ala-Phe-Leu-Leu-Tyr (Detmers *et al.*, 1998). However, in the periodontal pocket the relatively low affinity of *P. gingivalis* for peptides may not be a serious disadvantage because of the high proteolytic activity of this organism. In the periodontal pockets organisms like *P. micros* and *F. nucleatum* with much less potent proteolytic systems but with high affinity for peptides may thrive on peptides derived from *P. gingivalis*. Another highly proteolytic organism is *Bacteroides forsythus* and it is absolutely dependent on peptides for growth. It can salvage essential amino acids at 100-fold lower concentrations when present as peptides as compared to free amino acids (Wyss *et al.*, 1993). The peptides prepared by the highly proteolytic organisms are certainly a very rich nutritional source for many organisms and one can expect many physiological types of organisms to thrive where these peptides are produced.

Very little is known about the amino acid requirements of anaerobic oral bacteria, but a requirement for L-phenylalanine of *Capnocytophaga gingivalis, Eubacterium timidum, F. nucleatum, P. gingivalis, T. denticola,* and *Treponema vincentii* has been demonstrated (Wyss, 1993). A curious finding by Wyss (1992) was that a lecithin requirement by *T. vincentii* could be substituted for several months by adding the dipeptide valyl-lysine to the medium, but then the organism eventually stopped replicating.

Another ecological aspect of amino acid metabolism is that the excreted end products may be essential nutrients of other organisms. Thus *Treponema pectinovorum, T. socranskii, T. denticola* and *T. vincentii* are strictly dependent on either isobutyric or 2-methylbutyric acids for growth (Wyss, 1992). These acids are end products of valine and isoleucine metabolism and are excreted by most members of the genera *Porphyromonas* and *Prevotella* and also by some members of the genera *Peptostreptococcus* and *Eubacterium* (Collier *et al.*, 1998). The role of the proteolytic systems for the growth of the microbial consortia in periodontal pockets is also apparent when one studies the production of sulfides in these sites. Even in very shallow pockets (2 to 3 mm) there are high levels of sulfides (Persson, 1992). These microbial consortia have degraded the proteins, imported the peptides, and utilized the amino acids including L-cysteine and L-methionine (Claesson *et al.*, 1990; Persson *et al.*, 1990). Hydrogen sulfide formation is a common characteristic among oral anaerobic microorganisms, Gram-positive as well as Gram-negative. Of 75 tested oral species the majority formed significant amounts of hydrogen sulfide from L-cysteine, while only a few, mostly members of the genera *Porphyromonas* and *Fusobacterium,* formed methyl mercaptan from L-methionine (Persson *et al.*,

1990; Claesson *et al.*, 1990). It may be noted that these volatile sulfur compounds are significant components in oral malodor.

Nutrition and Energetics of Anaerobic Microbial Communities

In the oral cavity there may be microbial communities in all stages of development - from the very early colonization of bare surfaces to mature communities where all nutritional resources are fully exploited. The microorganisms obtain their energy for survival and growth by converting nutrients into compounds of lower free energy, which they then excrete. In microbial communities this conversion involves several physiological types of microorganisms, where the waste products from one type are sequentially used by others. In such systems there are also cross-feeding of vitamins and other essential nutrients. When such microbial communities run out of oxygen, methanogenic archaea often appear. They are the ultimate scavengers. They use hydrogen, formate, acetate and CO_2 and wring the last quantity of energy from these waste products of other microorganisms, and they produce methane gas (Gottschalk and Peinemann, 1992; Schink, 1997; Ferry, 1999). Methanogenic archaea have been demonstrated in dental biofilms (Brusa *et al.*, 1987; Belay *et al.*, 1988). The concept of 'waste chain' has been proposed by Dolfing and Prins (1996) for such waste handling in microbial communities, where a flow of nutrients involves several metabolic types of microorganisms and the waste products from one type are sequentially used by others. This should be distinguished from 'food chain', where the organisms of one trophic level, themselves, serve as food for the organisms of the next trophic level.

It may be noted that much less energy is available to a microbial community under anaerobic than under aerobic conditions.

$$C_6H_{12}O_6 + 6O_2 \rightarrow 6CO_2 + 6H_2O \ (\Delta G^{o'} = -2\ 870 \ kJ \cdot mol^{-1})$$

$$C_6H_{12}O_6 \rightarrow 3CO_2 + 3CH_4 \ (\Delta G^{o'} = -390 \ kJ \cdot mol^{-1})$$

In fact only 15% of the energy of the hexose will be available under anaerobic conditions (Schink, 1997).

The early colonizers on the bare oral surfaces are heterotrophic microorganisms often called primary fermenters, which obtain their nutrients by degrading organic polymers such as proteins and polysaccharides into oligomers and monomer (Figure 17). Some microorganisms use some of the monomers directly in the synthesis of new cell constituents. Such microorganisms are said to be auxotrophic for these compounds because they do not have the cellular machinery required for their synthesis. The other nutrients are degraded into precursor metabolites and during this degradation the cell obtains energy by substrate level phosphorylation or by building up a proton motive force that drives membrane-bound ATPase to produce ATP from ADP and inorganic phosphate. In an organism that grows on just sugar and inorganic salts only 12 precursor metabolites are required to synthesize all cell constituents. This can be achieved in *E. coli* with a genome of 4,639,221 base pairs (Blattner *et al.*, 1997). These precursor metabolites are glucose-6-phosphate, fructose-6-phosphate, pentose-5-phosphate, erythrose-4-phosphate, triose-phosphate, 3-

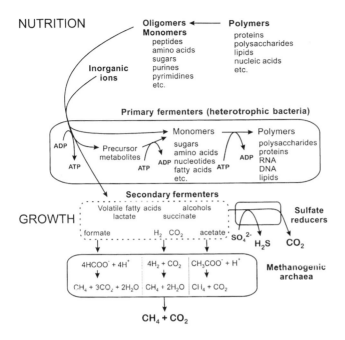

Figure 17. Flow of nutrients through an anaerobic microbial community where all nutrients are converted into CH_4 and CO_2 by the concerted action of primary and secondary fermenters, sulfate reducers, and methanogenic archaea.

phosphoglycerate, phosphoenolpyruvate, pyruvate, acetyl-CoA, α-ketoglutarate, and succinyl-CoA (Moat and Foster, 1995). Waste products excreted by the primary fermenters colonizing oral surfaces are volatile fatty acids, alcohols, succinate, lactate, H_2, and CO_2. In the case of CO_2 there may not be any net release, because many microorganisms readily use this gas. If sulfate-reducers are present in anaerobic microbial communities (Figure 17), they can use all types of waste products of the primary fermenters, oxidize them to CO_2 and simultaneously reduce sulfate to hydrogen sulfide (Schink, 1997). Sulfate-reducers of the genera *Desulfobacter* and *Desulfovibrio* have been isolated from periodontal pockets (Van der Hoeven, *et al.*, 1995). However, the significance of these microorganisms in the oral cavity has yet to be established. A key question is whether there are enough sulfates available for such conversions. The concentration of free sulfate in tissue fluid is less than 0.3 mmol·l^{-1} (Van der Hoeven, *et al.*, 1995). It is possible, however, that the high arylsulfatase activity of oral members of the genus *Campylobacter* (Wyss, 1989a) will mobilize significant amounts of free sulfate from tissue glycosaminoglycans in the periodontal pockets and in that way provide 'living-space' for the sulfate reducers.

Most of the microorganisms isolated from the oral cavity are primary fermenters. They easily grow in the laboratory. However, some of the oral microorganisms may have very specific nutritional requirements not satisfied by ordinary laboratory media. One example is *B. forsythus,* which is dependent on *N*-acetylmuramic acid for growth (Wyss, 1989). Before this requirement was known, *B. forsythus* only grew in laboratory media that had been unintentionally contaminated with this genuine bacterial product. To really understand the ecology of complex microbial communities one has to know the nutritional requirements of each population of the community, and in the best of these worlds one should also know the affinity for each nutrient by each population. Although oral microorganisms, compared to enteric bacteria, are considered to have very complex nutritional requirements, this is not always true. Most oral streptococci will grow in the presence of some vitamins and salts with sugars as an energy and carbon source, cysteine as a sulfur source, and ammonia as a nitrogen source. *S. salivarius* has the simplest requirements of oral streptococci. It grows luxuriantly in a medium containing glucose, ammonia, L-cysteine, inorganic salts and the vitamins nicotinic acid, biotin, thiamin, riboflavin, and pantothenic acid in an anaerobic atmosphere with CO_2 (Carlsson, 1971a). In this medium urea can substitute for the ammonia requirements. *S. mutans* and *S. sobrinus* have similar simple nutritional requirements under anaerobic conditions, but in addition they need the vitamins pyridoxine and *p*-aminobenzoic acid (Carlsson, 1970; Terleckyj *et al.*, 1975). Mutans streptococci also have an absolute requirement for Mn^{2+} (Bauer *et al.*, 1993), and it is probable that they share this requirement with the sanguis streptococci (Kolenbrander *et al.*, 1998). Of the sanguis streptococci *S. sanguinis* and *S. gordonii* have similar simple requirements as mutans streptococci under anaerobic conditions but do not require *p*-aminobenzoic acid while *S. oralis* is a little more fastidious (Carlsson, 1972). This species has the same vitamin requirements as *S. salivarius* but it also requires the amino acids arginine, glutamic acid and histidine in addition to cysteine for growth under anaerobic conditions (Carlsson, 1972). Under aerobic or microaerophilic conditions some of the other streptococci may also require arginine, glutamic acid and additional amino acids (Cowman *et al.*, 1975). When these studies were done *S. sanguinis, S. gordonii,* and *S. oralis* were all referred to as *S. sanguis*, but the studied strains have now been renamed according to Kilian and co-workers (1989) based upon previous observations (Carlsson, 1968). In mixed culture, *S. sanguinis* can satisfy the *p*-aminobenzoic acid requirement of *S. mutans* (Carlsson, 1971b). Of these streptococci *S. salivarius* colonizes the tongue (Krasse, 1954) and the mutans and sanguis streptococci the teeth (Carlsson, 1967). In 1970 Gibbons and co-workers suggested that the adherence of these streptococci to the tissue surfaces was the primary determinant of their colonization at various sites of the mouth (van Houte *et al.*, 1970; van Houte *et al.*, 1971). The similar nutritional requirements of these streptococci strongly support this suggestion.

Ammonia available in saliva will be an adequate nitrogen source for growth of streptococci under anaerobic conditions (Griffith and Carlsson, 1974). In addition, under aerobic conditions the mutans streptococci require glutamine or glutamic acid for growth (Carlsson, 1970) and they actually have a high-affinity glutamine transport system (Dashper *et al.*, 1995). Interestingly, in a study using signature-tagged mutagenesis, 126 virulence genes of *S. pneumoniae* were identified (Polissi *et al.*, 1998). Two genes, indispensable for *in-vivo* growth of *S. pneumoniae* in mice, coded for components of a high-affinity glutamine transport system. Another gene also indispensable for *in-vivo* growth showed striking similarity to the ATP binding component of the ABC transporter involved in multiple sugar uptake in *S. mutans* (Russell *et al.*, 1992).

In the studies of the nutritional requirements referred to above, minimal chemically defined media were used. Another way to study nutritional requirements is to use a chemically defined medium containing all possible nutrients. The nutritional requirements are then elucidated by deleting individual nutrients from the medium and determining their influences on growth. Wyss (1992) has successfully used this approach and his studies showed that the asaccharolytic *T. vincentii* and saccharolytic *T. maltophilum* required *N*-acetylglucosamine for growth (Bernet and Wyss, 1988; Wyss *et al.*, 1996). Hemin was required not only by *P. gingivalis* and *P. intermedia* (Gibbons and Macdonald, 1960) but also by *B. forsythus* (Wyss, 1992). The menadione requirement of *P. gingivalis* (Gibbons and Macdonald, 1960) was confirmed, and could be substituted by the menadione-precursor 1,4-dihydroxy-2-naphthoic acid (Wyss, 1992). In the oral cavity this menadione requirement may be satisfied by cross-feeding from *Veillonella* (Ramotar *et al.*, 1984). Polyamine, usually putrescine, was required by *T. pectinovorum, T. socranskii,* and *T. vincentii* (Wyss, 1992). It has to be noted, however, that these requirements may differ with growth conditions and combinations of available nutrients. In the chemically defined medium of Wyss (1992) adding coenzyme A to the medium could thus eliminate the *N*-acetylglucosamine requirement of *T. vincentii*.

As stated above the presence of methanogenic archaea in the oral cavity is strong evidence that there are anaerobic microbial communities in which the nutrients and most of the waste products are efficiently utilized. Methanogenic archaea can convert acetate, H_2, CO_2, and other one-carbon compounds into methane and CO_2 (Figure 17). For degradation of other waste products of the primary fermenters such

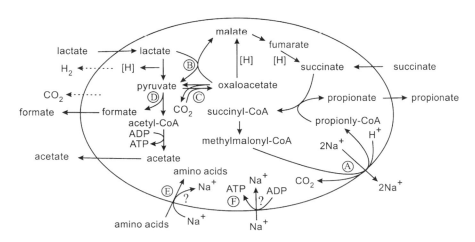

Figure 18. Energy metabolism of *Veillonella* (de Vries *et al.*, 1977; Denger and Schink, 1992). Oxidation of lactate to pyruvate is coupled to a reductive tricarboxylic acid cycle with a membrane-bound Na^+-pumping methylmalonyl-CoA decarboxylase (A) as a final step (Hilpert and Dimroth, 1991; Dimroth, 1997). Malic-lactic transhydrogenase (B) converts lactate in the presence of oxaloacetate into malate and pyruvate (Dolin *et al.*, 1965). The initial level of oxaloacetate for this reaction is produced from pyruvate by an ATP-independent oxaloacetate decarboxylase (C) (Ng *et al.*, 1982). Under anaerobic conditions pyruvate is degraded by pyruvate-formate lyase (D) (Distler and Kröncke, 1981). Veillonellae use amino acids (E) (Konings *et al.*, 1975; Hoshino, 1987; Durant *et al.*, 1997). The Na^+-gradient created by methylmalonyl-CoA decarboxylase may be involved in the transport of these amino acids (Denger and Schink, 1992). When starved cells are exposed to succinate, ATP levels increase (Janssen, 1992), but no Na^+-dependent ATPase has yet been found (F) (Hilpert and Dimroth, 1991).

as fatty acids longer than two carbon atoms and alcohols longer than one atom, a further group of microorganisms is needed, the secondary fermenters (Schink, 1997). Members of the genus *Veillonella* belong to this group. They have attracted interest in oral microbiology over the last 50 years because they use lactate and succinate as energy sources (Douglas, 1950; Rogosa, 1964; Ng and Hamilton, 1974; Delwiche *et al.*, 1985; Janssen, 1992) (Figure 18).

Veillonella may make up a significant part of the cultivable microorganisms from supragingival dental biofilm. Kilian and co-workers (1979) found a median around 30% veillonellae (2 – 67%) in biofilms from 12-year old Tanzanian children, and Milnes and Bowden (1985) found a median of 5% veillonellae (1 – 28%) in 1-year old Canadian Indian children. Veillonellae are thus very successful among secondary fermenters in competing for limiting nutrients in the supragingival biofilm. These microorganisms are also common in the intestines of humans and chickens and it has been suggested that veillonellae may have an antagonistic effect against enteropathogens by using up the amino acids in the intestine (Durant *et al.*, 1997). *Campylobacter sputorum* is another organism that uses lactate as an energy source under anaerobic conditions (de Vries *et al.*, 1980), while lactobacilli and members of the genus *Actinomyces* only use lactate under aerobic conditions (Murphy *et al.*, 1985; Takahashi and Yamada, 1999).

There are thus only small amounts of energy available in the waste products of the primary fermenters. This means that the secondary fermenters often have to use very sophisticated pathways to obtain this energy. For all known metabolic systems the minimal amount of energy required for the synthesis of one mole of ATP is 70 kJ (Schink, 1997). The decarboxylation of succinate to propionate in veillonellae (Figure 18:A) only produces a free energy change of one third of that ($\Delta G^{o'} = -25kJ \cdot mol^{-1}$) and a substrate level phosphorylation is not possible (Denger and Schink, 1992; Dimroth and Schink, 1998). The metabolic energy of this reaction is instead saved in the form of a Na^+ gradient across the cytoplasmic membrane and may be used in substrate uptake. *Veillonella* form ATP from succinate, but they are only able to grow on succinate in the presence of lactate (Janssen, 1992). The mechanism of ATP formation in this case is unknown. A membrane-bound Na^+-pumping ATPase as in *Propionigenum modestum* (Dimroth, 1997) has not been found in veillonellae (Hilpert and Dimroth, 1991, Dimroth and Schink, 1998). Lactate and succinate also stimulate the growth of *P. gingivalis* (Grenier, 1992; Wyss, 1992), but the mechanism of this stimulation has not been elucidated.

To extract the small amount of energy available in the waste products of the primary fermenters there is sometimes cooperation called syntrophy between metabolically different types of microorganisms. In such syntrophic cooperation between secondary fermenters both partners depend on each other to perform the required metabolic activities. In the laboratory this dependency cannot be overcome by simply adding a cosubstrate or any type of nutrient. An example is the syntrophy of two partner organisms initially thought to be one organism "*Methanobacillus omelianski*". One partner was responsible for

$$2CH_3CH_2OH + 2H_2O \rightarrow 2CH_3COO^- + 2H^+ + 4H_2 \ (\Delta G^{o'} = +19kJ \cdot 2 \text{ mol of ethanol}^{-1})$$

The other partner was a methanogenic archaea utilizing H_2

$$4H_2 + CO_2 \rightarrow CH_4 + 2H_2O \ (\Delta G^{o'} = -131kJ \cdot mol \ of \ methane^{-1})$$

and then the energetics of the syntrophy was

$$2CH_3CH_2OH + CO_2 \rightarrow 2CH_3COO^- + 2H^+ + CH_4 \ (\Delta G^{o'} = -112kJ \cdot mol \ of \ methane^{-1})$$

The degradation of ethanol to acetate and H_2 by the first organism is an endergonic reaction and can only provide energy for growth of this organism if the hydrogen partial pressure is kept low (<100Pa) by the second organism (Schink, 1997). Sulfate-reducing, homoacetogenic and fumarate-reducing microorganisms can also fulfill the hydrogen-consuming function of methanogenic archaea. The sulfate-reducers have been mentioned above. In homoacetogenic organisms the following exergonic reaction occurs:

$$4H_2 + CO_2 \rightarrow CH_3COO^- + H^+ + 2H_2O \ (\Delta G^{o'} = -95kJ \cdot mol \ of \ acetate^{-1})$$

Such organisms are present in the human colon (Leclerc *et al.*, 1997), but have not been reported to be present in the oral cavity. Spirochetes, closely related to human strains of the genus *Treponema* but isolated from termite guts, were recently shown to be homoacetogenic (Leadblatter *et al.*, 1999). This is of obvious interest since it has been shown that *T. denticola* significantly increases its acetogenesis in the presence of H_2 (Mikx, 1997). If this is a common characteristic among spirochetes, the dense spirochetal populations in deep periodontal pockets may play a significant role in keeping the partial pressure of H_2 low and in that way facilitate an efficient utilization of the nutritional resources in the pockets.

Fumarate reduction in the presence of formate or H_2 is a taxonomic characteristic of many oral members of the genus *Campylobacter* and these organisms may also play a significant role as hydrogen consumers in periodontal pockets. *C. rectus* displays H_2 metabolism (Gillespie and Barton, 1996) that is very similar to the paradigm organism of this metabolic type, *Wolinella succinogenes* (Gross *et al.*, 1998; Lancaster et al., 1999). Thus there is very efficient metabolic cooperation within an anaerobic microbial community, and this may result in methanogenesis, the final exergonic step in degradation of organic matter.

Perspective

During the 20th century investigators reduced the biofilm on teeth into the smallest possible functional pieces. There is today more knowledge about the indigenous microbial species in the oral cavity than in any other site of our body. We also have impressive knowledge about adherence mechanisms of these microorganisms, and some information on microbial physiology in relation to caries and periodontal disease. However, all of this is inadequate unless an appreciation is developed for interactions within the microbial communities colonizing the oral surfaces and between the microbial communities and their environment in the oral cavity. It is now time to use the knowledge, apply new technologies and consider biofilms as functional units.

This has to start with a more holistic view on the functions of the microbial cell as such. Biofilms develop until the environment no longer fully supports growth. In this chapter a few examples are presented describing how an organism prepares itself in an environment that is not optimal for growth. A picture emerges where it is clear that a significant part of the genome of an organism is devoted to handle such situations. It is the struggle for nutrients and protection against changes in temperature, osmolarity, pH, and oxygen levels that characterizes everyday life. Most of our present knowledge on the adaptation of an organism to its environment derives from studies where growth and survival of a wild-type microbial strain (mostly of *E. coli*) are compared to that of a strain with a knockout of a single gene. However, one does not have to take for granted that the product of this gene has the same effect on that organism when it is living in its natural 'real life' environment under multi-nutrient limitation. This calls for more holistic experimental approaches. The basis for such studies will be provided by the knowledge of the entire genome sequences of a representative number of organisms and the interpretation of these sequences into microbial physiology, *i.e.,* functional genomics. A further step to obtain a more holistic view of microbial physiology and ecological adaptation is to follow the changes in protein expression and concentrations of intracellular metabolites in response to alterations in the environment. Combining these approaches will allow a correlation of gene activity and metabolic activity in a cell at any one moment.

All the proteins in a translated microbial genome are called a proteome, and the actual goal of the science of proteomics is a quantitative description of protein expression and its changes under the influence of biological perturbations (Humphery-Smith *et al.*, 1997; Blackstock and Weir, 1999). Powerful analytical tools have paved the way for proteomics. A significant beginning was the introduction of two-dimensional polyacrylamide gel electrophoresis by O'Farrel (1975). The next step was when N-terminal amino acid sequencing could identify the separated polypeptides in the gel. A substantial increase in speed and sensitivity of peptide analysis has since been gained by the introduction of matrix assisted laser desorption and ionization in time of flight mass spectrometers (MALDI-TOF/MS) and triple-quadruple and ion trap mass spectrometers (Humphery-Smith *et al.*, 1997; James, 1997; Hochstrasser, 1998).

In the elucidation of microbial physiology under various environmental situations genomics and proteomics may fall short if not combined with information about concentrations of intracellular metabolites and kinetic properties of involved enzymes (Tweedale *et al.*, 1998). To analyze the metabolome (*i.e.,* the total intracellular pool of metabolites) there are basically two different approaches: non-invasive and invasive techniques. Nuclear magnetic resonance spectroscopy (NMR) is non-invasive and has high sensitivity. With ^{31}P-NMR and ^{13}C-NMR spectroscopy such high *in-vivo* sampling rates as twice per second can be obtained (Weuster-Botz and de Graaf, 1996). If all these techniques are highly standardized with scrupulous documentation of experimental data, microbial activities may be successfully analyzed by powerful bioinformatics in 'virtual laboratories' (Hochstrasser, 1998).

In oral microbiology the use of these techniques is in its infancy. The whole genome sequence of an oral microorganism has not yet been published but several of these projects are nearing completion. Changes in protein profiles using two-dimensional polyacrylamide gel electrophoresis have been studied in *Candida albicans* upon carbon-starvation at acid pH (Niimi *et al.*, 1996). Changes in protein profiles upon exposure to acid have also been studied in members of the genera *Lactobacillus, Streptococcus* and *Actinomyces* (Hamilton and Svensäter, 1998). For a long time, glycolytic intermediates of the metabolome in oral organisms have been studied with invasive techniques, and the findings have been very useful in elucidating the regulation of glycolysis and acid production in cariogenic organisms (Yamada and Carlsson, 1975; Iwami and Yamada, 1985: Takahashi *et al.*, 1991). With the rapid development of functional genomics (Martzen *et al.*, 1999) oral microbiologists can look into the future with confidence. However, there is one major obstacle in oral microbiology as well as in other areas of microbiology. Microbial biochemical physiology has not developed in pace with genomics. During the last few decades this field has actually been neglected in lieu of genomics. Without a thorough knowledge of the enzymes and their regulatory mechanisms at the protein level, genomics alone may be inadequate. Fortunately, many cellular functions are highly conserved among microorganisms. However, it is not too uncommon that one is surprised when one wants to confirm that an organism has solved a specific problem in the same way as others. For example, who would have imagined that GTP and pyrophosphate, and not ATP, is the 'energy currency' of glycolysis in oral members of *Actinomyces* (Takahashi and Yamada, 1999)?

Current methodologies can thus give us detailed information on global shifts in cellular functions of a microbial population under different environmental conditions (VanBogelen *et al.*, 1999). The real challenge of today is to extract similar information on microbial communities in biofilms. One has then to know the relationship between structure and function, and how the populations communicate and compete for nutrients (see "Microbial Recognition of the Environmental Situation" on page 74 and "Stress Handling by Microbial Communities" on page 93). In this approach of 'putting the pieces together', scanning confocal laser microscopy and various molecular methodologies are indispensable tools (Møller *et al.*, 1998). rRNA probes can be used to identify and quantify defined populations, and monitor the dynamics of the populations (Head *et al.*, 1998). By combining rRNA probes and confocal microscopy, syntrophic partners (see "Nutrition and Energetics of Anaerobic Microbial Communities" on page 104) in complex microbial communities can be identified (Harmsen *et al.*, 1996). With fluorescent molecular probes and confocal microscopy, variations in oxygen levels or pH within a biofilm can be analyzed (Caldwell *et al.*, 1992). New techniques are continuously being introduced and our lack of imagination may be the only limitation in unraveling the growth and nutrition of the microbial communities.

To understand how the microbial communities handle the situation on tooth surfaces, we will have to use the knowledge of the entire microbial world as our reference. In the present chapter it was discussed how microorganisms grow, how they inform themselves about the situation in their environment, and how they survive under starvation and shifts in aerobic and anaerobic conditions. The day we decipher the whole genomes of a representative number of oral microorganisms, we may have answers to some of these questions. The remaining questions about their physiology and ecology can then be explored from a more knowledgeable basis.

References

Abbe, K., J. Carlsson, S. Takahashi-Abbe, and T. Yamada. 1991. Oxygen and sugar metabolism in oral streptococci. Proc. Finn. Dent. Soc. 87: 477-487.

Adamson, M., and J. Carlsson. 1982. Lactoperoxidase and thiocyanate protect bacteria from hydrogen peroxide. Infect. Immun. 35: 20-24.

Aizenman, E., H. Engelberg-Kulka, and G. Glaser. 1996. An *Escherichia coli* chromosomal 'addiction module' regulated by 3',5'-bispyrophosphate: a model for programmed bacterial cell death. Proc. Natl. Acad. Sci. USA 93: 6059-6063.

Amano, A., H. Tamagawa, and S. Shizukuishi. 1986. Superoxide dismutase, catalase and peroxidases in oral anaerobic bacteria. J. Osaka Univ. Dent. Sch. 26: 187-192.

Andersson, C., M.-L. Sund, and L. Linder. 1984. Peptide utilization of oral streptococci. Infect. Immun. 43: 555-560.

Archibald, F. 1986. Manganese: its acquisition by and function in the lactic acid bacteria. Crit. Rev. Microbiol. 13: 63-109.

Åslund, F., and J. Beckwith. 1999a. The thioredoxin superfamily: redundancy, specificity, and gray-area genomics. J. Bacteriol. 18: 1375-1379.

Åslund, F., and J. Beckwith. 1999b. Bridge over troubled waters: sensing stress by disulfide bond formation. Cell 96: 751-753.

Aune, T.M., and E.L. Thomas. 1977. Accumulation of hypothiocyanite ion during peroxidase-catalyzed oxidation of thiocyanate ion. Eur. J. Biochem. 80: 209-214.

Aviv, M., H. Giladi, G. Schreiber, A.B. Oppenheim, and G. Glaser. 1994. Expression of the genes coding for the *Escherichia coli* integration host factor are controlled by growth phase, *rpoS,* ppGpp and by autoregulation. Mol. Microbiol. 14: 1021-1031.

Baba, T., and O. Schneewind. 1998. Instruments of microbial warfare: bacteriocin synthesis, toxicity and immunity. Trends Microbiol. 6: 66-71.

Banbula, A., P. Mak, M. Bugno, J. Silberring, A. Dubin, D. Nelson, J. Travis, and J. Potempa. 1999. Prolyl tripeptidase from *Porphyromonas gingivalis.* A novel enzyme with possible pathological implications for the development of periodontitis. J. Biol. Chem. 274: 9246-9252.

Barbieri, J.T., and C.D. Cox. 1979. Pyruvate oxidation by *Treponema pallidum.* Infect. Immun. 25: 157-163.

Barrett, J.F., and J.A. Hoch. 1998. Two-component signal transduction as a target for microbial anti-infective therapy. Antimicrob. Agent. Chemother. 42: 1529-1536.

Bauer, P.D., C. Trapp, D. Drake, K.G. Taylor, and R.J. Doyle. 1993. Acquisition of manganous ions by mutans group of streptococci. J. Bacteriol. 175: 819-825.

Becker, S., D. Vlad, S. Schuster, P. Pfeiffer, and G. Unden. 1997. Regulatory O_2 tensions for the synthesis of fermentation products in *Escherichia coli* and relation to aerobic respiration. Arch. Microbiol. 168: 290-296.

Beckers, H.J.A., and J.S. van der Hoeven. 1982. Growth rates of *Actinomyces viscosus* and *Streptococcus mutans* during early colonization of tooth surfaces in gnotobiotic rats. Infect. Immun. 35: 583-587.

Beighton, D., and H. Hayday. 1986. The influence of diet on the growth of streptococcal bacteria on the molar teeth of monkeys (*Macaca fascicularis*). Arch. Oral Biol. 31: 449-454.

Belay, N., R. Johnson, B.S. Rajagopal, E.D. de Macario, and L. Daniels. 1988. Methanogenic bacteria from human dental plaque. Appl. Environ. Microbiol. 54: 600-603.

Bender, G.R., S.V.W. Sutton, and R.E. Marquis. 1986. Acid tolerance, proton permeabilities, and membrane ATPases of oral streptococci. Infect. Immun. 53: 331-338.

Ben-Jacob, E., I. Cohen, and D.L. Gutnick. 1998. Cooperative organization of bacterial colonies: from genotype to morphotype. Annu. Rev. Microbiol. 52: 779-806.

Bergdahl, M. 2000. Salivary flow and oral complaints in adults from various age groups. Community Dent. Oral Epidemiol. 28: 59-66.

Berlett, B.S., and E.R. Stadtman. 1997. Protein oxidation in aging, disease, and oxidative stress. J. Biol. Chem. 272: 20313-20317.

Bernet, B., and C. Wyss. 1988. Identifizierung von 2-Acetamido-2-desoxy-α-D-glucopyranosyl-1-phosphat in Kuhmilch als Wachstumsfaktor für *Treponema vincentii*. Helv. Chim. Acta 71: 818-821.

Birkhed, D., K.-G. Rosell, and K. Granath. 1979. Structure of extracellular water-soluble polysaccharides synthesized from sucrose by oral strains of *Streptococcus mutans, Streptococcus salivarius, Streptococcus sanguis,* and *Actinomyces viscosus*. Arch. Oral Biol. 24: 53-61.

Björn, H., and J. Carlsson. 1964. Observations on a dental plaque morphogenesis. Odont. Rev. 15: 23-28.

Blackstock, W.P., and M.P. Weir. 1999. Proteomics: quantitative and physical mapping of cellular proteins. Trends Biotechnol. 17: 121-127.

Blattner, F.R., G. Plunkett III, C.A. Bloch, N.T. Perna, V. Burland, M. Riley, J. Collado-Vides, J.D. Glasner, C.K. Rode, G.F. Mayhew, J. Gregor, N.W. Davis, H.A. Kirkpatrick, M.A. Goeden, D.J. Rose, B. Mau, and Y. Shao. 1997. The complete genome sequence of *Escherichia coli* K-12. Science 277: 1453-1462.

Bloomquist, C.G., B.E. Reilly, and W.F. Liljemark. 1996. Adherence, accumulation, and cell division of a natural adherent bacterial population. J. Bacteriol. 178: 1172-1177.

Borg-Andersson, A., D. Birkhed, K. Berntorp, F. Lindgärde, and L. Matsson. 1998. Glucose concentration in parotid saliva after glucose/food intake in individuals with glucose intolerance and diabetes mellitus. Eur. J. Oral Sci. 106: 931-937.

Boss, A., A. Nussbaum-Shochat, and O. Amster-Choder. 1999. Characterization of the dimerization domain in BglG, an RNA-binding transcriptional antiterminator from *Escherichia coli*. J. Bacteriol. 181: 1755-1766.

Bouché, S., E. Klauck, D. Fischer, M. Lucassen, K. Jung, and R. Hengge-Aronis. 1998. Regulation of RssB-dependent proteolysis in *Escherichia coli*: a role for acetyl phosphate in a response regulator-controlled process. Mol. Microbiol. 27: 787-795.

Bradshaw, D.J., K.A. Homer, P.D. Marsh, and D. Beighton. 1994. Metabolic cooperation in oral microbial communities during growth on mucin. Microbiology 140: 3407-3412.

Bridges, B.A., and A. Timms. 1998. Effect of endogenous carotenoides and defective RpoS sigma factor on spontaneous mutation under starvation conditions in *Escherichia coli:* evidence for the possible involvement of singlet oxygen. Mut. Res. 403: 21-28.

Brusa, T., R. Conca, A. Ferrari, and A. Pecchioni. 1987. The presence of methanobacteria in human subgingival plaque. J. Clin. Periodontol. 14: 470-471.

Burne, R.A., Z.T. Wen, Y.-Y.M. Chen, and J.E.C. Penders. 1999. Regulation of expression of the fructan hydrolase gene of *Streptococcus mutans* GS-5 by induction and carbon catabolite repression. J. Bacteriol. 181: 2863-2871.

Button, D.K. 1985. Kinetics of nutrient-limited transport and microbial growth. Microbial. Rev. 49: 270-297.

Button, D.K. 1998. Nutrient uptake by microorganisms according to kinetic parameters from theory as related to cytoarchitecture. Microbiol. Mol. Biol. Rev. 62: 636-645.

Caldwell, C.E., and R.E. Marquis. 1999. Oxygen metabolism by *Treponema denticola*. Oral Microbiol. Immunol. 14: 66-72.

Caldwell, D.E., D.R. Korber, and J.R. Lawrence. 1992. Confocal laser microscopy and digital image analysis in microbial ecology. Adv. Microb. Ecol. 12: 1-67.

Calvo, J.M., and R.G. Matthews. 1994. The leucine-responsive regulatory protein, a global regulator of metabolism in *Escherichia coli*. Microbiol. Rev. 58: 466-490.

Carlsson, J. 1967. Presence of various types of non-hemolytic streptococci in dental plaque and in other sites of the oral cavity in man. Odont. Revy 18: 55-74.

Carlsson, J. 1968. A numerical taxonomic study of human oral streptococci. Odont. Revy 19: 137-160.

Carlsson, J. 1970. Nutritional requirements of *Streptococcus mutans*. Caries Res. 4: 305-320.

Carlsson, J. 1971a. Nutritional requirements of *Streptococcus salivarius*. J. Gen. Microbiol. 67: 66-76.

Carlsson, J., 1971b. Growth of *Streptococcus mutans* and *Streptococcus sanguis* in mixed culture. Arch. Oral Biol. 16: 963-965.

Carlsson. J. 1972. Nutritional requirements of *Streptococcus sanguis*. Arch. Oral Biol. 17: 1327-1332.

Carlsson, J., and V.S. Carpenter. 1980. The *recA⁺* gene product is more important than catalase and superoxide dismutase in protecting *Escherichia coli* against hydrogen peroxide toxicity. J. Bacteriol. 142: 319-321.

Carlsson, J., and J. Egelberg. 1965. Effect of diet on early plaque formation in man. Odont. Rev. 16: 112-125.

Carlsson, J., and M.-B.K. Edlund. 1987. Pyruvate oxidase in *Streptococcus sanguis* under various growth conditions. Oral Microbiol. Immunol. 2: 10-14.

Carlsson, J., M.-B.K. Edlund., and L. Hänström. 1984. Bactericidal and cytotoxic effects of hypothiocyanite-hydrogen peroxide mixtures. Infect. Immun. 44: 581-586.

Carlsson, J., Y. Iwami, and T. Yamada. 1983. Hydrogen peroxide excretion by oral

114

streptococci and effect of lactoperoxidase-thiocyanate-hydrogen peroxide. Infect. Immun. 40: 70-80.

Carlsson, J., and T. Johansson. 1973. Sugar and the production of bacteria in the human mouth. Caries Res. 7: 273-282.

Carlsson, J., J.T. Larsen, and M-B. Edlund. 1993. *Peptostreptococcus micros* has a uniquely high capacity to form hydrogen sulfide from glutathione. Oral Microbiol. Immunol. 8: 42-45.

Carlsson, J., and B. Sundström. 1968. Variation in composition of early dental plaque following ingestion of sucrose and glucose. Odont. Revy 19: 161-169.

Cassels, R., B. Oliva, and D. Knowles. 1995. Occurrence of the regulatory nucleotides ppGpp and pppGpp following induction of the stringent response in staphylococci. J. Bacteriol. 177: 5161-5165.

Chaillou, S., P.H. Pouwels, and P.W. Postma. 1999. Transport of of D-xylose in *Lactobacillus pentosus, Lactobacillus casei,* and *Lactobacillus plantarum:* evidence for a mechanism of facilitated diffusion via the phosphoenolpyruvate: mannose phosphotransferase system. J. Bacteriol. 181: 4768-4773.

Chen, Q., J.C. Arents, R. Bader, P.W. Postma, and O. Amster-Choder. 1997. BglF, the sensor of the *E. coli bgl* system, uses the same site to phosphorylate both a sugar and a regulatory protein. EMBO J. 16: 4617-4627.

Chen, Y.-Y.M., T.H. Hall, and R.A. Burne. 1998. *Streptococcus salivarius* urease expression: involvement of the phosphoenolpyruvate: sugar phosphotransferase system. FEMS Microbiol. Let. 165: 117-122.

Claesson, R., M-B. Edlund, S. Persson, and J. Carlsson. 1990. Production of volatile sulfur compounds by various *Fusobacterium* species. Oral Microbiol. Immunol. 5: 137-142.

Collier, L., A. Balows, and M. Sussman, eds. 1998. Topley & Wilson's Microbiology and Microbial Infections. Volume 2, Systematic Bacteriology. Arnold, London.

Condon, C., C. Squires, and C. L. Squires. 1995. Control of rRNA transcription in *Escherichia coli.* Microbiol. Rev. 59: 623-645.

Costerton, J.W., Z. Lewandowski, D.E. Caldwell, D.R. Korber, and H.M. Lappin-Scott. 1995. Microbial biofilms. Annu. Rev. Microbiol. 49: 711-745.

Costerton, J.W., P.S. Stewart, and E.P. Greenberg. 1999. Bacterial biofilms: a common cause of persistent infections. Science 284: 1318-1322.

Cowman, R.A., and S.S. Baron. 1997. Pathway for uptake and degradation of x-prolyl tripeptides in *Streptococcus mutans* VA-29R and *Streptococcus sanguis* ATCC 10556. J. Dent. Res. 76: 1477-1484.

Cowman, R.A., M.M. Perrella, B.O. Adams, and R.J. Fitzgerald. 1975. Amino acid requirements and proteolytic activity of *Streptococcus sanguis.* Appl. Microbiol. 30: 374-380.

Dashper, S.G., P.F. Riley, and E.C. Reynolds. 1995. Characterization of glutamine transport in *Streptococcus mutans.* Oral Microbiol. Immunol. 10: 183-187.

Dawes, C. 1983. A mathematical model of salivary clearance of sugar from the oral cavity. Caries Res. 17: 321-334.

Death, A., L. Notley, and T. Ferenci. 1993. Derepression of LamB protein facilitates outer membrane permeation of carbohydrates in *Escherichia coli* under conditions of nutrient stress. J. Bacteriol. 175: 1475-1483.

Debabov, D.V., M.P. Heaton, Q. Zhang, K.D. Steward, R.H. Lambalot, and F.C. Neuhaus. 1996. The D-alanyl carrier protein in *Lactobacillus casei*: cloning, sequencing, and expression of *dltC*. J. Bacteriol. 178: 3869-3876.

DeHaseth, P.L., M.L. Zupancic, and M.T. Record, Jr. 1998. RNA polymerase-promoter interactions: the comings and goings of RNA polymerase. J. Bacteriol. 180: 3019-3025.

Delwiche, E.A., J.J. Pestka, and M.L. Tortorello. 1985. The veillonellae: gram-negative cocci with a unique physiology. Annu. Rev. Microbiol. 39: 175-193.

Denger, K., and B. Schink. 1992. Energy conservation by succinate decarboxylation in *Veillonella parvula*. J. Gen. Microbiol. 138: 967-971.

DeRisi, J.L., V.R. Iyer, and P.O. Brown. 1997. Exploring the metabolic and genetic control of gene expression on a genomic scale. Science 278: 680-686.

Derré, I., G. Rapoport, and T. Msadek. 1999. CtsR, a novel regulator of stress and heat shock response, controls *clp* and molecular chaperone gene expression in Gram-positive bacteria. Mol. Microbiol. 31: 117-131.

Desai, J.D., and I. M. Banat. 1997. Microbial production of surfactants and their commercial potential. Microbiol. Mol. Biol. Rev. 61: 47-64.

Detmers, F.M., E.R.S. Kunji, F.C. Lanfermeijer, B. Poolman, and W.N. Konings. 1998. Kinetics and specificity of peptide uptake by the oligopeptide transport system of *Lactococcus lactis*. Biochemistry 37: 16671-16679.

De Vries, W., H.G.D. Niekus, M. Boellaard, and A.H. Stouthamer. 1980. Growth yields and energy generation by *Campylobacter sputorum* subspecies *bubulus* during growth in continuous culture with different hydrogen acceptors. Arch. Microbiol. 124: 221-227.

De Vries, W., T.R.M. Rietveld-Struijk, and A.H. Stouthamer. 1977. ATP formation associated with fumarate and nitrate reduction in growing cultures of *Veillonella alcalescens*. Antonie van Leeuenhoek 43: 153-167.

Dimroth, P. 1997. Primary sodium ion translocating enzymes. Biochim. Biophys. Acta 1318: 11-51.

Dimroth, P., and B. Schink. 1998. Energy conservation in the decarboxylation of dicarboxylic acids by fermenting bacteria. Arch. Microbiol. 170: 69-77.

Distler, W., and A. Kröncke. 1981. The lactate metabolism of the oral bacterium *Veillonella* from human saliva. Arch. Oral Biol. 26: 657-661.

Dmitriev, B.A., S. Ehlers, and E.T. Rietschel. 1999. Layered murein revisited: a fundamental new concept of bacterial cell wall structure, biogenesis and function. Med. Microbiol. Immunol. 187: 173-181.

Dolfing, J., and R.A. Prins. 1996. Methanogenic "food chains". ASM News 62: 117-118.

Dolin, M.I., E.F. Phares, and M.V. Long. 1965. Bound pyridine nucleotide of malic-lactic transhydrogenase. Biochem. Biophys. Res. Commun. 21: 303-310.

Doig, P., B. L. de Jonge, R. A. Alm, E. D. Brown, M. Uria-Nickelsen, B. Noonan, S. D. Mills, P. Tummino, G. Carmel, B. C. Guild, D. T. Moir, G. F. Vovis, and T. J. Trust. 1999. *Helicobacter pylori* physiology predicted from genomic comparison of two strains. Microbiol. Mol. Biol. Rev. 63: 675-707.

Dorman, C.J., J.C.D. Hinton, and A. Free. 1999. Domain organization and oligomerization among H-NS-like nucleoid-associated proteins in bacteria. Trends

116

Microbiol. 7: 124-128.

Douglas, H.C. 1950. On the occurrence of the lactate fermenting anaerobe, *Micrococcus lactilyticus,* in human saliva. J. Dent. Res. 29: 304-306.

Drapal, N., and G. Sawers. 1995. Purification of ArcA and analysis of its specific interaction with the *pfl* promoter-regulatory region. Mol. Microbiol. 16: 597-607.

Dukan, S., and T. Nyström. 1998. Bacterial senescence: stasis results in increased and differential oxidation of cytoplasmatic proteins leading to developmental induction of the heat shock regulon. Genes Dev. 12: 3431-3441.

Dunny, G., and B.A.B. Leonard. 1997. Cell-cell communication in gram-positive bacteria. Annu. Rev. Microbiol. 51: 527-564.

Durant, J.A., D.J. Nisbet, and S.C. Ricke. 1997. Comparison of batch culture growth and fermentation of a poultry *Veillonella* isolate and selected *Veillonella* species grown in a defined medium. Anaerobe 3: 391-397.

Durrett, R., and S. Levin. 1997. Allelopathy in spatially distributed populations. J. Theor. Biol. 185: 165-171.

Egger, L.A., H. Park, and M. Inouye. 1997. Signal transduction via histidyl-aspartyl phosphorelay. Genes to Cells 2: 167-184.

Fabret, C., V.A. Feher, and J.A. Hoch. 1999. Two-component signal transduction in *Bacillus subtilis:* how one organism sees its world. J. Bacteriol. 181: 1975-1983.

Fenno, J.C., A. Shaikh, G. Spatafora, and P. Fives-Taylor. 1995. The *fimA* locus of *Streptococcus parasanguis* encodes an ATP-binding membrane transport system. Mol. Microbiol. 15: 849-863.

Ferry, J.G., 1999. Enzymology of one-carbon metabolism in methanogenic pathways. FEMS Microbiol. Rev. 23: 13-38.

Franch, T., and K. Gerdes. 1996. Programmed cell death in bacteria: translational repression by mRNA end-pairing. Mol. Microbiol. 21: 1049-1060.

Fraser, C.M., S.J. Norris, G.M. Weinstock, O. White, G.G. Sutton, R. Dodson, M. Gwinn, E.K. Hickey, R. Clayton, K.A. Ketchum, E. Sodergren, J.M. Hardham, M.P. McLeod, S. Salzberg, J. Peterson, H. Khalak, D. Richardson, J.K. Howell, M. Chidambaram, T. Utterback, L. McDonald, P. Artiach, C. Bowman, M.D. Cotton, C. Fujii, S. Garland, B. Hatch, K. Horst, K. Roberts, M. Sandusky, J. Weidman, H. O. Smith, and J.C. Venter. 1998. Complete genome sequence of *Treponema pallidum*, the syphilis spirochete. Science 281: 375-388.

Fuqua, C., S.C. Winans, and E.P. Greenberg. 1996. Census and consensus in bacterial ecosystems: the LuxR-LuxI family of quorum-sensing transcriptional regulators. Annu. Rev. Microbiol. 50: 727-751.

Gerdes, K., A.P. Gultyaev, T. Franch, K. Pedersen, and N.D. Mikkelsen. 1997. Antisense RNA-regulated programmed cell death. Annu. Rev. Genet. 31: 1-31.

Gerhold, D., T. Rushmore, and C.T. Caskey. 1999. DNA chips: promising toys have become powerful tools. Trends Biochem. Sci. 24: 168-173.

Gibbons, R.J., and J.B. Macdonald. 1960. Hemin and vitamin K compounds as required factors for the cultivation of certain strains of *Bacteroides melaninogenicus.* J. Bacteriol. 80: 164-170.

Gibbons, R.J., and S.S. Socransky. 1962. Intracellular polysaccharide storage by organisms in dental plaques. Its relation to dental caries and microbial ecology of the oral cavity. Arch. Oral Biol. 7: 73-80.

Gillespie, M.J., and L.L. Barton. 1996. Hydrogenase coupled reactions in *Campylobacter rectus.* Anaerobe 2: 321-327.

Goffeau, A., B.G. Barrell, H. Bussey, R.W. Davis, B. Dujon, H. Feldmann, F. Galibert, J.D. Hoheisel, C. Jacq, M. Johnston, E.J. Louis, H.W. Mewes, Y. Murakami, P. Philippsen, H. Tettelin, and S.G. Oliver. 1996. Life with 6000 genes. Science 274: 546-567.

Gostick, D.O., J. Green, A.S. Irvine, M.J. Gasson, and J.R. Guest. 1998. A novel regulatory switch mediated by the FNR-like protein of *Lactobacillus casei.* Microbiology 144: 705-717.

Gotfredsen, M., and K. Gerdes. 1998. The *Escherichia coli relBE* genes belong to a new toxin-antitoxin gene family. Mol. Microbiol. 29: 1065-1076.

Gottesman, S. 1999. Regulation of proteolysis: developmental switches. Curr. Opin. Microbiol. 2: 142-147.

Gottschalk, G., and S. Peinemann. 1992. The anaerobic way of life. Vol. 1, p. 300-311. *In* A. Balows, H.G. Trüper, M. Dworkin, W. Harder, and K-H. Schleifer (ed.), The prokaryotes, 2nd ed. Springer-Verlag, New York.

Grant, R.A., D.J. Filman, S.E. Finkel, R. Kolter, and J.M. Hogle. 1998. The crystal structure of Dps, a ferritin homolog that binds and protects DNA. Nat. Struct. Biol. 5: 294-303.

Gray, K.M. 1997. Intercellular communication and group behavior in bacteria. Trends Microbiol. 5: 184-188.

Greene, S.R., and L.V. Stamm. 1999. Molecular characterization of a chemotaxis operon in the oral spirochete, *Treponema denticola.* Gene 232: 59-68.

Gregory, E.M., W.E.C. Moore, and L.V. Holdeman. 1978. Superoxide dismutase in anaerobes: survey. Appl. Environ. Microbiol. 35: 988-991.

Grenier, D. 1992. Nutritional interactions between two suspected periodontopathogens, *Treponema denticola* and *Porphyromonas gingivalis.* Infect. Immun. 60: 5298-5301.

Griffith, C.J., and J. Carlsson. 1974. Mechanism of ammonia assimilation in streptococci. J. Gen. Microbiol. 82: 253-260.

Grönroos, L., M. Saarela, J. Mättö, U. Tanner-Salo, A. Vuorela, and S. Alaluusua. 1998. Mutacin production by *Streptococcus mutans* may promote transmission of bacteria from mother to child. Infect. Immun. 66: 2595-2600.

Gross, R., J. Simon, F. Theis, and A. Kröger. 1998. Two membrane anchors of *Wolinella succinogenes* hydrogenase and their function in fumarate and polysulfide respiration. Arch. Microbiol. 170: 50-58.

Guggenheim, B. 1970. Extracellular polysaccharides and microbial plaque. Int. Dent. J. 20: 657-678.

Hamilton, I.R., and G. Svensäter. 1998. Acid-regulated proteins induced by *Streptococcus mutans* and other oral bacteria during acid shock. Oral Microbiol. Immunol. 13: 292-300.

Harmsen, H.J.M., H.M.P. Kengen, A.D.L. Akkermans, A.J.M. Stams, and W.M. de Vos. 1996. Detection and localization of syntrophic propionate-oxidizing bacteria in granular sludge by in situ hybridization using 16S rRNA-based oligonucleotide probes. Appl. Environ. Microbiol. 62: 1656-1663.

Håverstein, L.S. 1998. Identification of a competence regulon in *Streptococcus pneumoniae* by genomic analysis. Trends Microbiol. 6: 297-299.

Håverstein, L.S., P. Gaustad, I.F. Nes, and D.A. Morrison. 1996. Identification of the streptococcal competence-pheromone receptor. Mol. Microbiol. 21: 863-869.

Håverstein, L.S., R. Hakenbeck, and P. Gaustad. 1997. Natural competence in the genus *Streptococcus:* evidence that streptococci can change pherotype by interspecies recombinational exchange. J Bacteriol. 179: 6589-6594.

Head, I.M., J.R. Saunders, and R.W. Pickup. 1998. Microbial evolution, diversity, and ecology: a decade of ribosomal RNA analysis of uncultivated microorganisms. Microb. Ecol. 35: 1-21.

Hecker, M., and U. Völker. 1998. Non-specific, general and multiple stress resistance of growth-restricted *Bacillus subtilis* cells by the expression of the σ^B regulon. Mol. Microbiol. 29: 1129-1136.

Helmann, J.D. 1999. Anti-sigma factors. Curr. Opin. Microbiol. 2: 135-141.

Hengge-Aronis, R. 1999. Interplay of global regulators and cell physiology in the general stress response of *Escherichia coli*. Curr. Opin. Microbiol. 2: 148-152.

Hillman, J.D., A.L. Dzuback, and S.W. Andrews. 1987. Colonization of the human oral cavity by a *Streptococcus mutans* mutant producing increased bacteriocin. J. Dent. Res. 66: 1092-1094.

Hillman, J.D., J. Novák, E. Sagura, J.A. Gutierrez, T.A. Brooks, P.J. Crowley, M. Hess, A. Azizi, K.-P. Leung, D. Cvitkovitch, and A.S. Bleiweis. 1998. Genetic and biochemical analysis of mutacin 1140, a lantibiotic from *Streptococcus mutans*. Infect. Immun. 66: 2743-2749.

Hilpert, W., and P. Dimroth. 1991. On the mechanism of sodium ion translocation by methylmalonyl-CoA decarboxylase from *Veillonella alcalescens*. Eur. J. Biochem. 195: 79-86.

Hinode, D., R. Nakamura, D. Grenier, and D. Mayrand. 1998. Cross-reactivity of specific antibodies directed to heat shock proteins from periodontopathogenic bacteria of human origin. Oral Microbiol. Immunol. 13: 55-58.

Hochman, A. 1997. Programmed cell death in prokaryotes. Crit. Rev. Microbiol. 23: 207-214.

Hochschild, A., and S.L. Dove. 1998. Protein-protein contacts that activate and repress prokaryotic transcription. Cell 92: 597-600.

Hochstrasser, D.F. 1998. Proteome in perspective. Clin. Chem. Lab. Med. 36: 825-836.

Holme, T., and H. Palmstierna. 1956. Changes in glycogen and nitrogen-containing compounds in *Escherichia coli B* during growth in deficient media. I. Nitrogen and carbon starvation. Acta Chem. Scand. 10: 578-586.

Holmstrøm, K., T. Tolker-Nielsen, and S. Molin. 1999. Physiological states of individual *Salmonella typhimurium* cells monitored by in situ reverse transcription-PCR. J. Bacteriol. 181: 1733-1738.

Hoshino, E. 1987. L-Serine enhances the anaerobic lactate metabolism of *Veillonella dispar* ATCC 17745. J. Dent. Res. 66: 1162-1165.

Hoshino, E., F. Frölander, and J. Carlsson. 1978. Oxygen and the metabolism of *Peptostreptococcus anaerobius* VPI4330-1. J. Gen. Microbiol. 107: 235-248.

Hughes, K.T., and K. Mathee. 1998. The anti-sigma factors. Annu. Rev. Microbiol. 52: 231-286.

Humphery-Smith, I., S.J. Cordwell, and W.P. Blackstock. 1997. Proteome research: complementarity and limitations with respect to the RNA and DNA worlds. Electrophoresis 18: 1217-1242.

Igarashi, T., A. Yamamoto, and N. Goto. 1998. Detection of dextranase-producing gram-negative oral bacteria. Oral Microbiol. Immunol. 13: 382-386.

Iuchi, S., and E.C.C. Lin. 1991. Adaptation of *Escherichia coli* to respiratory conditions: regulation of gene expression. Cell 66: 5-7.

Iwami, Y., and T. Yamada. 1985. Regulation of glycolytic rate in *Streptococcus sanguis* grown under glucose-limited and glucose-excess conditions in a chemostat. Infect. Immun. 50: 378-381.

James, P. 1997. Breakthroughs and views: of genomes and proteomes. Biochem. Biophys. Res. Commun. 231: 1-6.

Janssen, P.H. 1992. Growth yield increase and ATP formation linked to succinate decarboxylation in *Veillonella parvula*. Arch. Microbiol. 157: 442-445.

Jenkinson, H.F., R.A. Baker, and G.W. Tannock. 1996. A binding-lipoprotein-dependent oligopeptide transport system in *Streptococcus gordonii* essential for uptake of hexa- and heptapeptides. J. Bacteriol. 178: 68-77.

Jenney Jr., F.E., M.F.J.M. Verhagen, X. Cul, and M.W.W. Adams. 1999. Anaerobic microbes: oxygen detoxification without superoxide dismutase. Science 286: 306-309.

Juarez, Z.E., and M.W. Stinson. 1999. An extracellular protease of *Streptococcus gordonii* hydrolyzes type IV collagen and collagen analogues. Infect. Immun. 67: 271-278.

Kakinuma, Y. 1998. Inorganic cation transport and energy transduction in *Enterococcus hirae* and other streptococci. Microbiol. Mol. Biol. Rev. 62: 1021-1045.

Kawamura, Y., X-G. Hou, F. Sultana, H. Miura, and T. Ezaki. 1995. Determination of 16S rRNA sequences of *Streptococcus mitis* and *Streptococcus gordonii* and phylogenetic relationships among members of the genus *Streptococcus*. Int. J. Syst. Bacteriol. 45: 406-408.

Kelstrup, J., and R.J. Gibbons. 1969. Bacteriocins from human and rodent streptococci. Arch. Oral Biol. 14: 251-258.

Kiley, P.J., and H. Beinert. 1999. Oxygen sensing by global regulator, FNR: the role of the iron-sulfur cluster. FEMS Microbiol. Rev. 22: 341-352.

Kilian, M., L. Mikkelsen, and J. Henrichsen. 1989. Taxonomic study of viridans streptococci: description of *Streptococcus gordonii* sp. nov. and emended description of *Streptococcus sanguis* (White and Niven 1946), *Streptococcus oralis* (Bridge and Sneath 1982), and *Streptococcus mitis* (Andrewes and Horder 1906). Int. J. Syst. Bacteriol. 39: 471-484.

Kilian, M., A. Thylstrup, and O. Fejerskov. 1979. Predominant plaque flora of Tanzanian children exposed to high and low water fluoride concentrations. Caries Res. 13: 330-343.

Kleanthous, C., A.M. Hemmings, G.R. Moore, and R. James. 1998. Immunity proteins and their specificity for endonuclease colicins: telling right from wrong in protein-protein recognition. Mol. Microbiol. 28: 227-233.

Knappe, J., and G. Sawers. 1990. A radical-chemical route to acetyl-CoA: the anaerobically induced pyruvate formate-lyase system in *Escherichia coli*. FEMS Microbiol. Rev. 75: 383-398.

Koch, A.L. 1997. Microbial physiology and ecology of slow growth. Microbiol. Mol. Biol. Rev. 61: 305-318.

Koga, T., T. Kusuzaki, H. Asakawa, H. Senpuku, T. Nishihara, and T. Noguchi. 1993. The 64-kilodalton GroEL-like protein of *Actinobacillus actinomycetemcomitans*. J. Periodont. Res. 28: 475-477.

Kolenbrander, P.E., R.N. Andersen, and N. Ganeshkumar. 1994. Nucleotide sequence of the *Streptococcus gordonii* PK488 coaggregation adhesin gene, *scaA,* and ATP-binding cassette. Infect. Immun. 62: 4469-4480.

Kolenbrander, P.E., R.N. Andersen, R.A. Baker, and H.F. Jenkinson. 1998. The adhesion-associated *sca* operon in *Streptococcus gordonii* encodes an inducible high-affinity ABC transporter for Mn^{2+} uptake. J. Bacteriol. 180: 290-295.

Konings, W.N., J. Boonstra, and W. de Vries. 1975. Amino acid transport in membrane vesicles of obligately anaerobic *Veillonella alcalescens*. J. Bacteriol. 122: 245-249.

Konings, W.N., B. Poolman, and H.W. van Veen. 1994. Solute transport and energy translocation in bacteria. Antonie van Leeuwenhoek 65: 369-380.

Konings, W.N., J.S. Lolkema, H. Bolhuis, H.W. van Veen, B. Poolman, and A.J.M. Driessen. 1997. The role of transport processes in survival of lactic acid bacteria. Energy transduction and multidrug resistance. Antonie van Leeuwenhoek 71: 117-128.

Kovárová-Kovar, K., and T. Egli. 1998. Growth kinetics of suspended microbial cells: from single-substrate-controlled growth to mixed-substrate kinetics. Microbiol. Mol. Biol. Rev. 62: 646-666.

Krasse, B. 1954.The proportional distribution of *Streptococcus salivarius* and other streptococci in various parts of the mouth. Odont. Revy 5: 203-211.

Kroos, L., B. Zhang, H. Ichikawa, and Y.-T.N. Yu. 1999. Control of σ factor activity during *Bacillus subtilis* sporulation. Mol. Microbiol. 31: 1285-1294.

Kunji, E.R.S., G. Fang, C.M. Jeronimus-Stratingh, A.P. Bruins, B. Poolman, and W.N. Konings. 1998. Reconstruction of the proteolytic pathway for use of β-casein by *Lactococcus lactis*. Mol. Microbiol. 27: 1107-1118.

Kunji, E.R.S., I. Mierau, A. Hagting, B. Poolman, and W.N. Konings. 1996. The proteolytic systems of lactic acid bacteria. Antonie van Leeuwenhoek 70: 187-221.

Kunst, F., Kunstl, F., N. Ogasawara, I. Moszer, A. M. Albertini, G. Alloni, V. Azevedo, M.G. Bertero, P. Bessières, A. Bolotin, S. Borchert, R. Borriss, L. Boursier, A. Brans, M. Braun, S.C. Brignell, S. Bron S. Brouillet, C.V. Bruschi, B. Caldwell, V. Capuano, N.M. Carter, S-K. Choi, J-J. Codani, I. F. Connerton, N.J. Cummings, R.A. Daniel, F. Denizot, K.M. Devine, A. Düsterhöft, S.D. Ehrlich, P.T. Emmerson, K.D. Entian, J. Errington, C. Fabret, E. Ferrari, D. Foulger, C. Fritz, M. Fujita, Y. Fujita, S. Fuma, A. Galizzi, N. Galleron, S-Y. Ghim, P. Glaser, A. Goffeau, E.J. Golightly, G. Grandi, G. Guiseppi, B.J. Guy, K. Haga, J. Haiech, C.R. Harwood, A. Hénaut, H. Hilbert, S. Holsappel, S. Hosono, M-F. Hullo, M. Itaya, L. Jones, B. Joris, D. Karamata, Y. Kasahara, M. Klaerr-Blanchard, C. Klein, Y. Kobayashi, P. Koetter, G. Koningstein, S. Krogh, M. Kumano, K. Kurita, A. Lapidus, S. Lardinois,

J. Lauber, V. Lazarevic, S-M. Lee, A. Levine, H. Liu, S. Masuda, C. Mauël, C. Médigue, N. Medina, R.P. Mellado, M. Mizuno, D. Moestl, S. Nakai, M. Noback, D. Noone, M. O'Reilly, K. Ogawa, A. Ogiwara, B. Oudega, S-H. Park, V. Parro, T.M. Pohl, D. Portetelle, S. Porwollik, A. M. Prescott, E. Presecan, P. Pujic, B. Purnelle, G. Rapoport, M. Rey, S. Reynolds, M. Rieger, C. Rivolta, E. Rocha, B. Roche, M. Rose, Y. Sadaie, T. Sato, E. Scanlan, S. Schleich, R. Schroeter, F. Scoffone, J. Sekiguchi, A. Sekowska, S.J. Seror, P. Serror, B-S. Shin, B. Soldo, A. Sorokin, E. Tacconi, T. Takagi, H. Takahashi, K. Takemaru, M. Takeuchi, A. Tamakoshi, T. Tanaka, P. Terpstra, A. Tognoni, V. Tosato, S. Uchiyama, M. Vandenbol, F. Vannier, A. Vassarotti, A. Viari, R. Wambutt, E. Wedler, H. Wedler, T. Weitzenegger, P. Winters, A. Wipat, H. Yamamoto, K. Yamane, K. Yasumoto, K. Yata, K. Yoshida, H-F. Yoshikawa, E. Zumstein, H. Yoshikawa, and A. Danchin. 1997. The complete genome sequence of the gram-positive bacterium *Bacillus subtilis*. Nature 390: 249-256.

Kuramitsu, H.K. 1998. Proteases of *Porphyromonas gingivalis*: what don't they do? Oral Microbiol. Immunol. 13: 263-270.

Lagerlöf, F., and C. Dawes. 1984. The volume of saliva in the mouth before and after swallowing. J. Dent. Res. 63: 618-621.

Lamont, R.J., and H.F. Jenkinson. 1998. Life below the gum line: pathogenic mechanisms of *Porphyromonas gingivalis*. Microbiol. Mol. Biol. Rev. 62: 1244-1263.

Lancaster, C.R.D., A. Kröger, M. Auer, and H. Michel. 1999. Structure of fumarate reductase from *Wolinella succinogenes* at 2.2 Å resolution. Nature 402: 377-385.

Lange, R., and R. Hengge-Aronis. 1994. The cellular concentration of the σ^S subunit of RNA polymerase in *Escherichia coli* is controlled at the level of transcription, translation, and protein stability. Genes Dev. 8: 1600-1612.

Lazazzera, B.A., and A.D. Grossman. 1998. The ins and outs of peptide signaling. Trends Microbiol. 6: 288-294.

Lazdunski, C.J., E. Bouveret, A. Rigal, L. Journet, R. Lloubès, and H. Bénédetti. 1998. Colicin import into *Escherichia coli* cells. J. Bacteriol. 180: 4993-5002.

Leadbetter, J.R., T.M. Schmidt, J.R. Graber, and J.A. Breznak. 1999. Acetogenesis from H_2 plus CO_2 by spirochetes from termite guts. Science 283: 686-689.

Leclerc, M., A. Bernalier, G. Donadille, and M. Lelait. 1997. H_2/CO_2 metabolism in acetogenic bacteria isolated from human colon. Anaerobe 3: 307-315.

Li, H., S. Arakawa, Q.-D. Deng, and H. Kuramitsu. 1999. Characterization of a novel methyl-accepting chemotaxis gene, *dmcB,* from the oral spirochete *Treponema denticola.* Infect. Immun. 67: 694-699.

Liljemark, W.F., C.G. Bloomquist, B.E. Reilly, C.J. Bernards, D.W. Townsend, A.T. Pennock, and J.L. LeMoine. 1997. Growth dynamics in a natural biofilm and its impact on oral disease management. Adv. Dent. Res. 11: 14-23.

Linton, K.J., and C.F. Higgins. 1998. The *Escherichia coli* ATP-binding cassette (ABC) proteins. Mol. Microbiol. 28: 5-13.

Listgarten, M.A., H.E. Mayo, and R. Tremblay. 1975. Development of dental plaque on epoxy resin crowns in man. A light and electron microscopic study. J. Periodontol. 46: 10-26.

Liu, X., and T. Ferenci. 1998. Regulation of porin-mediated outer membrane permeability by nutrient limitation in *Escherichia coli.* J. Bacteriol. 180: 3917-3922.

Loesche, W.J., F. Gusberti, G. Mettraux, T. Higgins, and S. Syed. 1983. Relationship between oxygen tension and subgingival bacterial flora in untreated human periodontal pockets. Infect. Immun. 42: 659-667.

Loewen, P.C., and R. Hengge-Aronis. 1994. The role of the sigma factor σ^S (KatF) in bacterial global regulation. Annu. Rev. Microbiol. 48: 53-80.

Loewen, P.C., B. Hu, J. Strutinsky, and R. Sparling. 1998. Regulation in the *rpoS* regulon of *Escherichia coli.* Can. J. Microbiol. 44: 707-717.

Lu, B., and B.C. McBride. 1994. Stress response of *Porphyromonas gingivalis.* Oral Microbiol. Immunol. 9: 166-173.

Marais, A., G.L. Mendz, S.L. Hazell, and F. Mégraud. 1999. Metabolism and genetics of *Helicobacter pylori:* the genome era. Microbiol. Mol. Biol. Rev. 63: 642-674.

Martínez-Costa, O.H., M.A. Fernández-Moreno, and F. Malpartida. 1998. The *relA/spoT*-homologous gene in *Streptomyces coelicolor* encodes both ribosome-dependent (p)ppGpp-synthesizing and –degrading activities. J. Bacteriol. 180: 4123-4132.

Martzen, M.R., S.M. McCraith, S.L. Spinelli, F.M. Torres, S. Fields, E.J. Grayhack, and E.M. Phizicky. 1999. A biochemical genomics approach for identifying genes by the activity of their products. Science 286: 1153-1155.

Matsumoto, J., M. Higuchi, M. Shimada, Y. Yamamoto, and Y. Kamio. 1996. Molecular cloning and sequence analysis of the gene encoding the H_2O-forming NADH-oxidase from *Streptococcus mutans.* Biosci. Biotechnol. Biochem. 60: 39-43.

Michán, C., M. Manchado, G. Dorado, and C. Pueyo. 1999. In vivo transcription of the *Escherichia coli oxyR* regulon as a function of growth phase and in response to oxidative stress. J. Bacteriol. 181: 2759-2764.

Mierau, I., E.R.S. Kunji, K.J. Leehouts, M.A. Hellendoorn, A.J. Haandrikman, B. Poolman, W.N. Konings, G. Venema, and J. Kok. 1996. Multiple-peptidase mutants of *Lactococcus lactis* are severely impaired in their ability to grow in milk. J. Bacteriol. 178: 2794-2803.

Mikx, F.H.M. 1997. Environmental effects on the growth and proteolysis of *Treponema denticola* ATCC 33520. Oral Microbiol. Immunol. 12: 249-253.

Milnes, A.R., and G.H.W. Bowden. 1985. The microflora associated with developing lesions of nursing caries. Caries Res. 19: 289-297.

Missiakas, D., and S. Raina. 1998. The extracytoplasmic function sigma factors: role and regulation. Mol. Microbiol. 28: 1059-1066.

Moat, A.G., and J.W. Foster (ed.). 1995. Microbial physiology. 3rd ed. New York. Wiley-Liss.

Møller, S., C. Sternberg, J.B. Andersen, B.B. Christensen, J.L. Ramos, M. Givskov, and S. Molin. 1998. In situ gene expression in mixed-culture biofilm: evidence of metabolic interactions between community members. Appl. Environ. Microbiol. 64: 721-732.

Monchois, V., R.-M. Willemot, and P. Monsan. 1999. Glucansucrases: mechanism of action and structure-function relationships. FEMS Microbiol. Rev. 23: 131-151.

Monedero, V., M.J. Gosalbes, and G. Pérez-Martínez. 1997. Catabolite repression in *Lactobacillus casei* ATCC 393 is mediated by ccpA. J. Bacteriol. 179: 6657-6664.

Monod, J. 1949. The growth of bacterial cultures. Annu. Rev. Microbiol. 3: 371-393.

Mooney, R.A., I. Artsimovitch, and R. Landick. 1998. Information processing by RNA polymerase: recognition of regulatory signals during RNA chain elongation. J. Bacteriol. 180: 3265-3275.

Moskovitz, J., B.S. Berlett, J.M. Poston, and E.R. Stadtman. 1999. Methionine sulfoxide reductase in antioxidant defense. Meth. Enzymol. 300: 239-244.

Msadek, T.1999. When the going gets tough: survival strategies and environmental signaling networks in *Bacillus subtilis*. Trends Microbiol. 7: 201-207.

Murphy, M.G., L. O'Connor, D. Walsh, and S. Condon. 1985. Oxygen dependent lactate utilization by *Lactobacillus plantarum*. Arch. Microbiol. 141: 75-79.

Nakano, M.M., and P. Zuber. 1998. Anaerobic growth of a "strict aerobe" (*Bacillus subtilis*). Annu. Rev. Microbiol. 52: 165-190.

Naqui, A., B. Chance, and E. Cadenas. 1986. Reactive oxygen intermediates in biochemistry. Annu. Rev. Biochem. 55: 137-166.

Narberhaus, F. 1999. Negative regulation of bacterial heat shock genes. Mol. Microbiol. 31: 1-8.

Nes, I.F., and J.R. Tagg. 1996. Novel lantibiotics and their pre-peptides. Antonie van Leeuwenhoek 69: 89-97.

Ng, S.K.C., and I.R. Hamilton. 1974. Gluconeogenesis by *Veillonella parvula* M_4: evidence for the indirect conversion of pyruvate to P-enolpyruvate. Can. J. Microbiol. 20: 19-28.

Ng, S.K.C., M. Wong, and I.R. Hamilton. 1982. Properties of oxaloacetate decarboxylase from *Veillonella parvula*. J. Bacteriol. 150: 1252-1258.

Niimi, M., M.G. Shepherd, and B.C. Monk. 1996. Differential profiles of soluble proteins during the initiation of morphogenesis in *Candida albicans*. Arch. Microbiol. 166: 260-268.

Nissen-Meyer, J., and I.F. Nes. 1997. Ribosomally synthesized antimicrobial peptides: their function, structure, biogenesis, and mechanism of action. Arch. Microbiol. 167: 67-77.

Nyström, T. 1994. The glucose-starvation stimulon of *Escherichia coli*: induced and repressed synthesis of enzymes of central metabolic pathways and role of acetyl phosphate in gene expression and starvation survival. Mol. Microbiol. 12: 833-843.

Nyström, T. 1998. To be or not to be: the ultimate decision of the growth-arrested bacterial cell. FEMS Microbiol. Rev 21: 283-290.

Nyström, T. 1999. Starvation, cessation of growth and bacterial aging. Curr. Opin. Microbiol. 2: 214-219.

Nyström, T., C. Larsson, and L. Gustafsson. 1996. Bacterial defense against aging: role of the *Escherichia coli* ArcA regulator in gene expression, readjusted energy flux and survival during stasis. EMBO J. 15: 2319-3228.

Nyvad, B. 1993. Microbial colonization of human tooth surfaces. APMIS 101: Suppl. No 32, 1-45.

O'Farrel, P.H. 1975. High resolution two-dimensional electrophoresis of proteins. J Biol. Chem. 250: 4007-4021.

Ogawa, T., K. Tomita, T. Ueda, K. Watanabe, T. Uozumi, and H. Masaki. 1999. A cytotoxic ribonuclease targeting specific transfer RNA anticodons. Science 283: 2097-2100.

Palmer, R.J., and D.C. White. 1997. Developmental biology of biofilms: implication for treatment and control. Trends Microbiol. 5: 435-440.

Parker, R.B., and H.R. Creamer. 1971. Contribution of plaque polysaccharides to growth of cariogenic microorganisms. Arch. Oral Biol. 16: 855-862.

Perego, M. 1998. Kinase-phosphatase competition regulates *Bacillus subtilis* development. Trends Microbiol. 6: 366-370.

Perego, M., P. Glaser, A. Minutello, M.A. Strauch, K. Leopold, and W. Fischer. 1995. Incorporation of D-alanine into lipoteichoic acid and wall teichoic acid in *Bacillus subtilis*. Identification of genes and regulation. J. Biol. Chem. 270: 15598-15606.

Perraud, A.-L., V. Weiss, and R. Gross. 1999. Signalling pathways in two-component phosphorelay systems. Trends Microbiol. 7: 115-120.

Perry, D., and H.K. Kuramitsu. 1981. Genetic transformation of *Streptococcus mutans*. Infect. Immun. 32: 1295-1297.

Persson, S., 1992. Hydrogen sulfide and methyl mercaptan in periodontal pockets. Oral Microbiol. Immunol. 7: 378-379.

Persson, S., M.-B. Edlund, R. Claesson, and J. Carlsson. 1990. The formation of hydrogen sulfide and methyl mercaptan by oral bacteria. Oral Microbiol. Immunol. 5: 195-201.

Peypoux, F., J.M. Bonmatin, and J. Wallach. 1999. Recent trends in the biochemistry of surfactin. Appl. Microbiol. Biotechnol. 51: 553-563.

Polissi, A., A. Pontiggia, G. Feger, M. Altieri, H. Mottl, L. Ferrari, and D. Simon. 1998. Large-scale identification of virulence genes from *Streptococcus pneumoniae*. Infect. Immun. 66: 5620-5629.

Postle, K. 1999. Active transport by customized β-barrels. Nat. Struct. Biol. 6: 3-6.

Postma, P.W., J.W. Lengeler, and G.R. Jacobson. 1993. Phosphoenolpyruvate: carbohydrate phosphotransferase systems of bacteria. Microbiol. Rev. 57: 543-594.

Pruitt, K.M., B. Mansson-Rahemtulla, and J. Tenovuo. 1983. Detection of hypothiocyanite (OSCN⁻) ion in human parotid saliva and the effect of pH on OSCN⁻ generation in the salivary peroxidase antimicrobial system. Arch. Oral Biol. 28: 517-525.

Pütsep, K., C.-I. Brändén, H.G. Boman, and S. Normark. 1999. Antibacterial peptide from *H. pylori*. Science 398: 671-672.

Qi, F., P. Chen, and P.W. Caufield. 1999. Functional analysis of the promoters in the lantibiotic mutacin II biosynthetic locus in *Streptococcus mutans*. Appl. Environ. Microbiol. 65: 652-658.

Rafay, A.M., K.A. Homer, and D. Beighton. 1996. Effect of mucin and glucose on proteolytic and glycosidic activities of *Streptococcus oralis*. J. Med. Microbiol. 44: 409-417.

Ramotar, K., J.M. Conly, H. Chubb, and T.J. Louie. 1984. Production of menadione by intestinal anaerobes. J. Infect. Dis. 150: 213-218.

Rauhut, R. and G. Klug. 1999. mRNA degradation in bacteria. FEMS Microbiol. Rev. 23: 353-370.

Reizer, J., C. Hoischen, F. Titgemeyer, C. Rivolta, R. Rabus, J. Stülke, D. Karamata, M.H. Saier Jr, and W. Hillen. 1998. A novel protein kinase that controls carbon catabolite repression in bacteria. Mol. Microbiol. 27: 1157-1169.

Riley, M.A. 1998. Molecular mechanisms of bacteriocin evolution. Annu. Rev. Genet. 32: 255-278.

Riley, M.A., and D.M. Gordon. 1999. The ecological role of bacteriocins in bacterial competition. Trends Microbiol. 7: 129-133.

Rogers, A.H., A.L. Pfennig, N.J. Gully, and P.S. Zilm. 1991. Factors affecting peptide catabolism by oral streptococci. Oral immunol. Microbiol. 6: 72-75.

Rogers, A.H., J. van der Hoeven, and F.H.M. Mikx. 1979. Effect of bacteriocin production by *Streptococcus mutans* on the plaque of gnotobiotic rats. Infect. Immun. 23: 571-576.

Rogosa, M. 1964. The genus *Veillonella*, I. General cultural, ecological, and biochemical considerations. J. Bacteriol. 87: 162-170.

Rojo, F. 1999. Repression of transcription initiation in bacteria. J. Bacteriol. 181: 2987-2991.

Romeo, T. 1998. Global regulation by the small RNA-binding protein CrsA and the non-coding RNA molecule CsrB. Mol. Microbiol. 29: 1321-1330.

Ross, K.F., C.W. Ronson, and J.R. Tagg. 1993. Isolation and characterization of the lantibiotic salivaricin A and its structural gene *salA* from *Strepococcus salivarius* 20P3. Appl. Environ. Microbiol. 59: 2014-2021.

Ross, R.P., and A. Claiborne. 1997. Evidence for regulation of the NADH peroxidase gene (*npr*) from *Enterococcus faecalis* by OxyR. FEMS Microbiol. Lett. 151: 177-183.

Russell, R.R.B., J. Aduse-Opoku, I.C. Sutcliffe, L. Tao, and J.J. Ferretti. 1992. A binding protein-dependent transport system in *Streptococcus mutans* responsible for multiple sugar metabolism. J. Biol. Chem. 267: 4631-4637.

Ruthberg, B., 1997. Antitermination of transcription of catabolic operons. Mol. Microbiol. 23: 413-421.

Sahl, H.-G., and G. Bierbaum. 1998. Lantibiotics: biosynthesis and biological activities of uniquely modified peptides from gram-positive bacteria. Annu. Rev. Microbiol. 52: 41-79.

Sawers, G. 1999. The aerobic/anaerobic interface. Curr. Opin. Microbiol. 2: 181-187.

Schink, B. 1997. Energetics of syntrophic cooperation in methanogenic degradation. Microbiol. Mol. Biol. Rev. 61: 262-280.

Sedewitz, B., K.H. Schleifer, and F. Götz. 1984. Purification and biochemical characterization of pyruvate oxidase from *Lactobacillus plantarum*. J. Bacteriol. 160: 273-278.

Shapiro, J.A. 1998. Thinking about bacterial populations as multicellular organisms. Annu. Rev. Microbiol. 52: 81-104.

Sherrill, C., and R.C. Fahey. 1998. Import and metabolism of glutathione by *Streptococcus mutans*. J. Bacteriol. 180: 1454-1459.

126

Shockman, G.D., M.L. Higgins, L. Daneo-Moore, S.J. Mattingly, J.R. Di Persio, and B. Terleckyj. 1976. Studies of balanced and unbalanced growth of *Streptococcus mutans*. J. Dent. Res. 55: A10-A18.

Sigler, P.B., Z. Xu, H.S. Rye, S.G. Burston, W.A. Fenton, and A.L. Horwich. 1998. Structure and function in GroEL-mediated protein folding. Annu. Rev. Biochem. 67: 581-608.

Silversmith, R.E., and R.B. Bourret. 1999. Throwing the switch in bacterial chemotaxis. Trends Microbiol. 7: 16-22.

Simpson, C.L, and R.R.B. Russell. 1998a. Identification of a homolog of ccpA catabolite repressor protein in *Streptococcus mutans*. Infect. Immun. 66: 2085-2092.

Simpson, C.L., and R.R.B. Russell. 1998b. Intracellular α-amylase of *Streptococcus mutans*. J. Bacteriol. 180: 4711-4717.

Smith, A.J., R.G. Quivey, Jr., and R.C. Faustoferri. 1996. Cloning and nucleotide sequence analysis of the *Streptococcus mutans* membrane-bound, proton-translocating ATPase operon. Gene 183: 87-96.

Smith, V.H., and D.J. Pippin. 1998. Implications of resource-ratio theory for oral microbial ecology. Eur. J. Oral Sci. 106: 605-615.

Sokatch, J.R. 1969. Bacterial physiology and metabolism. Academic Press, London, p. 163-193.

Socransky, S.S., A.D. Manganiello, D. Propas, V. Oram, and J. van Houte. 1977. Bacteriological studies of developing supragingival dental plaque. J. Periodont. Res. 12: 90-106.

Spatafora, G.A., M. Sheets, R. June, D. Luyimbazi, K. Howard, R. Hulbert, D. Barnard, M. El Janne, and M. C. Hudson. 1999. Regulated expression of the *Streptococcus mutans dlt* genes correlates with intracellular polysaccharide accumulation. J. Bacteriol. 181: 2363-2372.

Spiess, C., A. Beil, and M. Ehrmann. 1999. A temperature-dependent switch from chaperone to protease in a widely conserved heat shock protein. Cell 97: 339-347.

Spiro, S. 1994. The FNR family of transcriptional regulators. Antonie van Leeuenhoek 66: 23-36.

Spiro, S., and J.R. Guest. 1990. FNR and its role in oxygen-regulated gene expression in *Escherichia coli*. FEMS Microbiol. Rev. 75: 399-428.

Stadtman, E.R. 1992. Protein oxidation and aging. Science 257: 1220-1224.

Stock, J. 1999. Sensitivity, cooperativity and gain in chemotaxis signal transduction. Trends Microbiol. 7: 1-4.

Storz, G., and J.A. Imlay. 1999. Oxidative stress. Curr. Opin. Microbiol. 2: 188-194.

Struhl, K. 1999. Fundamentally different logic of gene regulation in eukaryotes and prokaryotes. Cell 98: 1-4.

Stülke, J., M. Arnaud, G. Rapoport, and I. Martin-Verstraete. 1998. PRD – a protein domain involved in PTS-dependent induction and carbon catabolite repression of catabolic operons in bacteria. Mol. Microbiol. 28: 865-874.

Stülke, J., and W. Hillen. 1998. Coupling physiology and gene regulation in bacteria: the phosphotransferase sugar uptake system delivers the signals. Naturwissenschaften 85: 583-592.

Stülke, J., and W. Hillen. 1999. Carbon catabolite repression in bacteria. Curr. Opin. Microbiol. 2: 195-201.

Sturr, M.G., and R.E. Marquis. 1992. Comparative acid tolerances and inhibitor sensitivities of isolated F-ATPases of oral lactic acid bacteria. Appl. Environ. Microbiol. 58: 2287-2291.

Sund, M.-L., C. Branting, and L.E. Linder. 1989. Water-soluble glucans from *Streptococcus mutans* strain Ing-Britt as an energy source for bacterial growth. Caries Res. 23: 256-260.

Sze, C.C., and V. Shingler. 1999. The alarmone (p)ppGpp mediates physiological-responsive control at the σ^{54}-dependent Po promoter. Mol. Microbiol. 31: 1217-1228.

Takahashi, N., Y. Iwami, and T. Yamada. 1991. Metabolism of intracellular polysaccharide in the cells of *Streptococcus mutans* under strictly anaerobic conditions. Oral Microbiol. Immunol. 6: 299-304.

Takahashi, N., and T. Yamada. 1999. Glucose and lactate metabolism by *Actinomyces naeslundii*. Crit. Rev. Oral Biol. Med. 10: 487-503.

Tang-Larsen, J., R. Claesson, M.-B. Edlund, and J. Carlsson. 1995. Competition for peptides and amino acids among periodontal bacteria. J. Periodont. Res. 30: 390-395.

Tappan, H. 1974. Molecular oxygen and evolution, p. 81-135. *In* O. Hayaishi (ed.), Molecular oxygen in biology: topics in molecular oxygen research. North-Holland Publishing Company, Amsterdam.

Taylor, B.L., and I.B. Zhulin. 1999. PAS domains: internal sensors of oxygen, redox potential, and light. Microbiol. Mol. Biol. Rev. 63: 479-506.

Tenovuo, J., K.M. Pruitt, and E.L. Thomas. 1982. Peroxidase antimicrobial system of human saliva: hypothiocyanite levels in resting and stimulated saliva. J. Dent. Res. 61: 982-985.

Terleckyj, B., N.P. Willett, and G.D. Shockman. 1975. Growth of several cariogenic strains of oral streptococci in a chemical defined medium. Infect. Immun. 11: 649-655.

Thauer, R.K., K. Jungermann, and K. Decker. 1977. Energy conservation in chemotrophic anaerobic bacteria. Bacteriol. Rev. 41: 100-180.

Thomas, E.L., and T.M. Aune. 1978. Lactoperoxidase, peroxide, thiocyanate antimicrobial systems: correlation of sulfhydryl oxidation with antimicrobial action. Infect. Immun. 20: 456-463.

Tolker-Nielsen, T., K. Holmstrøm, and S. Molin. 1997. Visualization of specific gene expression in individual *Salmonella typhimurium* cells by in situ PCR. Appl. Environ. Microbiol. 63: 4196-4203.

Tweedale, H., L. Notley-McRobb, and T. Ferenci. 1998. Effect of slow growth on metabolism of *Escherichia coli*, as revealed by global metabolite pool ("metabolome") analysis. J. Bacteriol. 180: 5109-5116.

Unden, G., S. Becker, J. Bongaerts, J. Schirawski, and S. Six. 1994. Oxygen regulated gene expression in facultatively anaerobic bacteria. Antonie van Leeuwenhoek 66: 3-23.

Unden, G., and J. Bongaerts. 1997. Alternative respiratory pathways of *Escherichia coli*: energetics and transcriptional regulation in response to electron acceptors.

Biochim. Biophys. Acta 1320: 217-234.

Unden, G., and J. Schirawski. 1997. The oxygen-responsive transcriptional regulator FNR of *Escherichia coli:* the search for signals and reactions. Mol. Microbiol. 25: 205-210.

Vadeboncoeur, C., and M. Pelletier. 1997. The phosphoenolpyruvate: sugar phosphotransferase system of oral streptococci and its role in the control of sugar metabolism. FEMS Microbiol. Rev. 19: 187-207.

VanBogelen, R.A., K.D. Greis, R.M. Blumenthal, T.H. Tani, and R.G. Matthews. 1999. Mapping regulatory networks in microbial cells. Trends Microbiol. 7: 320-328.

Van der Hoeven, J.S., and P.J.M. Camp. 1993. Mixed continuous culture of *Streptococcus mutans* with *Streptococcus sanguis* or with *Streptococcus oralis* as a model to study the ecological effects of the lactoperoxidase system. Caries Res. 27: 26-30.

Van der Hoeven, J.S., C.W.A. van den Kieboom, and P.J.M. Camp. 1990. Utilization of mucin by oral *Streptococcus* species. Antonie van Leeuwenhoek 57: 165-172.

Van der Hoeven, J.S., C.W.A. van den Kieboom, and M.J.M. Schaeken. 1995. Sulfate-reducing bacteria in the periodontal pocket. Oral Microbiol. Immunol. 10: 288-290.

Van Houte, J., R.J Gibbons, and S.B. Banghart. 1970. Adherence as a determinant of the presence of *Streptococcus salivarius* and *Streptococcus sanguis* in the human tooth surface. Arch. Oral Biol. 15: 1025-1034.

Van Houte, J., R.J. Gibbons, and A.J. Pulkkinen. 1971. Adherence as an ecological determinant for streptococci in the human mouth. Arch. Oral Biol. 16: 1131-1141.

Van Houte, J., and C.A. Saxton. 1971. Cell wall thickening and intracellular polysaccharide in microorganisms of the dental plaque. Caries Res. 5: 30-43.

Wassarman, K.M., A. Zhang, and G. Storz. 1999. Small RNAs in *Escherichia coli.* Trends Microbiol. 7: 37-45.

Weiger, R., L. Netuschil, C. von Ohle, U. Schlagenhauf, and M. Brecx. 1995. Microbial generation time during the early phases of supragingival dental plaque formation. Oral Microbiol. Immunol. 10: 93-97.

Weisberg, R.A., and M.E. Gottesman. 1999. Processive antitermination. J. Bacteriol. 181: 359-367.

Weuster-Botz, D., and A.A. de Graaf. 1996. Reaction engineering methods to study intracellular metabolite concentrations. Adv. Biochem. Eng. Biotechnol. 54: 75-108.

Willcox, M.D.P., M. Patrikakis, and K.W. Knox. 1995. Degradative enzymes of oral streptococci. Aust. Dent. J. 40: 121-128.

Williams, R.M., and S. Rimsky. 1997. Molecular aspects of the *E. coli* nucleoid protein, H-NS: a central controller of gene regulatory networks. FEMS Microbiol. Lett. 156: 175-185.

Winzeler, E.A., D.D. Shoemaker, A. Astromoff, H. Liang, K. Anderson, B. Andre, R. Bangham, R. Benito, J.D. Boeke, H. Bussey, A. M. Chu, C. Connelly, K. Davis, F. Dietrich, S. W. Dow, M. El Bakkoury, F. Foury, S.H. Friend, E. Gentalen, G. Giaever, J.H. Hegemann, T. Jones, M. Laub, H. Liao, N. Liebundguth, D.J. Lockhart, A. Lucau-Danila, M. Lussier, N. M'Rabet, P. Menard, M. Mittmann, C.

Pai, C. Rebischung, J.L. Revuelta, L. Riles, C. J. Roberts, P. Ross-MacDonald, B. Scherens, M. Snyder, S. Sookhai-Mahadeo, R.K. Storms, S. Véronneau, M. Voet, G. Votckaert, T.R. Ward, R. Wysocki, G.S. Yen, K. Yu, K. Zimmermann, P. Philippsen, M. Johnston, R. W. Davis. 1999. Functional characterization of the *S. cerevisiae* genome by gene deletion and parallel analysis. Science 285: 901-906.

Visick, J.E., and S. Clarke. 1995. Repair, refold, recycle: how bacteria can deal with spontaneous and environmental damage to proteins. Mol. Microbiol. 16: 835-845.

Visick, J.E., H. Cai, and S. Clarke. 1998. The L-isoaspartyl protein repair methyltransferase enhances survival of aging *Escherichia coli* subjected to secondary environmental stresses. J. Bacteriol. 180: 2623-2629.

Wösten, M.M.S.M. 1998. Eubacterial sigma-factors. FEMS Microbiol. Rev. 22: 127-150.

Wyss, C. 1989a. *Campylobacter-Wolinella* group organisms are the only oral bacteria that form arylsulfatase-active colonies on a synthetic indicator medium. Infect. Immun. 57: 1380-1383.

Wyss, C. 1989. Dependence of proliferation of *Bacteroides forsythus* on exogenous *N*-acetylmuramic acid. Infect. Immun. 57: 1757-1759.

Wyss, C. 1992. Growth of *Porphyromonas gingivalis, Treponema denticola, T. pectinovorum, T. socranskii,* and *T. vincentii* in a chemically defined medium. J. Clin. Microbiol. 30: 2225-2229.

Wyss, C. 1993. Aspartame as a source of essential phenylalanine for the growth of oral anaerobes. FEMS Microbiol. Lett. 108: 255-258.

Wyss, C., B.K. Choi, P. Schüpbach, B. Guggenheim, and U.B. Göbel. 1996. *Treponema maltophilum* sp. nov., a small oral spirochete isolated from periodontal lesions. Int. J. Syst. Bacteriol. 46: 745-752.

Wyss, C., P. Hunziker, and S. Klauser. 1993. Support of peptide-dependent growth of *Bacteroides forsythus* by synthetic fragments of haemoglobin or fetuin. Arch. Oral Biol. 38: 979-984.

Yamada, T., and J. Carlsson. 1975. Regulation of lactate dehydrogenase and change of fermentation products in streptococci. J. Bacteriol. 124: 55-61.

Zhang, A., S. Altuvia, A. Tiwari, L. Argaman, R. Hengge-Aronis, and G. Storz. 1998. The OxyS regulatory RNA represses *rpoS* translation and binds the Hfq (HF-I) protein. EMBO J. 17: 6061-6068.

Zheng, M., F. Åslund, and G. Storz. 1998. Activation of the oxyR transcription factor by reversible disulfide bond formation. Science 279: 1718-1721.

From: *Oral Bacterial Ecology: The Molecular Basis*
ISBN 1-898486-22-0 ©2000 Horizon Scientific Press, Wymondham, U.K.

3

ADHESION AS AN ECOLOGICAL DETERMINANT IN THE ORAL CAVITY

Richard J. Lamont and Howard F. Jenkinson

Contents

Introduction

The oral cavity harbors a diverse, abundant and complex microbial community. Bacteria accumulate on both the hard and soft oral tissues in a sessile biofilm. These organisms engage the host in an intricate cellular and molecular dialogue, the outcome of which normally serves to constrain the bacteria in a state of commensal harmony. Under certain circumstances, however, the oral microbiota can be directly or indirectly responsible for disease (see Chapters 5 and 6). An understanding of the basis of microbial colonization thus provides insight into both oral ecology and the underlying pathogenic mechanisms of oral bacteria. This chapter will focus on oral microbial adhesion, which, as depicted in Figure 1, is the underlying process that drives colonization and ultimately disease progression.

Acquisition of the Oral Bacterial Microbiota

The oral cavity presents a habitat that is paradoxically at once inviting and challenging to potential bacterial colonizers. A warm, moist, generally nutrient rich environment is a situation that would tend to favor colonization. Conversely, the mechanical shearing forces of salivary flow and tongue movement would tend to dislodge and expel bacteria. The importance of salivary flow in controlling microbial colonization in the oral cavity is well illustrated by the finding that individuals with xerostomia (low salivary flow rate) suffer from rapid overgrowth of the plaque biofilm on the tooth surface and rampant caries, and are highly susceptible to mucosal lesions. A fundamental attribute of a successful colonizer is thus the ability to adhere to the available surfaces in the mouth and resist the cleansing action of shear forces. In general, bacterial adhesion mechanisms comprise a thermodynamic component providing surface-surface attraction, and a higher affinity adhesin-receptor component involving complementary molecules that interact stereochemically and impart specificity on the reaction. As first observed by Gibbons and colleagues in the 1970's specificity of adhesion can in turn bestow specificity of colonization. Thus, the tissue specific tropisms displayed by oral bacteria are driven to a large extent by the complex recognition system that exists between adhesion conferring molecules on bacterial and host surfaces (Gibbons and van Houte, 1975). Indeed, bacteria that lack specialized adherence mechanisms are either lost from the oral cavity or found in sites that are highly retentive. Before describing the molecular basis of oral bacterial adhesion, it is necessary to consider the milieu in which oral bacteria must function, and how the oral environment can impinge upon adhesion.

Environmental Factors Influencing Adhesion
The mouth is awash with salivary fluid that continually bathes and coats the oral tissues. Whole saliva contains not only the secretory products of the salivary glands, but also bacterial products, dietary components, and serum molecules that originate as an exudate in gingival crevicular fluid, as well as other compounds derived from eukaryotic cells or from gastric or respiratory reflux. Bacteria that enter the mouth, either from contaminated objects or directly from other infected individuals, become suspended in saliva. Specific molecules within saliva, such as histatins, cystatins,

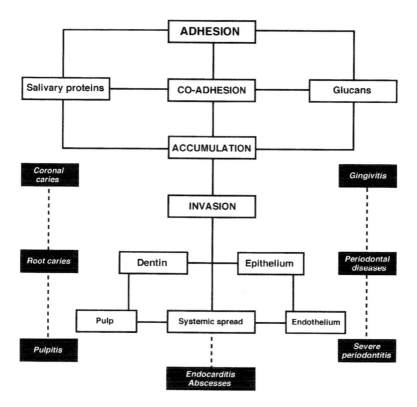

Figure 1. Adhesion: the pivotal event in oral microbial colonization. Examples of diseases associated with microbial activities are noted in filled boxes; dotted lines indicate that the lower may progress from the upper, but not necessarily. Adhesion of bacteria to hard or soft tissues surfaces is promoted by salivary components such as proteins, glycoproteins, and glucans. Co-adhesion of bacteria leads to the accumulation of microbial cells and their products, and of host-derived products, to form the complex biofilm known as plaque. These accumulations may result in the production of excess acid (promoting demineralization of enamel and onset of caries), or in inflammation and tissue destruction leading to gingivitis and more severe forms of periodontal disease. Invasion of tissues such as dentin or gingival epithelium promotes deep-seated infections and the possibility of systemic spread and infection at distant body sites.

lysozyme, lactoferrin and salivary peroxidase, can inhibit growth and metabolic activities of susceptible bacteria (Scannapieco, 1994). Salivary neuraminidase and other glycosidic enzymes, along with lysosomal enymes derived from neutrophils also have the potential to modulate adhesion depending on the involvement of the substrate in the interaction. Serum-derived molecules such as fibronectin or fibrinogen can block binding by saturating receptors or by sterically hindering the interaction. Furthermore, saliva contains specific aggregating molecules that will agglutinate bacteria, facilitating their removal by expectoration or swallowing before adhesion can occur (Scannapieco, 1994; Jenkinson and Lamont, 1997). By way of redress, however, the salivary coating or pellicle on oral surfaces contains molecules such as proline rich proteins (PRP) and glycoproteins, statherin, mucins, and α-amylase that function as the receptors for bacterial adhesins. Thus, initial bacterial adhesion to tooth surfaces, and to a lesser extent epithelial cell surfaces, is mediated by interactions with deposited salivary molecules. Several strains of *Streptococcus gordonii* and *Actinomyces naeslundii*, for example, adhere to acidic PRP deposited in salivary pellicle. The binding sites within PRP-1 reside in the carboxy-terminal dipeptide PQ (residues 149 and 150) (Li *et al.*, 1999). The existence of pellicle receptors in the fluid phase of saliva presents bacteria with the problem of distinguishing between the two forms and avoiding aggregation prior to adhesion. There are several mechanisms by which this may be accomplished. Binding of *A. naeslundii* with PRP will only occur after the reactive region is exposed by a conformation change when the protein adsorbs to the surface (Gibbons and Hay, 1988). The interaction between proteins of the streptococcal antigen I/II family and salivary agglutinin glycoprotein (SAG) is mediated by physically distinct domains within the antigen I/II polypeptides. Differing affinity of these domains for the immobilized or soluble forms of SAG would allow adhesion of cells to immobilized glycoproteins despite the presence of excess fluid-phase receptors in saliva (Jenkinson and Demuth, 1997). Another means by which bacteria may be able to bind to more than one receptor simultaneously is by spatially separating adhesins on the cell surface. For example, α-amylase bound from solution localizes to the polar and septal regions of *S. gordonii,* and hence may not impede the activities of other surface adhesins (Scannapieco, 1994). This arrangement may also prevent amylase-mediated cross-linking of bacteria and subsequent aggregation. Amylase deposited in the enamel salivary pellicle can thus promote the adhesion of streptococci possessing amylase-specific adhesins (Scannapieco, 1994).

Despite their close proximity in the oral cavity, the hard, calcified, tissues of the teeth and the soft tissues of the oral epithelia, support distinct microbiota. This is in part related to differences in retention, as epithelial cells desquamate; and in part related to surface properties such as the presence of higher levels of matrix proteins on epithelial cells. Oral bacteria can demonstrate exquisite specificity for different oral tissues, a tropism that is consequent to specificity of adhesion. *S. sanguis* and related streptococci along with *Actinomyces spp.* are the predominant colonizers of enamel salivary pellicle (Carlsson, 1967; Ellen, 1976), and a variety of adhesins for salivary molecules have been identified in these organisms. Indeed, organisms such as *S. sanguis* are so well adapted to live on the tooth surface that they are not found in pre-dentate infants and are rapidly lost from the mouth in edentulous patients. The actinomyces and streptococci thus exemplify the Gibbons principle that the relative resistance or susceptibility of an oral tissue to colonization by a particular bacterial species is associated with the ability of that species to attach to the tissue (Gibbons, 1984).

The presence of bacteria in the oral cavity stimulates the host immune response. In addition to secretory IgA (S-IgA) in salivary secretions, oral fluid also contains IgG and neutrophils derived from the gingival crevicular fluid. S-IgA and IgG can inhibit bacterial adhesion and promote bacterial agglutination (Williams and Gibbons, 1972). Many of the oral streptococci produce IgA1 proteases (Cole *et al.*, 1994) that cleave specifically at the hinge region of human IgA1, a process that has the potential to provide an ecological advantage and promote colonization (Kilian *et al.*, 1988).

The accumulation of bacteria on oral surfaces irrevocably alters the local environment. Thus the microbial community is continually in flux, with colonization and succession of bacterial species occurring as conditions evolve. Mutans streptococci, for example, will preferentially colonize the tooth surface during a window of infectivity that occurs about 20 months following tooth eruption (Caufield *et al.*, 1993). An interesting corollary is that elimination of mutans streptococci in later life by temporary control agents such as monoclonal antibodies can inhibit re-emergence by these organisms (Ma *et al.*, 1990). Furthermore, as oral surfaces rapidly become colonized with the early streptococcal- and actinomyces-rich microbiota, later colonizers, including gram negative anaerobes such as *Porphyromonas gingivalis*, encounter and adhere to surfaces comprised of antecedent bacteria and their products rather than to salivary pellicle (Gibbons and Nygaard, 1970; Slots and Gibbons, 1978). Such bacteria-bacteria binding may not only favor colonization but also promotes nutritional interrelationships and intercellular signaling mechanisms as discussed in more detail later in the chapter.

Mechanisms of Oral Bacterial Adhesion

In order to adhere to a surface, bacteria must first come in close proximity. Many factors can assist microorganisms approach surfaces. In flowing systems, convective transport by fluid dynamic forces causes accumulation of bacteria at solid-liquid interfaces where there is a viscous boundary layer. Frictional drag and turbulent downsweeps will also contribute to contact with the surface (Gilbert *et al.*, 1993). Surface irregularities present micro-niches where the bacteria can be protected from shear forces. In non-flowing systems, diffusive transport resulting from Brownian motion can serve, albeit more slowly, to concentrate suspended bacteria at a surface. Bacteria, of course, do not always behave like inert particles and can actively participate in their surface localization by chemotactically-driven motile responses.

Following arrival at a surface, the next challenge is to avoid displacement. As bacterial surfaces and the enamel pellicle are generally negatively charged, these two surfaces would tend to repel each other. Attraction can still occur, however, as predicted by the DLVO theory of energetic interactions in biocolloidal systems. In the classical DLVO interaction (Israelachvili and McGuiggan, 1988; van Loosdrecht *et al.*, 1990), the double layer electrostatic repulsive force is roughly exponential in distance dependence with its strength dependent on the surface charge density and its range dependent on the electrolyte concentration of the suspending medium. This repulsive force can be overcome by attractive van der Waals forces that arise from fluctuating molecular dipoles and prevail at small separations. The DLVO theoretical framework thus provides for two separation distances at which attraction will occur: the primary minimum at a separation distance of <1 nm, and the secondary minimum at a separation distance of 10-20 nm. Initial bacterial adhesion tends to occur at the secondary minimum and is a reversible process with continuous exchange between free and adherent cells. To remain at the surface for a longer period of time, bacteria form higher affinity bonds utilizing specific surface molecules that interact stereochemically with cognate receptors via intermolecular van der Waals forces, electrostatic interactions, and hydrogen bonds. Formation of these stronger, essentially irreversible, short-range associations requires energy to propel the bacteria from the secondary to primary minimum. This can be provided by Brownian motion or thermal energy possessed by the bacteria on their approach to the surface. Moreover, the expression of filamentous fimbriae or fibrils, and surface polysaccharides can bridge the gap between the secondary and primary minima (Marshall, 1980; van Loosdrecht *et al.*, 1990; Gilbert *et al.*, 1993) and facilitate the adhesive interactions.

Complex biological systems rarely conform strictly to simple mathematical models, and bacterial adhesion is no exception. There exist a variety of types of interaction that can modify the basic DLVO theory. Electrostatic repulsive forces can also be overcome by cation bridging and hydrophobic interactions, whereas repulsion can be enhanced by hydration forces (Israelachvili and McGuiggan, 1988). Vicinal effects of other organisms are also important in the adhesion process (van der Mei *et al.*, 1993). In depth kinetic studies bear witness to the complexity of bacterial adhesion. Adhesion of *S. sanguis* to saliva-coated hydroxyapatite (the base mineral of enamel), for example, involves multiple co-operatively interacting molecules (Hasty *et al.*, 1992). Two kinetically distinct steps are discernible: first, a reversible interaction mediated by electrostatic and hydrophobic forces, followed by a time dependent shift to a higher affinity binding involving multiple, non-hydrophobicity dependent adhesins (Cowan *et al.*, 1987). Salivary receptors containing sialic acid at the active site can be involved in the higher affinity association, although different stains of *S. sanguis* possess distinct adhesins. Positive cooperativity is manifest in the adhesion kinetics of a variety of streptococci and may be explained by antecedent adherent organisms modifying receptors in the salivary pellicle or excreting compounds to which additional cells can adhere (van der Mei *et al.*, 1993).

The physiological and kinetic studies of adhesion that were first undertaken 15-20 years ago have been developed further by molecular studies. These have provided much detailed information about individual adhesin function, and in some instances have identified the primary amino acid sequences involved in ligand-binding (Jenkinson and Demuth, 1997). However, such analyses have not generally been integrated with ecological studies, so an appreciation is lacking of the role and importance of specific adhesin proteins in oral colonization *in vivo*. In this regard, generation of isogenic streptococcal mutants has led to identification of surface proteins influencing colonization and virulence in rodent models (Jenkinson and Lamont, 1997). In the future it will be prudent to revise physiological considerations, and to combine the molecular knowledge gained about adhesin structure and function with new and more sophisticated techniques for analysis of binding kinetics, such as those based on atomic force microscopy. With that as the backdrop, the next section considers in more detail two components of the adhesion process that have received much attention, namely hydrophobicity and lectin-like binding.

Cell Surface Hydrophobicity

The concept of bacterial cell surface hydrophobicity (CSH) has influenced strongly ideas about the mechanisms of attachment of primary colonizers to salivary pellicle.

Although there are some positive correlations between the CSH values of bacterial cells and their corresponding affinities of binding experimental salivary pellicles (Doyle *et al.*, 1990), the suggestion that CSH is an adhesion determinant is a contentious issue. Techniques for assessing CSH, such as hexadecane-partitioning or contact angle measurements, at best provide composite CSH values across surfaces that are inherently heterogeneous. Thus, while hydrophobic forces may indeed stabilize individual adhesin-receptor interactions (Doyle *et al.*, 1990), assessments of bacterial CSH are not generally predictive of adhesion levels. CSH comparisons can, however, detect major differences in bacterial cell surface composition. For example, novobiocin-resistant mutants of streptococci have altered surface hydrophobicity (Jenkinson, 1987) as well as altered adhesion properties. Bacterial cell surface proteins are major determinants of CSH (Doyle *et al.*, 1990). Isogenic mutants of *S. mutans* that are deficient in production of SpaP or AgB polypeptide (antigen I/II family) show reduced CSH and reduced adhesion to experimental pellicle (Harrington and Russell, 1993; Lee *et al.*, 1989). Although common genetic determinants for CSH in oral gram-positive bacteria are not apparent, emerging evidence for *S. gordonii*, *S. sanguis* and closely-related streptococci suggests that their hydrophobic properties may to a greater extent be determined by expression of a high molecular mass cell-wall linked protein designated CshA (McNab *et al.*, 1994). This 259-kDa polypeptide carries a number of features that are common to gram-positive cell-wall anchored polypeptides. It comprises a leader peptide that directs secretion across the cytoplasmic membrane, an extensive amino acid repeat block region, and a C-terminal cell wall anchorage sequence, including the motif LPxTG that is cleaved and becomes covalently linked via the Thr residue to peptidogylcan (Navarre and Schneewind, 1994). The amino acid repeat block region comprises 13 blocks of a 101 amino acid repeat, dominated by the amino acids Gly, Pro, Thr and Val. It is this region that is believed to be involved in determining cell surface hydrophobicity. Cells that lack this polypeptide are less hydrophobic (McNab *et al.*, 1994) and the presence and amount of CshA at the cell surface correlates well with CSH measurements of *S. gordonii* and *S. sanguis* strains. Moreover, expression of the CshA protein on the surface of the heterologous species *Enterococcus faecalis* demonstrated that it confers hydrophobic properties on these otherwise hydrophilic cells (McNab *et al.*, 1999). However, isogenic mutants of *S. gordonii* deficient in CshA production are unaffected in their ability to adhere to experimental salivary pellicles. This reinforces the notion that CSH values are not predictive of adhesion levels to salivary pellicle.

Lectin-Like Adhesion
While hydrophobic reactions are involved in oral bacterial adhesion to pellicle receptors, and in coaggregation of different bacterial types (Jenkinson, 1987), the site-specificity of colonization within the oral ecosystem is believed to be due to, at least in part, bacterial adhesin-receptor recognition. Specificity is characteristically determined by stereochemical interactions. Oral bacteria have acquired multiple adhesins, and some of these adhesins recognize more than one ligand (Jenkinson and Demuth, 1997). As has been demonstrated in other adhesin-receptor systems, it is likely that variations in primary sequence of oral bacterial adhesins affect their

receptor specificities and thus modulate bacterial colonization. However, in the ecological context, it is not known whether differing affinities of individual adhesins present on the surface of bacteria are of greater or lesser significance than the combined binding activity of the total adhesin complement. It is certain, though, that adhesion to pellicle is maximized when several adhesin epitopes are co-expressed on different surface antigens (Gong and Herzberg, 1997).

Multiple adherent interactions occur between oral bacteria and salivary pellicle, epithelial cells, other oral bacteria, matrix proteins such as fibronectin and collagen, and platelets. A common example is binding of carbohydrate (glycosidic) receptors by bacterial polypeptide adhesins (lectins). Three major classes of receptors have been identified for these lectin-based interactions. The most widely-recognized ligands are galactosyl moieties that are present on host cells and tissue proteins, and on the surfaces of oral bacteria (Whittaker *et al.*, 1996). Sialic acid-containing receptors, present on salivary glycoproteins and host cells are also bound by many oral microbes. The third class of receptor, for mutans-group streptococci, includes bacterial or dietary glucans within plaque. Glucan receptors are recognized by glucan-binding proteins. These are special bacterial lectins that carry characteristic blocks of amino acid repeats that bind polymeric glycosidic structures (Giffard and Jacques, 1994).

Recognition of galactosyl moieties appears to be a common theme in oral bacterial adhesion. The type 2 fimbriae of *A. naeslundii* recognize both Galβ1→3GalNAc and GalNAcβ1→3Gal glycosidic linkages present in the cell wall polysaccharides of a range of oral streptococci (Whittaker *et al.*, 1996) thus promoting interbacterial adhesion within plaque. These motifs are also present within host cell receptors, such as glycoproteins or glycosphingolipids on oral epithelial cells (Cisar *et al.*, 1997). Different strains of *A. naeslundii* appear to express fimbrial adhesins with slightly different sugar specificities (Strömberg and Borén, 1992) that may determine tissue tropism in colonization. Furthermore, *A. naeslundii* can prime target glycoproteins by cleaving the terminal sialic acid to expose the penultimate galactosides (Costello *et al.*, 1979; Ellen *et al.*, 1980). Enzymatic exposure of cryptic receptors (cryptitopes) has been proposed as a common property of oral bacteria that confers a strong selection advantage for any organism that colonizes a mucosal or tooth surface (Gibbons, 1989). *Fusobacterium nucleatum*, a major component of mature dental plaque, expresses at least one major lectin-like adhesin. These outer membrane proteins mediate adhesion to *P. gingivalis* and galactose-sensitive attachment to mammalian cells (Kinder and Holt, 1989; Shaniztki *et al.*, 1997). *Prevotella loescheii* also bears a galactoside-specific adhesin associated with fimbriae that binds host cells and oral streptococci (Weiss *et al.*, 1989). The recognition of galactosyl receptors may confer an adhesive advantage given that galactosyl linkages are core components in a wide range of host oligosaccharides. Moreover, the presence of host-like glycoconjugate motifs within the streptococcal cell wall polysaccharides may reduce the immunogenicity of the bacterial cell surface and influence the ecological distribution of these bacteria within the oral cavity (Cisar *et al.*, 1997). Many oral streptococci, along with *Actinobacillus actinomycetemcomitans* and *P. gingivalis*, also possess the sialyl-Lewis(x) (sLe(x); Neu5Ac $\alpha2\rightarrow3$ Gal ß1→4) (Fuc $\alpha1\rightarrow3$) (GlcNAc-R) antigen on the cell surface (Hirota *et al.*, 1995). In addition to assisting immune evasion, this molecule might bind to host antigens of the selectin family which could promote binding to endothelial cells. This may be a factor in the initiation of the events leading to infective endocarditis once the organisms have gained access systemically.

Potential receptors for sialic acid-binding bacterial lectins include O-linked oligosaccharides of salivary mucins, S-IgA1, and surface glycoproteins e.g. leukosialin on polymorphonuclear leukocytes. The binding activities of the SspA and SspB (antigen I/II family) surface proteins of *S. gordonii* are sensitive to sialic-acid inhibition (Demuth *et al.*, 1996). However, these proteins do not appear to represent the major sialic acid-binding lectin on *S. gordonii* that is associated with hemagglutination and the adhesion of bacterial cells to $\alpha2\rightarrow3$-linked sialic acid on O-linked oligosaccharides of IgA1 (Ruhl *et al.*, 1996). Evidence suggests that a fibrillar surface glycoprotein antigen of molecular mass >200-kDa mediates hemagglutination by *S. gordonii* and related strains (Takahashi *et al.*, 1997).

Given the enormous diversity of oligosaccharide chains present on host and bacterial cells within the oral cavity, and the large number of microbial carbohydrate-binding specificities, a major concern is the lack of experimental support for biological relevance of specific adhesin-receptor interactions. Nonetheless, it is apparent from studies of a range of oral bacteria, including streptococci and *P. gingivalis*, that the presence of multiple adhesins is a common configuration. Such multimodal adhesion mechanisms may facilitate attachment to a variety of surfaces presenting differing receptors, and also increase the avidity of binding to individual substrates as desorption would require simultaneous adhesin detachment. In addition, adhesion to host cells can be a prelude to the modulation of eukaryotic intracellular signal transduction pathways. Thus, the specificity or strength of the adhesive interaction, or the sequence in which different adhesins engage their receptors, may be important in the manipulation of the biological activity of the host cell.

Bacterial Surface Structures and Adhesion

It has long been recognised that bacteria elaborate a variety of different surface structures or appendages that are involved in colonization of the host. These structures may mediate bacterial adhesion to host surfaces, promote invasion of host cells, allow bacterial cell-cell contact and communication, or protect bacteria from host immune defences. Since bacteria isolated from the oral environment usually possess one or more kinds of surface appendage, this is taken to indicate the importance of surface structures to bacterial growth and survival *in vivo*. During laboratory sub-culture in the absence of host environmental selective pressures, some oral bacteria may turn-off expression of their cell surface structures (see, for example, Fine *et al.*, 1999). This is frequently accompanied by loss of bacterial adhesion properties. It is not always clear, however, if these adhesion properties are directly attributable to the surface appendages themselves or if adhesins are simply co-regulated and subject to similar genetic or environmental controls on expression. Attempts to correlate surface structures with adhesive functions for many oral microorganisms have been unsatisfactory. Even with the development of genetic transfer systems for generating isogenic mutants, many issues remain unresolved, primarily because of the multifactorial nature of the oral bacterial adhesion processes. Nevertheless, there are instances of clear correlation between adhesive ability and fimbriae production, in particular for *A. naeslundii* and *P. gingivalis*. On the other hand, much of the data purporting to establish adhesive functions to streptococcal surface structures are somewhat equivocal.

Fimbriae and Fibrils

Fimbriae (or pili) are thread- (or hair-) like appendages (2 - 8 nm in diameter) on the surfaces of bacteria that are composed of helically-assembled protein subunits termed fimbrillins (or pilins). Many types of fimbriae are known to present adhesins as minor protein components either along their sides or at their tip. Structures that fit the morphological description of fimbriae are found on *A. naeslundii, P. gingivalis* and some species of oral streptococci (Figure 2, C and D). These oral bacterial fimbriae appear to form thin, flexible structures, 3 to 5 nm wide and between 1 and 3 µm long, and on this basis are generally reminiscent of *E. coli* K fimbriae. A second type of surface structure, which is morphologically distinct from a fimbria, has been detected on a wide range of oral bacteria (Handley, 1990). These structures, termed fibrils (Figure 2A), are usually much shorter than fimbriae and extend only 50-200 nm from the cell surface. They have tapered ends and it is difficult to measure their individual widths as they are often densely packed and clumped. Fibrils may be peritrichous, densely or sparsely distributed according to strain, or in the streptococci may be localised as a lateral crest or polar tuft composed of fibrils of more than one length (Handley *et al.*, 1991; Jameson *et al.*, 1995).

Streptococcus Fibrils

S. gordonii is one of the predominant oral colonizers and strains of this species carry peritrichous, relatively sparse, fibrils. Recent evidence suggests that the cell wall-anchored protein CshA (259-kDa, described above) is the structural and functional component of *S. gordonii* fibrils (McNab *et al.*, 1999) that project some 60.7 $\pm$ 14.5 nm from the cell surface. When the *cshA* gene is cloned and expressed in a strain of *Enterococcus faecalis* which does not normally produce fibrils, enterococcal cells produce fibrils of identical morphology to those on *S. gordonii*. Hence, the *cshA* gene product appears to contain all the information necessary for fibril production in an appropriate host. The CshA polypeptide is also a major adhesin in *S. gordonii*, participating in adhesive interactions with oral microorganisms (*A. naeslundii, S. oralis*, and *Candida albicans*) and fibronectin (McNab *et al.*, 1996). Thus CshA, like the fibrillar antigens of *S. salivarius*, is capable of forming an adhesive appendage. The adhesive epitopes are localized to within the N-terminal 93-kDa segment of mature CshA polypeptide that is present at the globular end or tip of the fibril (McNab *et al.*, 1996). These observations have led to the hypothesis that the extensive amino acid repeat block region of CshA polypeptide may act as a scaffold to present the adhesive domain distal from the cell surface (McNab *et al.*, 1999). Moreover, the repeat block region is highly antigenic, and it seems that antibodies may be directed mainly towards this region of the polypeptide. This would confer an ecological advantage to the bacteria in that adhesive functions of the N-terminal region of the polypeptide might not be blocked by mucosal antibodies.

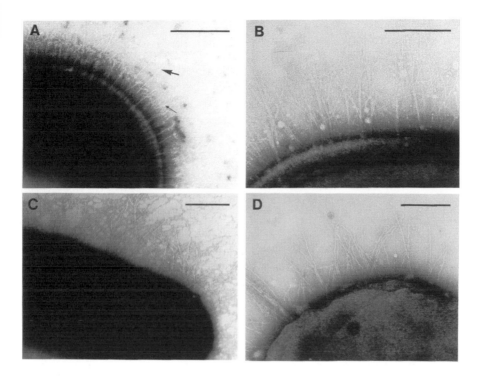

Figure 2. Surface structures of some oral bacteria that promote adhesion. Cells were negatively stained and visualized by electron microscopy. (A), *Streptococcus salivarius* HB showing peritrichous fibrils of two lengths. The large arrow indicates a longer fibril (159-209 nm), a number of which emanate through a fringe of shorter fibrils (70-111 nm) indicated by the smaller arrow. (B), *Streptococcus parasanguis* FW213 showing peritrichous fimbriae of length 198-244 nm. (C), *Actinomyces naeslundii* T14V producing thin flexible fimbriae of variable lengths and some in excess of 0.5 μm. (D), *Porphyromonas gingivalis* WPH35 showing individual fimbriae (up to 400 nm in length), some of which have a curly appearance, as well as small clumps or bundles of fimbriae. Bar markers are 0.2 μm. Photographs in panels A, B and D were kindly provided by P. S. Handley (University of Manchester, Manchester, UK), and that in panel C by M. K. Yeung (University of Texas, San Antonio, USA).

Streptococcus crista is a secondary colonizer forming 'corn-cob' coaggregates with *Corynebacterium matruchotii* and *F. nucleatum* (Lancy *et al.*, 1983) that are prevalent in mature plaque (Figure 3A). *S. crista* is 'tufted' with long (420±39.3nm) fibrils projecting outwards through a dense crest of shorter fibrils (242±13.9nm) on the side of the cell (Handley *et al.*, 1991). Although *S. crista* strains adhere specifically to *C. matruchotii* and *F. nucleatum* (Lancy *et al.*, 1983) via their tufts of fibrils, the fibril components mediating this interaction have not been identified. The reaction may be mediated by several different molecules, since heat or trypsin treatments reduce corncob formation (Lancy *et al.*, 1983), and the tuft fibrils react with antiserum specific for the backbone of lipoteichoic acid (LTA) (Mouton *et al.*,1980). Attempts to identify the corn-cob adhesins on *S. crista* strains have been largely unsuccessful.

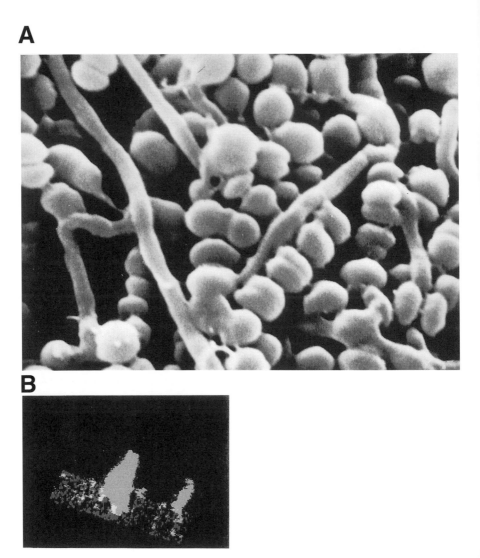

Figure 3. Examples of co-adhering bacteria. (A) Scanning electron micrograph of "corn-cob" formation between *Fusobacterium nucleatum* and *Streptococcus crista*. These structures are commonly observed in mature plaque. Kindly provided by Drs B. Rosan and J. DiRienzo (University of Pennsylvania, Philadelphia, USA). Originally published in *Infection and Immunity*, volume 40, page 305, reproduced with permission. (B) Scanning laser confocal image of *Porphyromonas gingivalis* (light gray) biofilm formation on cells of *S. gordonii* (dark gray). Co-localized bacteria appear white. *P. gingivalis* bind to the streptococci, in preference to the saliva-coated substrate on which the streptococci are attached, and rapidly form towering microcolonies. Kindly provided by Drs J.W. Costerton and G.S. Cook (Montana State University, Bozeman, USA). Originally published in the *Journal of Periodontal Research*, volume 33, page 325, reproduced with permission.

Fibrillar strains of *S. salivarius* generally produce densely-packed fibrils (length 70 to 111 nm) as well as more sparsely distributed fibrils (159 to 209 nm) projecting through the shorter structures (Figure 2A). In *S. salivarius* HB the shorter fibrils may be further sub-divided into at least two sub-classes of lengths 91 nm and 72 nm. The 91 nm-long fibrils bind *Veillonella* and are composed of AgB (VBP), a 320-kDa protein with covalently-bound carbohydrate. The 72 nm fibrils are composed of a 220-280-kDa protein (AgC) that is more heavily glycosylated and mediates binding to host cells (Weerkamp *et al.*, 1986a,b). The purified fibrillar proteins adopt rod-like structures with globular portions at their ends. Since AgB and AgC are co-expressed on the bacterial surface, it is not clear how AgC-mediated binding of streptococci to host cells occurs in the presence of the longer fibrils.

Streptococcus Fimbriae

While it is convenient to distinguish between fibrils and fimbriae present on the surface of streptococci on the basis of their differing structural morphologies, there is no biochemical evidence to suggest their protein components are fundamentally different. Indeed, to the extent that they have been characterized, protein components of both types of structure possess conventional wall anchor motifs and are composed of repetitive amino acid blocks. The peritrichous fimbriae of *S. parasanguis* FW213 express as a dense surface array which is unusual for oral streptococci (Figure 2B). These fimbrial arrays are composed of a novel protein, designated Fap1, with a highly repetitive amino acid residue structure (Wu and Fives-Taylor, 1999). The polypeptide precursor of 2,552 amino acid residues carries a 50 aa residue N-terminal leader peptide, a cell wall-anchorage sequence at the C-terminus, and 1,000 repeats of the dipeptide (E/V/I)S, the functional significance of which is not understood at present. It is conceivable that Fap1 might become assembled into fimbriae by a process analogous to curli formation in *E. coli* (Hammar *et al.*, 1996) whereby subunits nucleate upon a surface-anchored polypeptide. Since each Fap1 molecule has a putative cell-wall-anchorage region, this proposed assembly mechanism requires that the C-terminally cleaved Fap1 molecule is protected from cell-wall-anchorage and instead polymerized into fimbriae.

Actinomyces naeslundii Fimbriae

A. naeslundii fimbriae (5 nm wide and approximately 1.5 μm long) are the best characterized of all gram-positive oral bacterial fimbriae (Figure 2C) and two major types have been identified. Some strains carry both type 1 and type 2 fimbriae, while others carry only type 2 fimbriae (Yeung and Cisar, 1990). Type 1 fimbriae are associated with adhesion of *A. naeslundii* to salivary acidic PRP and to statherin deposited within salivary pellicles on oral surfaces (Clark *et al.*, 1989). Since, however, antibodies raised to the type 1 fimbrial subunit protein FimP do not block adhesion of *A. naeslundii* cells to salivary pellicle (Cisar *et al.*, 1991), it is currently thought that FimP may not be the primary adhesin for PRP. Type 2 fimbriae are involved in the adhesion of *A. naeslundii* to glycosidic receptors on epithelial cells, polymorphonuclear leukocytes, and oral streptococcal cells. The lectin-like adhesion of *A. naeslundii* to these substrates is inhibited by galactose and N-acetylgalactosamine (see Whittaker *et al.*, 1996), and is associated with a 95-kDa

143

fimbrial polypeptide that is distinct from the structural FimA subunit (Klier *et al.*, 1997).

The genes encoding the structural subunits of type 1 and type 2 fimbriae encode proteins of molecular masses 54-kDa to 59-kDa (Yeung and Cisar, 1990). The subunit protein precursors contain an N-terminal leader peptide and a C-terminal wall-anchorage region containing the LPxTG cleavage recognition sequence, although the proteins themselves do not contain internal amino acid repeat blocks. The presence of a C-terminal cell wall-anchorage region within the fimbrial subunit proteins implies that, in order to form fimbriae, individual subunits are not bound to cell wall peptidoglycan but must become linked to one another. Two hypotheses for fimbrial assembly have been suggested. One proposes that FimA and FimP are processed by the general gram-positive wall-protein cleavage (sortase) machinery (Navarre and Schneewind, 1994), linked to peptidoglycan precursors and then, instead of being incorporated into cell wall, become linked to other respective fimbrial subunits. An alternative hypothesis, like that proposed for assembly of Fap fimbriae in *S. parasanguis* (Wu and Fives-Taylor, 1999), is that cell-wall-linked molecules act as nucleators upon which processed subunits become polymerized. Presumably, signals within the fimbrial polypeptides, in addition to the LPxTG motif, must be recognized in order to direct subunit assembly as opposed to cell-wall linkage.

The type 1 fimbrial gene cluster in *A. naeslundii* T14V contains seven genes, amongst which are *fimP* and a downstream gene *orf4*. The product of this gene shows 40% amino acid identity to the *orf365* peptide that is found dowstream of *fimA* (Yeung and Ragsdale 1997). The *orf365* gene product is essential for production of type 2 fimbriae (Yeung *et al.*, 1998) suggesting that fimbrial assembly in *Actinomyces* requires accessory proteins.

Porphyromonas gingivalis Fimbriae
Fimbriae are the major adhesion-mediating determinants in *P. gingivalis*. At least three different types have been identified, but all are peritrichous, up to 3 μm long and 5 nm in width (Figure 2D). The major class of fimbriae is composed of a protein subunit designated fimbrillin (FimA), of approximately 43-kDa (Lee *et al.*, 1991; Hamada *et al.*, 1994). The region downstream of the *fimA* gene contains four open reading frames, two of which encode 50-kDa and 80-kDa polypeptides that appear to be minor structural components of fimbriae (Yoshimura *et al.*, 1993). Fimbrial assembly in *P. gingivalis* is under complex genetic and environmental controls (Xie and Lamont, 1999), and requires proteolytic processing mediated by the arginine-specific proteinases RgpA or RgpB (Nakayama *et al.*, 1996).

Fimbrillin binds PRP, statherin, lactoferrin, oral epithelial cells, oral streptococci and *A. naeslundii*, fibrinogen and fibronectin (see Lamont and Jenkinson, 1998). Binding to salivary proteins involves multiple regions on the molecule, with the most active domains being located within the C-terminal half of the protein between residues 266 and 337 (Amano *et al.*, 1996b). The combined activities of binding sites are important in enhancing binding of fimbriae to salivary protein receptors, thus establishing adhesion of cells to saliva-coated surfaces. Statherin is bound by fimbrillin only when the former is deposited upon a surface such as hydroxyapatite (Amano *et al.*, 1996a). It is believed that upon binding to a surface, conformational

144

changes occur within the C-terminal regions of statherin molecules, thus exposing hidden residues (cryptitopes) to which fimbrillin binds. Salivary PRP are also believed to change conformation upon adsorption to a surface, and the sequence PQGPPQ, which occurs four times within PRP-1 is recognized by fimbrillin (Kataoka *et al.*, 1997). Fimbriae also show chemotactic properties, bind a β2 integrin on mouse macrophages (Takeshita *et al.*, 1998), induce cytokine production, and are necessary for *P. gingivalis* invasion of epithelial cells (Weinberg *et al.*, 1997) and endothelial cells (Deshpande *et al.*, 1998).

Actinobacillus actinomycetemcomitans Fimbriae
A. actinomycetemcomitans is a gram-negative capnophile closely associated with localized juvenile periodontitis. Freshly isolated strains possess fimbriae but these are rapidly lost upon laboratory subculture (Fine *et al.*, 1999). The presence of fimbriae imparts an internal star-shaped appearance to colonies on solid media (Rosan *et al.*, 1988), usually designated the rough colony type. The fimbriae demonstrate a peritrichous, bundle-forming arrangement, are approximately 5 nm wide and can reach several μm in length. They are involved in adhesion to saliva-coated surfaces, epithelial cells and fibroblasts (Rosan *et al.*, 1988; Harano *et al.*, 1995; Fine *et al.*, 1999). The fimbrial subunit appears to be a 6.5 kDa protein (Flp) that bears homology to the *Neisseria gonorrhoeae* type 4 fimbriae, which are synthesized via the general protein secretion pathway for fimbrial assembly (Hultgren *et al.*, 1996; Inoue *et al.*, 1998). The *flp* gene comprises part of an operon that includes genes that may be involved in protein secretion and fimbrial assembly (Haase *et al.*, 1999). Whether the fimbrial structural component *per se* mediates adhesion, or whether fimbrial-associated proteins (Ishihara *et al.*, 1997) act as adhesins, remains to be determined.

Repertoire of Adhesin Functions

Fimbrial Adhesins
The functions of filamentous surface structures on oral bacteria have, in many cases, been associated with adhesion. Models for fimbrial adhesion predict that the long-range associations of bacteria with host cells mediated by fimbriae may help overcome the electrostatic repulsion of two negatively-charged cell surfaces. Moreover, fimbriae or fimbriae associated proteins often possess domains that interact directly with cognate receptors on other cells. However, as described above, there are many fimbriae or fibrils described on gram-positive bacteria for which specific binding activities have not been identified. For example, recent evidence suggests that aggregative fimbrial-like structures of up to 4 μm in length produced by rough-colony type *Peptostreptococcus micros* strains are not necessary for adhesion to epithelial cells (Kremer *et al.*, 1999) and may in fact be obstructive to adhesion. These kinds of observations suggest that fimbriae could serve essential *in vivo* roles that are not necessarily related to primary adhesion, such as functions associated with nutrition, protection or cell-cell communication.

Non-Fimbrial Adhesins

Significantly, some of the best characterized oral bacterial adhesins do not appear to be associated with surface fimbriae or fibrils. They are held at the cell surface of gram-positive bacteria by covalent linkage to peptidoglycan, or though binding to other surface molecules; and at the gram-negative cell surface through intercalation with the outer membrane. The amylase-binding protein (AbpA) of *S. gordonii* is a small (20-kDa) cell-wall-linked polypeptide that forms no discernible surface structures. It may mediate binding of cells to α-amylase in salivary pellicle, and provides a means by which bacteria can sequester host metabolic enzymic activity from saliva for their own nutrient acquistion (Rogers *et al.*, 1998). The antigen I/II family proteins, found on mutans-group and mitis-group streptococci, are also cell-wall-anchored polypeptides and bind a wide range of substrates including salivary glycoproteins, collagen, *A. naeslundii* and *P. gingivalis* (Jenkinson and Demuth, 1997). These proteins also do not appear to form surface structures, although they may contribute to the surface fringe on *S. mutans* as visualized by electron microscopy (Lee *et al.*, 1989). Multiple binding functions of the antigen I/II family polypeptides may be attributed to different regions of primary sequence, some possessing broad-specificity receptor recognition and others unique-receptor specificities.

Adhesins with Alternate Functions

Multi-functional (or oligospecific) adhesins, like the antigen I/II family proteins, are now appreciated as being part of the adhesive repertoire of most gram-positive bacteria. However, it is only now becoming apparent just how many surface proteins with assigned enzymic or transport functions, based on primary sequence homologies, may also act as adhesins. The most significant cases in point are the lipoproteins of gram-positive bacteria that are modified post-translationally at the N-terminal cysteine with covalently-linked fatty acid, and found associated with the outer leaflet of the cytoplasmic membrane (Sutcliffe and Russell, 1995). The lipoproteins of gram-positive bacteria that have been characterized to date are solute-binding proteins that have their counterparts as the periplasmic proteins of gram-negative bacteria that deliver substrates for uptake by ATP-binding cassette (ABC) type transport systems. These high-affinity substrate-capture lipoproteins on the surfaces of gram-positive bacteria are free to interact with environmental substrates. It would not require much imagination to envisage that such proteins could function as adhesins if they were able to bind their substrates, or analogs, when immobilized. There is convincing evidence that the oligopeptide-binding proteins of *S. gordonii* and *S. pneumoniae*, in addition to facilitating binding and transport via the Hpp or Ami permeases, respectively, of oligopeptides up to 7 amino acid residues in length, are involved also in mediating streptococcal adhesion (Cundell *et al.*, 1995; Jenkinson *et al.*, 1996). The HppA protein in *S. gordonii* also behaves as an environmental sensor of oligopeptides that regulate the development of competence and of surface protein adhesin production (McNab and Jenkinson, 1998). It is not unexpected that oligopeptide mediated cell-communication functions are linked with adhesion, since the regulated expression of adhesins would be important for the development and maintenance of plaque communities responsive to intercellular signals.

A second family of ABC transporter-associated proteins, designated LraI

(Jenkinson, 1994), is found amongst the streptococci and enterococci, and has been consistently implicated in adhesion. These proteins are proposed to form a new cluster 9 group of solute-binding proteins (Dintilhac *et al.*, 1997), also present in a wide range of organisms, including *Salmonella, Staphylococcus, Treponema* and *Yersinia*, that are involved in the uptake of metal ions (Fe^{2+}, Mn^{2+}, Zn^{2+}). In *S. gordonii* and *S. pneumoniae*, the LraI polypeptides are lipoproteins (ScaA and PsaA) that function in the uptake of Mn^{2+} ions (Dintilhac *et al.*, 1997; Kolenbrander *et al.*, 1998). Mutations in the *sca* or *psa* operons affect growth of bacteria in manganese-depleted environments, the development of competence for transformation, and the production of other cell-surface proteins (Novak *et al.*, 1998). In addition, inactivation of the lipoprotein genes *fimA* (in *S. parasanguis*) or *lmb* (in *S. agalactiae*), affects adhesion of bacterial cells to fibrin (Burnette-Curley *et al.*, 1995) or laminin (Spellerberg *et al.*, 1999), respectively. These effects on cell adhesion, caused by mutations in LraI protein genes, could, at least in part, be pleiotropic as a result of disturbing an important metabolic control circuit. However, purified FimA protein binds to salivary glycoproteins and inhibits adhesion of *S. parasanguis* to experimental pellicles (Oligino and Fives-Taylor, 1993), implicating a direct role in adhesion. A current hypothesis, that takes into account most of the experimental evidence, is that under conditions of depleted metal ions (Mn^{2+}, Fe^{2+}) such as might exist in the host environment, expression of the LraI family cell-surface lipoproteins is up-regulated (Kolenbrander *et al.*, 1998). These proteins would act to scavenge metal ions and concentrate them within the bacterial community, as well as enhance adhesion of bacteria to host tissues and to other bacteria.

Adhesins and Nutrition

The idea that nutrient acquisition and adhesion functions in oral bacteria are intimately linked is clearly demonstrated in *P. gingivalis* by the production of proteinases. Extracellular (cell-free) or cell-surface-bound forms of arginine-specific (Rgp) or lysine-specific (Kgp) proteinases act on a variety of host substrates, including fibronectin, hemoglobin, and immunoglobulins to generate peptides for bacterial growth. In addition, the cell-bound protease complexes carry adhesion (hemagglutinin) domains that are thought to bind host cells and thus target the bacterial proteinases to their substrates (Progulske-Fox *et al.*, 1989; Lantz *et al.*, 1991; Li *et al.*, 1991; Pavloff *et al.*, 1997). The proteinases are necessary for processing of fimbrillin, the major structural component of *P. gingivalis* adhesion fimbriae, and they also modify host substrates such as fibronectin to uncover arginine-peptides to which the fimbriae adhere (Kontani *et al.*, 1997). The co-evolution of nutritional and adhesin functions ensures that the colonization process is nutritionally-sensitive. Moreover, for *P. gingivalis*, survival is also nutritionally-directed since the proteinases destroy a wide range of host defense mechanisms.

There are numerous other hydrolytic enzymes produced by oral bacteria that may be retained associated with the cell surface and that could act as adhesins (as summarized in Figure 4). Although not formally demonstrated, it is reasonable to suggest that enzymes such as sialidase (neuraminidase), IgA1 protease, and collagenase, that are secreted by oral streptococci could act as adhesins in the absence of, or prior to, substrate cleavage. Indeed, other enzymes with well-characterized

SUBSTRATE

BACTERIAL NUTRIENT

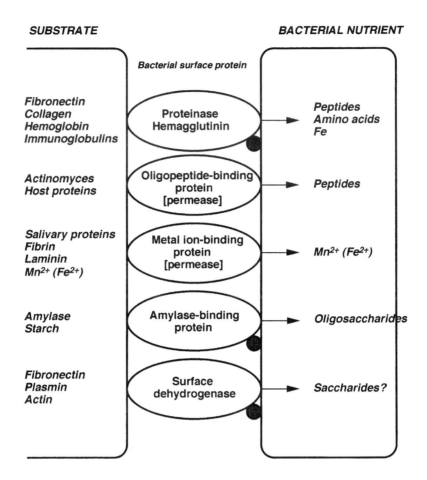

Figure 4. The intimate links between oral bacterial adhesion and nutrition. Hydrolytic enzymes such as the proteinases of *Porphyromonas gingivalis* promote adhesion of bacteria to host cells and matrix components. The breakdown products of proteolysis are taken-up by the bacterial cells via specialized transport systems (filled circles). In streptococci, the binding-protein components of oligopeptide and metal ion permeases have been shown to possess adhesive activities. Permeases may therefore serve dual functions in adhesion and nutrient uptake. The α-amylase-binding protein in *Streptococcus gordonii* is an example of a mechanism by which bacteria can acquire a host-derived hydrolytic function for their nutritional benefit. Oligosaccharides derived from starch breakdown will be taken-up by a multiple sugar permease. Streptococci, and other organisms such as *Candida albicans*, also appear to present on their cell surfaces some of the glycolytic pathway enzymes. The surface dehydrogenase (glyceraldehyde-3-phosphate dehydrogenase) functions as an adhesin, and in concert with other glycolytic enzymes may provide a cell-surface energy-generating system that could charge, hypothetically, modification of other surface molecules or the uptake of saccharides.

activities have been demonstrated to act as adhesins. The glucosyltransferase (GtfG) of *S. gordonii* binds to human endothelial cells (Vacca-Smith *et al.*, 1994), while a key glycolytic pathway enzyme glyceraldehyde-3-phosphate dehydrogenase is found on the surfaces of streptococci and staphylococci where it binds a range of human tissue proteins including fibronectin, plasmin and transferrin (Jenkinson and Lamont, 1997; Modun and Williams, 1999).

In addition to the direct binding of enzymes, the products of enzyme activity can also be involved in adherence. It has long been recognized that there is an association between the glucose polymers (glucans) ensuing from glucosyltransferase activity and adherence of streptococci (see Gibbons, 1984). Glucans play a key role in the formation of dental plaque because they adhere to smooth surfaces and mediate coadhesion of bacterial cells. The glucan-mediated accumulation of mutans streptococci is enhanced through the activities of glucan-binding proteins (GBPs) as described below. In addition, water-soluble forms of extracellular glucans and fructans are a form of bacterial storage polymer that may be degraded by glucanases or fructanases and utilized as a source of metabolizable energy under conditions of nutrient deprivation (see Kuramitsu, 1993).

Glucosyltransferase (GTF) and fructosyltransferase (FTF) enzymes produced by streptococci are responsible for the production of glucans and fructans respectively. In *S. mutans*, GTFs are encoded by three genes denoted *gtfB*, *gtfC* and *gtfD*. The GtfB enzyme (162-kDa) synthesizes a water-insoluble glucan made up of $\alpha-1,3$-linked glucose residues; GtfC (149-kDa) synthesizes a low-molecular-mass and partly water-soluble glucan; and GtfD (155-kDa) produces a water-soluble $\alpha-1,6$-linked glucan. Enzymes with high sequence identities to GtfB and synthesizing water-insoluble glucans are produced by *S. sobrinus* (GtfI) and *S. downei* (GtfI). Some of these enzymes (GtfD, GtfI) are primer-stimulated (or primer-dependent) while others (GtfB, GtfC) do not apparently require a glucan primer for polysaccharide synthesis.

The importance of extracellular polysaccharide production to cariogenesis has been well-established in animal models of experimental caries. All three GTFs and the product of the *ftf* gene (83-kDa) in *S. mutans* are involved in eliciting caries, the severity of which depends upon the amount and type of polysaccharide produced and on the rate of degradation of the polymers (Munro *et al.*, 1995). The GTFs all contain a common four-domain structure consisting of a signal peptide, a poorly-conserved N-terminal region of about 200 amino acid residues, a catalytic domain (~800 aa residues) which is well-conserved, and a C-terminal domain of ~500 aa residues that binds glucans. The glucan-binding domains are composed of series of repeated aa residue blocks originally designated A, B or C repeats (Ferretti *et al.*, 1987; Banas *et al.*, 1990). A detailed comparison of all the sequenced GTFs has further suggested the presence of a fundamental repeated conserved YG motif running through the A, B and C repeats within the C-terminal regions (Giffard and Jacques, 1994). These repeats act as receptors for glucans, but it is generally considered that the GBPs, not GTFs, mediate cell adhesion to glucans.

Other streptococci, such as *S. gordonii* and *S. salivarius*, which are not considered amongst the more cariogenic species, also produce extracellular glucans and fructans. The mixed-linked glucan product of the *S. gordonii* GTF enzyme (GtfG) promotes accumulation of *S. gordonii* cells on surfaces (Vickerman *et al.*, 1991) despite the current lack of evidence that *S. gordonii* produces GBPs. *S. salivarius* on the other hand has been shown to produce at least four GTF activities (Simpson *et al.*, 1995). One of the main sources of GTF in saliva may come from *S. salivarius*. When incorporated into salivary pellicle these GTFs synthesize glucans to which other

149

streptococci such as *S. mutans* may then adhere (Schilling and Bowen, 1992). Thus the synthesis and binding of glucan by oral streptococci is a major ecological factor in the development of complex plaque.

As techniques for genetic analysis of oral bacteria become more widely-developed, and methods for determining adhesin structure and activity more sophisticated, a fuller appreciation of the adhesin repertoire of oral microorganisms will be achieved. It is already evident that past concepts of adhesion being mediated solely through the expression of defined surface structures are outdated. Although fimbriae and fibrils on oral bacteria frequently demonstrate multiple binding functions, there are many instances of surface appendages for which no adhesive functions have been identified. In fact, a wide range of binding functions have been now attributed to proteins that are not composed of discernible surface structures. A simplistic molecular viewpoint is that adhesion is mediated through a combination of the activities of specialized adhesins, together with other surface proteins with physiological (metabolic) functions. In the oral ecosystem, expression of the specialized adhesins may be chiefly under cell-cycle or genetic (phase) control, while expression of the alternate adhesins may be environmentally (nutritionally)-regulated.

Bacterial Cell-Cell Interactions

Dental plaque biofilms are complex multicellular entities comprising numerous bacterial species and their products along with host and dietary components. In order to maintain the cohesive integrity of this biofilm it is necessary that the fundamental building blocks, the bacteria, are firmly anchored to each other (Gibbons and Nygaard, 1970). The ability of bacteria to attach to other bacteria is a common attribute that has been demonstrated for over 700 strains representing 14 genera and is readily visible in undisturbed plaque (Listgarten, 1976). Generally termed coaggregation, or coadhesion (as aggregates are not always formed *in vitro*), this occurs through specific interactions between complementary surface molecules on partner cells. It is also mediated through bridging of cells by multivalent salivary molecules such as salivary mucins, agglutinin glycoproteins, and glucans (Lamont *et al.*, 1991; Jenkinson and Lamont, 1997). Most interactions occur between different genera, with *Streptococcus*, *Actinomyces* and *Fusobacterium* species demonstrating the greatest number of coaggregation partners. Streptococci also participate in intra-generic coaggregations with other streptococcal species in binding reactions that are usually inhibitable by galactose (Whittaker *et al.*, 1996). Bacterial recognition and attachment to specific partner organisms then drives the temporal development of the supragingival and subgingival plaque biofilms. It will also influence greatly the microbial composition of plaque and therefore disease etiology. Although the ecological significance of bacterial coadhesion has yet to be conclusively demonstrated *in vivo*, there is overwhelming evidence from *in vitro* studies, which include simulated oral environments, and from correlative *in vivo* studies, that the specificity of bacterial cell-cell adhesion is critical in the development and maintenance of the complex oral microbial ecosystem.

Mechanisms of Bacterial Coadhesion

Bacterial cell-cell coadhesion is no different from other specific adhesion mechanisms and involves, in simplest form, noncovalent, stereochemical interactions typified by a lectin-like protein on one partner cell recognizing a carbohydrate (receptor) moiety on the other partner cell. For example, the PlaA adhesin of *Prevotella loescheii* (London and Allen, 1990) and the type 2 fimbrial lectins of *A. naeslundii* (Cisar *et al.*, 1997) recognize the antigenically-diverse linear cell wall polysaccharides of *S. oralis*, *S. sanguis* and related streptococci that contain the host-like recognition motifs GalNAcβ1→3Gal or Galβ1→3GalNAc. It is apparent, though, that for many inter-bacterial binding reactions, multiple cell surface components are involved in reciprocal adhesin-receptor bindings, and these may also include direct protein-protein interactions. The oral yeast *Candida albicans*, which colonizes mucosal surfaces, denture acrylic, and which can be found in subgingival plaque, binds to oral streptococci by at least three mechanisms. A proteinaceous *C. albicans* adhesin recognizes streptococcal cell wall carbohydrate, while two adhesin families on the streptococcal cell surface, comprising CshA and antigen I/II polypeptides, bind the yeast cell surface mannoproteins (Holmes *et al.*, 1996). Similarly, coadhesion between *P. gingivalis* and *S. gordonii* is mediated by the *P. gingivalis* fimbriae and a surface protein of 35 kDa, along with the streptococcal Ssp surface proteins (Lamont *et al.* 1993, 1994). The active region of the SspB protein is located towards the C-terminus within aa residues 1167-1250 (Brooks *et al.*, 1997). In Figure 5 some of the bacterial cell-cell interactions between different oral species are depicted. This is not a comprehensive diagram of all the various interactions that have been identified. Instead, it has been designed to demonstrate the potential adhesive networks that can be generated amongst ecologically and pathogenically relevant species. The cell-interactions involve not only the primary colonizing bacteria (such as *Streptococcus* and *Actinomyces*), and the secondary colonizing organisms (*Fusobacterium*, *Porphyromonas*, *Prevotella*) binding to the primary colonziers, but also specific interactions between the later colonizers, such as *Porphyromonas*, *Treponema* and *Bacteroides forsythus*. This emphasizes that coadhesion is potentially relevant at all stages of development of the plaque biofilm. Moreover, it seems not insignificant that the coadhering pathogenic bacteria *P. gingivalis*, *T. denticola* and *B. forsythus*, that are associated primarily with adult periodontitis, are found as a discrete complex *in vivo* and are co-isolated from the periodontal pocket (Socransky *et al.*, 1998).

Role of Bacterial Coadhesion

Most oral microorganisms such as mitis-group streptococci, *A. naeslundii*, *F. nucleatum* and *P. gingivalis* (see Figure 5) seem to have the ability to bind to surfaces in the oral cavity that are coated with salivary molecules. In particular, PRP and statherin components of saliva become rapidly deposited onto enamel surfaces and support the adhesion of the above bacteria to the extent that they can survive in early plaque. Other species, such as *T. denticola*, *B. forsythus* and *Prevotella loescheii*, do not appear to have a high affinity for binding deposited salivary proteins. These organisms, and others with lower affinities of binding to pellicle, may depend therefore upon adhesion to antecedent bacteria in order to colonize successfully. In

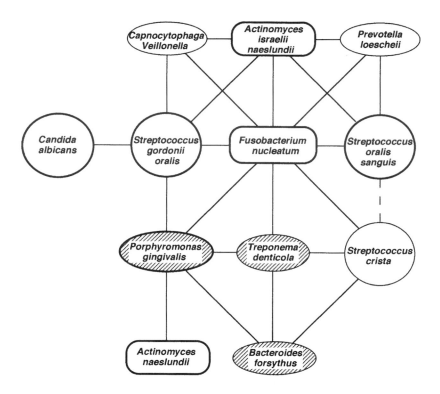

Figure 5. Coadhesion of oral microorganisms involves multiple interacting complexes. Solid lines represent inter-genus binding, dashed line represents intra-genus binding. Individual interactions often involve multiple adhesin-receptor pairs. Early colonizers of oral surfaces such as streptococci and *Actinomyces* spp. exhibit multiple coaggregating (coadhering) partners, as does *Fusobacterium nucleatum*. Bacteria such as *Prevotella loescheii*, *Treponema denticola*, *Bacteroides forsythus* and *Streptococcus crista* do not appear to bind with high affinity the major enamel salivary pellicle components such as PRPs and statherin. Therefore, these bacteria may rely on those organisms such as streptococci, *A. naeslundii*, *F. nucleatum*, and *Porphyromonas gingivalis* that do bind PRPs and statherin (depicted by bold outlines). As *F. nucleatum*, and *P. gingivalis* are strict anaerobes and later colonizers of plaque, they may rely more on coadhesion than binding to salivary proteins for colonization. Binding of some species is dependent upon an intermediary: for example, *B. forsythus* will bind to *F. nucleatum* through *S. crista*, that forms 'corn-cob' structures in mature plaque. *P. gingivalis*, *T. denticola* and *B. forsythus* (hatched) form a complex in the periodontal pocket that is most strongly implicated in adult periodontitis.

some instances, however, it is unclear whether the formation of large aggregates in suspension will enhance colonization or, by analogy to saliva-mediated aggregation of bacteria, facilitate clearance of organisms from the oral cavity. For example, *Actinomyces* species that coaggregate with streptococci *in vitro*, do not colonize streptococcal-coated surfaces *in vivo* (Skopek *et al.*, 1993). Thus it seems likely that coadhesion mechanisms between bacteria can differ depending upon whether the partner cells are in suspension or bound to a surface and exhibiting a biofilm phenotype. Indeed, *P. gingivalis* cells bind to *Streptococcus*-rich plaque when introduced into the mouths of human volunteers (Slots and Gibbons, 1978), yet large inter-species coaggregates are not formed consistently. *P. gingivalis* and oral streptococci such as *S. gordonii* do participate in a multi-modal adhesion event; however, this binding is of low affinity when both species are in suspension, hence the failure to form large coaggregates. In contrast, *P. gingivalis* adheres avidly to sessile *S. gordonii* and will preferentially localize on streptococcal cells rather than onto saliva-coated surfaces. Once attached to *S. gordonii*, *P. gingivalis* rapidly forms a biofilm comprising towering microcolonies separated by fluid-filled channels (Figure 3B) (Cook *et al.*, 1998). As *P. gingivalis* is an obligate anaerobe it will only survive at sites of reduced oxygen tension. Regions of plaque containing streptococci thus provide attractive targets for *P. gingivalis* colonization since available oxygen will be removed by the facultative anaerobes. The ability to form a biofilm of "classical" morphology implies the existence of density sensing communication systems as has been documented in other biofilm-forming gram-negative organisms (Costerton, *et al.*, 1995; Davies *et al.*, 1998), although more work is required to clarify this point.

Nutritional and metabolic relationships between bacteria in the developing biofilm may be facilitated by coadhesion. The most-cited example of this is the utilization by *Veillonella* of organic acids, formed by streptococci, as a carbon source for growth. Removal of lactate from the environment is predicted to raise the pH, but at the same time drives lactate flux with increased utilization of fermentable carbohydrates. A multitude of nutritional benefits could be theorized to result from bacterial coadhesion between organisms that secrete specific hydrolytic enzyme activities and those that do not, thereby enhancing the latter organisms' growth by co-opting a new metabolic function. The advantages to these kinds of coadhesion-based associations may be one reason that organisms such as *V. alcalescens* and *S. salivarius* are frequently isolated together on the tongue (Weerkamp and McBride, 1981).

Glucan-Mediated Coadhesion

Extracellular polysaccharides produced by mutans-group streptococci, some mitis-group bacteria, and *A. naeslundii* also play a key role in the formation of plaque (see Chapter 1). Glucan polymers in particular adhere to smooth surfaces and mediate bacterial cell-cell adhesion. The glucan-mediated accumulation of mutans-group streptococci in dental biofilms is enhanced by the activities of a group of cell-surface-associated proteins designated the glucan-binding proteins (GBPs). Many proteins with glucan-binding activity have been identified in *S. mutans* and *S. sobrinus*. Three distinct GBPs are known for *S. mutans* and at least three GBPs are produced by *S. sobrinus*. The three GBPs from *S. sobrinus* (GBP-2, GBP-3 and GBP-4) are neither antigenically nor structurally related to each other, and none has significant antigenic relationship with the *S. mutans* GBPs (Smith *et al.*, 1998).

The major GBP (GbpA) of *S. mutans* Ingbritt is a 56-kDa polypeptide with considerable sequence identity (60%) to the C-terminal amino acid residue repeat block regions of the glucan-synthesizing enzymes (glucosyltransferases, GTFs) GtfB, GtfC and GtfD (Banas *et al.*, 1990). These repeat block regions contain a conserved 'YG' repeat of approximately 21 residues containing one (or more) aromatic residues (usually tyrosine, Y) followed by glycine (G), usually 4 residues downstream (Giffard and Jacques, 1994). The YG repeats are present in highly regular arrays not only within the glucosyltransferase enzymes and GbpA of *S. mutans*, but also within the C-terminal domains of *Clostridium difficile* toxin A, and a class of *Streptococcus pneumoniae* surface proteins designated the choline-binding proteins (Rosenow *et al.*, 1997). Pneumococci produce a cell-wall teichoic acid that is phosphorylcholine-substituted, and the C-terminal YG repeat blocks are proposed to interact non-covalently with choline, thereby anchoring the proteins to the cell surface. In GTF, the repeat blocks bind the glucan template onto which the enzyme polymerizes glucose residues. The YG repeat then is currently thought to represent a general structural theme for binding of polypeptides to carbohydrate polymers (von Eichel-Streiber *et al.*, 1992). The binding specificities of these proteins could be determined by the lengths of the repeated units and by the spacing of conserved residues within these units, as well as by the identities of less well-conserved residues.

While evidence suggests that the YG blocks provide a common mechanism for gram-positive bacterial protein binding to repeating carbohydrate polymers, only GbpA of *S. mutans* has been shown to bind glucan in this way. It seems likely that the *S. sobrinus* GBPs identified to date do not utilize this mechanism since there is no evidence for any structural relationships of the GBPs to the GTFs produced by this organism (Smith *et al.*, 1998). Although it is clear that the GBPs are necessary for glucan-binding and glucan-mediated aggregation of mutans-group streptococci *in vitro* (Ma *et al.*, 1996), functional roles for the GBPs in the development of plaque and dental caries have yet to be experimentally proven. Recent work from Hazlett *et al.* (1998) has demonstrated that an isogenic GbpA mutant of *S. mutans* colonized rat molars to a similar degree to the wild-type parent strain, but was more cariogenic. This somewhat surprising result was attributed to the fact that GbpA mutants form a softer and less-cohesive biofilm. Under these conditions, the greater plaque porosity might facilitate more rapid nutrient influx and fermentation, with subsequent increased acidogenicity and demineralization. These experimental results illustrate the complex, and sometimes paradoxical, associations between microbial colonization and etiology of oral microbial disease.

Consequences of Adhesion

Adhesion is not a passive "hand-holding" event. Rather, evidence is accumulating that adhesion involves sensing and response reactions whereby bacteria and oral surfaces form an interactive interface. There thus exists an elaborate communications network that generates and transmits signals among bacteria, epithelial cells and the underlying cells in the periodontal tissues. Bacteria possess a variety of adhesins with differing receptor specificities and affinities, that can potentially impinge to varying degrees upon a diversity of receptor-dependent host cell biochemical pathways. The resulting cellular products and fate make a significant impact on the local innate host defense mechanisms that serve to control the microbial challenge. One of the visible and most dramatic outcomes of this cross-talk is bacterial internalization within eukaryotic cells. Such bacterial invasion is thought to be a pathogenic mechanism of *P. gingivalis* and *A. actinomycetemcomitans* and is shared by a variety of important pathogens including *Salmonella, Shigella, Listeria,* and

Yersinia. An intracellular environment may benefit bacteria by providing a nutritionally rich milieu, largely protected from the ravages of the host immune system. In order to invade non-professional-phagocytic cells, such as epithelial cells, bacteria must be able to first adhere to the surface. They then induce a signal that ultimately causes the cytoskeletal rearrangements that underlie the membrane invaginations that engulf the bacteria and effect their internalization.

Invasion

P. gingivalis can invade primary cultures of epithelial cells (Lamont *et al.*, 1992), multilayered pocket epithelium (Sandros *et al.*, 1994), transformed epithelial cells (Duncan *et al.*, 1993) and endothelial cells (Deshpande *et al.*, 1998; Dorn *et al.*, 1999). Invasion of primary epithelial cells (Lamont *et al.*, 1995), involves both microfilament and microtubule reconstruction (Figure 6). Fimbriae mediated adhesion, possibly in concert with other adhesin-receptor pairs, can stimulate the

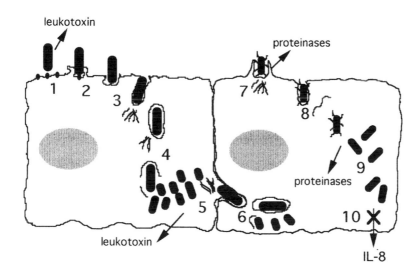

Figure 6. Patterns of epithelial cell invasion for two periodontal bacteria associated with severe forms of periodontal disease. *Actinobacillus actinomycetemcomitans* cells (left) bind to the surface of cultured epithelial cells via specific receptors (1) causing the membrane to ruffle and efface (2). The cytoplasmic membrane invaginates (3) and the bacteria become engulfed within a membrane vesicle. Actin filaments (3,4) appear associated with those vesicles that contain bacteria. *A. actinomycetemcomitans* then destroys the membrane vesicles (possibly by secretion of Phospholipase C) releasing the bacteria into the cytoplasm where they grow and divide rapidly (5). Bacteria become localized at membrane protrusions through which they enter adjoining epithelial cells (6). A major virulence factor of *A. actinomycetemcomitans* is a powerful leukotoxin that destroys immune cell function. *Porphyromonas gingivalis* cells (right) bind through their surface fimbriae (7) to a surface receptor on primary gingival cells. Microtubules and microfilaments are rearranged to facilitate invagination of the membrane that results in the engulfment of bacterial cells (8). *P. gingivalis* cells rapidly locate in the cytoplasm where they replicate freely and can produce proteinases that destroy host proteins (9). *P. gingivalis* has multiple effects on host cells, one of which is to block release of IL-8 that normally stimulates leukocytes (10). Bacterial lipopolysaccharide stimulates release of other cytokines from macrophages and promotes bone resorption.

signal transduction that leads to cytoskeletal remodeling. *P. gingivalis* cells are also capable of secreting a novel set of proteins when in contact with epithelial cells (Park and Lamont, 1998). Proteins secreted under these conditions are often translocated directly into the host cell cytoplasm. These translocated proteins can then impinge upon eukaryotic signaling pathways. The extent to which the *P. gingivalis* secreted proteins possess intracellular effector functions remains to be determined. Internalized *P. gingivalis* rapidly become located in the cytoplasm (Figure 7) where they can survive for extended periods and replicate. Cell-to-cell spread has not been observed and the ultimate fate of internal *P. gingivalis* is uncertain. The process of the cellular information flow that is disrupted by *P. gingivalis* is being unraveled. *P. gingivalis* induces a transient increase in epithelial cell cytosolic $[Ca^{2+}]$, as a result of release of Ca^{2+} from intracellular stores (Izutsu *et al.*, 1996). Such calcium ion fluxes are likely to be important in many signaling events and may converge on calcium gated ion channels in the cytoplasmic membrane, cytoskeletal remodeling, or nuclear transcription factors. The collective action of the subversion of epithelial cell intracellular pathways by *P. gingivalis* can have phenotypic effects with immediate relevance to the disease process. Regulation of matrix metalloproteinase (MMP) production by gingival epithelial cells is disrupted following contact with the organism (Fravalo *et al.*, 1996), thus interfering with extracellular matrix repair and reorganization. Invasion of *P. gingivalis* also has implications for innate host immunity. Secretion of interleukin (IL)-8 by gingival epithelial cells is inhibited following *P. gingivalis* invasion (Figure 6). *P. gingivalis* is also able to antagonize IL-8 secretion following stimulation of epithelial cells by common plaque commensals (Darveau *et al.*, 1998). Inhibition of IL-8 accumulation by *P. gingivalis* at sites of bacterial invasion could have a debilitating effect on innate host defense in the periodontium where bacterial exposure is constant. The host would no longer be able to detect the presence of bacteria and direct leukocytes for their removal. The ensuing overgrowth of bacteria would then contribute to a burst of disease activity.

A. *actinomycetemcomitans* invades epithelial cells by a dynamic multistep process (Fives-Taylor *et al.*, 1999). Initial attachment, primarily to the transferrin receptor, induces effacement of the microvilli and the bacteria enter through ruffled apertures in the cell membrane (Figure 6). Actin microfilament rearrangements are required for the entry of some strains while other strains enter by actin-independent mechanisms. Internal bacteria are initially constrained within a host-derived membrane vacuole, but this membrane is soon broken down and the bacteria are present in the cytoplasm where they can replicate. What follows is a remarkable example of bacterial orchestration of host cell function. A. *actinomycetemcomitans* induces the formation of surface membrane protrusions through which the organism can migrate and enter into adjacent cells. The formation of these protrusions is consequent to bacterial interaction with the plus-ends of microtubules and movement through them may depend on bacterial cell division.

Both *P. gingivalis* and A. *actinomycetemcomitans* can demonstrably invade cells in tissue culture. The questions then arise: to what extent does this happen *in vivo*, and what is its significance? Neither of these questions are readily tractable. Immunofluorescence studies have revealed the presence of *P. gingivalis* and A.

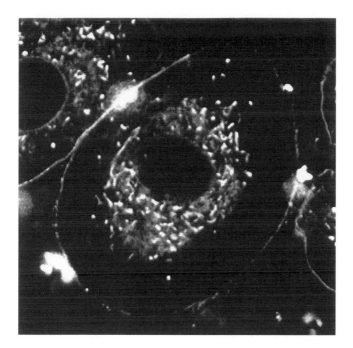

Figure 7. Fluorescent image analysis of a 0.2 μm thick optical section of a gingival epithelial cell exposed to *P. gingivalis*. Image was obtained by widefield deconvolution microscopy using a lipid dye to highlight the plasma membrane and fluorescently labeled *P. gingivalis*. The bacteria have invaded the epithelial cell in high numbers and are concentrated in the perinuclear region. Kindly provided by Drs Izutsu and Belton (University of Washington, Seattle, USA).

actinomycetemcomitans within cells from biopsy tissue (Saglie *et al.*, 1988); however, more modern imaging techniques that can optically section through cells would be required to definitively answer the question. As to significance, one possibility is that the bacteria use an intracellular location as a pied-à -terre which serves as a site for recrudescence of active disease episodes. Alternatively, uptake of bacteria by epithelial cells may be a means to sequester pathogenic organisms which are then eliminated following apoptotic epithelial cell death. In either event, the outcome of overall association between bacteria and epithelial cells has importance for the ecological balance in the oral cavity. Moreover, adherent bacteria need not internalize for attempted manipulation of information flow to occur, as is well illustrated by *T. denticola*.

Host-Signaling without Invasion

T. denticola is a spirochete that is associated with the advancing periodontal lesion. Epithelial cells in contact with *T. denticola in vitro* rapidly undergo depolymerization and rearrangement of actin microfilaments and, concomitantly, experience shrinkage and loss of cell-cell adhesion and substratum detachment (Ellen, 1999). Expression of desmoplakin II, a desmosome-associated protein, and cytokeratins are also reduced. Potential effector molecules of *T. denticola* appear to be an outer-membrane chymotrypsin-like protease along with other outer membrane proteins including a major surface protein (Msp) of 53-kDa. The chymotrypsin-like enzyme can degrade junctional complexes and, following transport into the cell, disturb the actin cytoskeleton (Uitto *et al.*, 1995). The 53-kDa Msp can be translocated into the epithelial cell membrane where it forms a conductance ion channel and depolarizes the membrane (Mathers *et al.*, 1996). The actin rearrangements may be induced by impinging on IP_3-dependent calcium ion signaling pathways. *T. denticola* outer membrane proteins cause an immediate and long-lasting suppression of normal intracellular calcium responses, by inhibiting Ca^{2+} release from intracellular stores and interfering with calcium release-activated channels (Ellen, 1999).

Bacterial Cell Signaling
To come full circle, not only do eukaryotic cells respond to the presence of bacteria, but the bacteria themselves have mechanisms to sense their location and respond accordingly. Although this is better documented for non-oral pathogens, evidence is accumulating for similar genetic control systems in important oral bacteria (Burne *et al.*, 1997; Jenkinson and Lamont, 1997; Vickerman and Clewell, 1997; Lamont and Jenkinson, 1998). Transcription of the *S. mutans gtf* genes can be upregulated in response to contact with saliva coated hydroxyapatite (Hudson and Curtis, 1990), and in *S. gordonii* a gene designated *rgg* positively regulates expression of glucosyltransferase (Vickerman and Clewell, 1997). Other adhesins that can show differential expression include: the streptococcal antigen I/II family; the oligopeptide permeases that are associated, either directly or indirectly, with streptococcal adhesion; the CshA adhesin of *S. gordonii*; the platelet aggregation-associated protein of *S. sanguis,* and a laminin-binding protein in *S. gordonii*, that are upregulated in the presence of collagen; and *P. gingivalis* fimbrillin that may be regulated by cell density or contact with surfaces. As adhesion is an important ecological determinant, the array of adhesin expression can be expected to be a tightly controlled process. The ideal number and types of adhesins expressed at, for example, the mucous membranes may not be the most advantageous configuration if the bacteria gain access to deeper tissues where such adhesion might promote uptake by the host's professional phagocytic cells. Bacteria and host cells thus participate in a highly choreographed interchange which will establish the nature of their long term relationship, be it expulsion or persistence with or without disease.

Concluding Remarks

Bacterial adhesion can be seen as a defining event for oral bacteria. Once firmly anchored to oral surfaces, bacteria can be expected to remain at that site for a long period (in relation to bacterial doubling time) and may alter their phenotypic properties accordingly. The multiplicity of adhesins that bacteria produce is testimony to the ecological importance of adhesion and the opportunities available to sessile organisms. The specificity and expression of these adhesins determines, to a large extent, the ecological niche of individual species within the oral cavity. Accumulation into a complex plaque biofilm provides for nutritional and protective advantages not afforded by planktonic growth. These accumulations are responsible for superficial tissue damage, such as that associated with gingivitis, and if unchecked can lead to permanent loss of enamel and dentin (caries), or periodontal tissues (periodontal diseases) (see Figure 1). For some species of bacteria, adhesion to host tissues is the prelude to internalization within host cells, and the development of systemic disease conditions (Figure 1). Since adhesion is of paramount ecological significance in oral infections, strategies aimed at blocking adhesin functions and/ or subsequent signaling through immunization or drug therapies may receive greater attention in oral disease prevention in the future.

References

Amano, A., K. Kataoka, P. A. Raj, R. J. Genco, and S. Shizukuishi. 1996a. Binding sites of salivary statherin for *Porphyromonas gingivalis*. Infect. Immun. 64: 4249-4254.

Amano, A., A. Sharma, J-Y. Lee, H. T. Sojar, P. A. Raj, and R. J. Genco. 1996b. Structural domains of *Porphyromonas gingivalis* recombinant fimbrillin that mediate binding to salivary proline-rich protein and statherin. Infect. Immun. 64: 1631-1637.

Banas, J. A., R. R. B. Russell, and J. J. Ferretti. 1990. Sequence analysis of the gene for the glucan-binding protein of *Streptococcus mutans* Ingbritt. Infect. Immun. 58: 667-673.

Brooks, W., D. R. Demuth, S. Gil, and R. J. Lamont. 1997. Identification of a *Streptococcus gordonii* SspB domain that mediates adhesion to *Porphyromonas gingivalis*. Infect. Immun. 65: 3753-3758.

Burne, R. A., Y-Y. M. Chen, and J. E. C. Penders. 1997. Analysis of gene expression in *Streptococcus mutans* in biofilms *in vitro*. Adv. Dent. Res. 11: 100-109.

Burnette-Curley, D., V. Wells, H. Viscount, C. L. Munro, J. C. Fenno, P. Fives-Taylor, and F. L. Macrina. 1995. FimA, a major virulence factor associated with *Streptococcus parasanguis* endocarditis. Infect. Immun. 63: 4669-4674.

Carlsson, J. 1967. Presence of various types of non-haemolytic streptococci in dental plaque and in other sites of the oral cavity in man. Odontol Revy 18: 55-74.

Caufield, P. W., G. R. Cutter, and A. P. Dasanayake. 1993. Initial acquisition of mutans streptococci by infants: evidence for a discrete window of infectivity. J. Dent. Res. 72: 37-45.

Cisar, J. O., E. L. Barsumian, R. P. Siraganian, W. B. Clark, M. K. Yeung, S. D. Hsu, S. H. Curl, A. E. Vatter, and A. L. Sandberg. 1991. Immunochemical and functional studies of *Actinomyces viscosus* T14V type 1 fimbriae with monoclonal and polyclonal antibodies directed against the fimbrial subunit. J. Gen. Microbiol. 137: 1971-1979.

Cisar, J. O., A. L. Sandberg, G. P. Reddy, C. Abeygunawardana, and C. A. Bush. 1997. Structural and antigenic types of cell wall polysaccharides from viridans group streptococci with receptors for oral actinomyces and streptococcal lectins. Infect. Immun. 65: 5035-5041.

Clark, W. B., J. E. Beem, W. E. Nesbitt, J. O. Cisar, C. C. Tseng, and M. J. Levine. 1989. Pellicle receptors for *Actinomyces viscosus* type 1 fimbriae *in vitro*. Infect. Immun. 57: 3003-3008.

Cook, G. S., J. W. Costerton, and R. J. Lamont. 1998. Biofilm formation by *Porphyromonas gingivalis* and *Streptococcus gordonii*. J. Periodont. Res. 33: 323-327.

Cole, M. F., M. Evans, S. Fitzsimmons, J. Johnson, C. Pearce, M. J. Sheridan, R. Wientzen, and G. Bowden. 1994. Pioneer oral streptococci produce immunoglobulin A1 protease. Infect. Immun. 62: 2165-2168.

Costello, A. H., J. O. Cisar, P. E. Kolenbrander, and O. Gabriel. 1979. Neuraminidase-dependent hemagglutination of human erythrocytes by human strains of *Actinomyces viscosus* and *Actinomyces naeslundii*. Infect. Immun. 26: 563-572.

Costerton, J. W., Z. Lewandowski, D. E. Caldwell, D. R. Korber, and H. M. Lappin-Scott. 1995. Microbial Biofilms. Annu. Rev. Microbiol. 49: 711-745.

Cowan, M. M., K. G. Taylor, and R. J. Doyle. 1987. Energetics of the initial phase of adhesion of *Streptococcus sanguis* to hydroxylapatite. J. Bacteriol. 169: 2995-3000.

Cundell, D. R., B. J. Pearce, J. Sandros, A. M. Naughton, and H. R. Masure. 1995. Peptide permeases from *Streptococcus pneumoniae* affect adherence to eukaryotic cells. Imfect. Immun. 63: 2493-2498.

Darveau, R. P., C. M. Belton, R. A. Reife, and R. J. Lamont. 1998. Local chemokine paralysis: a novel pathogenic mechanism of *Porphyromonas gingivalis*. Infect. Immun. 66: 1660-1665.

Davies, D. G., M. R. Parsek, J. P. Pearson, B. H. Iglewski, J. W. Costerton, and E. P. Greenberg. 1998. The involvement of cell-to-cell signals in the development of a bacterial biofilm. Science 280: 295-298.

Demuth, D. R., Y. Duan, W. Brooks, A. R. Holmes, R. McNab, and H. F. Jenkinson. 1996. Tandem genes encode cell-surface polypeptides SspA and SspB which mediate adhesion of the oral bacterium *Streptococcus gordonii* to human and bacterial receptors. Mol. Microbiol. 20: 403-413.

Deshpande, R.G., M. B. Khan, and C. A. Genco. 1998. Invasion of aortic and heart endothelial cells by *Porphyromonas gingivalis*. Infect. Immun. 66: 5337-5343.

Dintilhac, A., G. Alloing, C. Granadel, and J-P. Claverys. 1997. Competence and virulence of *Streptococcus pneumoniae*: Adc and PsaA mutants exhibit a requirement for Zn and Mn resulting from inactivation of putative ABC metal permeases. Mol. Microbiol. 25: 727-739.

Dorn, B. R., W. A. Dunn, and A. Progulske-Fox. 1999. Invasion of human coronary artery cells by periodontal pathogens. Infect. Immun. 67: 5792-5798.

Doyle, R. J., M. Rosenberg, and D. Drake. 1990. Hydrophobicity of oral bacteria, p. 387-419. *In* R. J. Doyle and M. Rosenberg (ed.), Microbial cell surface hydrophobicity. ASM Press, Washington, DC.

Duncan, M. J., S. Nakao, Z. Skobe, and H. Xie. 1993. Interactions of *Porphyromonas gingivalis* with epithelial cells. Infect. Immun. 61: 2260-2265.

Ellen, R. P. 1976. Establishment and distribution of *Actinomyces viscosus* and *Actinomyces naeslundii* in the human oral cavity. Infect. Immun. 14: 1119-1124.

Ellen, R. P. 1999. Perturbation and exploitation of host cell cytoskeleton by periodontal pathogens. Microbes Infect. 1: 621-632.

Ellen, R. P., E. D. Fillery, K. H. Chan, and D. A. Grove. 1980. Sialidase-enhanced lectin-like mechanism for *Actinomyces viscosus* and *Actinomyces naeslundii* hemagglutination. Infect Immun 27: 336-343.

Ferretti, J.J., M.L. Gilpin, and R.R.B. Russell. 1987. Nucleotide sequence of a glucosyltransferase gene from *Streptococcus sobrinus* Mfe28. J. Bacteriol. 169: 4271-4278.

Fine, D. H., D. Furgang, H. C. Schreiner, P. Goncharoff, J. Charlesworth, G. Ghazwan, P. Fitzgerald-Bocarsly, and D. H. Figurski. 1999. Phenotypic variation in *Actinobacillus actinomycetemcomitans* during laboratory growth: implications for virulence. Microbiol. 145: 1335-1347.

Fives-Taylor, P. M., D. Hutchins Meyer, K. P. Mintz, and C. Brissette. 1999. Virulence factors of *Actinobacillus actinomycetemcomitans*. Periodontol. 2000 20: 136-167.

Fravalo, P., C. Menard, and M. Bonnaure-Mallet. 1996. Effect of *Porphyromonas gingivalis* on epithelial cell MMP-9 type IV collagenase production. Infect. Immun. 64: 4940-4945.

Gibbons, R. J. 1984. Microbial ecology. Adherent interactions which may affect microbial ecology in the mouth. J. Dent. Res. 63: 378-385.

Gibbons, R. J. 1989. Bacterial adhesion to oral tissues: a model for infectious diseases. J. Dent. Res. 68: 750-760.

Gibbons, R. J., and J. van Houte. 1975. Bacterial adherence in oral microbial ecology. Annu. Rev. Microbiol. 29: 19-44.

Gibbons, R. J., and D. I. Hay. 1988. Human salivary acidic proline-rich proteins and statherin promote the attachment of *Actinomyces viscosus* LY7 to apatitic surfaces. Infect. Immun. 56: 2990-2993.

Gibbons, R. J., and M. Nygaard. 1970. Interbacterial aggregation of plaque bacteria. Arch. Oral Biol 15: 1397-1400.

Giffard, P. M., and N. A. Jacques. 1994. Definition of a fundamental repeating unit in streptococcal glucosyltransferase glucan-binding regions and related sequences. J. Dent. Res. 73: 1133-1141.

Gilbert, P., D. J. Evans, and M. R. Brown. 1993. Formation and dispersal of bacterial biofilms *in vivo* and *in situ*. J. Appl. Bacteriol. 74: 67S-78S.

Gong, K., and M. C. Herzberg. 1997. *Streptococcus sanguis* expresses a 150-kilodalton two-domain adhesin: characterization of several independent adhesin epitopes. Infect. Immun. 65: 3815-3821.

Haase, E. M., J. L. Zmuda, and F. A. Scannapieco. 1999. Identification and molecular analysis of rough-colony-specific outer membrane proteins of *Actinobacillus actinomycetemcomitans*. Infect. Immun. 67: 2901-2908

Hamada, S., T. Fujiwara, S. Morishima, I. Takahashi, I. Nakagawa, S. Kimura, and T. Ogawa. 1994. Molecular and immunological characterization of the fimbriae of *Porphyromonas gingivalis*. Microbiol. Immunol. 38: 921-930.

Hammar, M., Z. Bian, and S. Normark. 1996. Nucleator-dependent intercellular assembly of adhesive curli organelles in *Escherichia coli*. Proc. Natl. Acad. Sci. USA 93: 6562-6566.

Handley, P. S. 1990. Structure, composition, and functions of surface structures on oral bacteria. Biofouling 2: 239-264.

Handley, P., A. Coykendall, D. Beighton, J. M. Hardie, and R. A. Whiley. 1991. *Streptococcus crista* sp. nov., a viridans streptococcus with tufted fibrils, isolated from the human oral cavity and throat. Int. J. Syst. Bacteriol. 41: 543-547.

Harano, K., A. Yamanaka, and K. Okuda. 1995. An antiserum to a synthetic fimbrial peptide of *Actinobacillus actinomycetemcomitans* blocked adhesion of the microorganism. FEMS Microbiol. Lett. 130: 279-286.

Harrington, D. J., and R. R. B. Russell. 1993. Multiple changes in cell wall antigens of isogenic mutants of *Streptococcus mutans*. J. Bacteriol. 175: 5925-5933.

Hasty, D. L., I. Ofek, H. S. Courtney, and R. J. Doyle. 1992. Multiple adhesins of streptococci. Infect. Immun. 60: 2147-2152.

Hazlett, K. R. O., S. M. Michalek, and J. A. Banas. 1998. Inactivation of the *gbpA* gene of *Streptococcus mutans* increases virulence and promotes *in vivo* accumulation of recombinations between glucosyltransferase B and C genes. Infect. Immun. 66: 2180-2185.

Hirota, K., H. Kanitani, K. Nemoto, T. Ono, and Y. Miyake. 1995. Cross-reactivity between human sialyl Lewis(x) oligosaccharide and common causative oral bacteria of infective endocarditis. FEMS Immunol. Med. Microbiol. 12: 159-164.

Holmes, A. R., R. McNab, and H. F. Jenkinson. 1996. *Candida albicans* binding to the oral bacterium *Streptococcus gordonii* involves multiple adhesin-receptor interactions. Infect. Immun. 64: 4680-4685.

Hudson, M. C., and R. Curtiss. 1990. Regulation of expression of *Streptococcus mutans* genes important to virulence. Infect. Immun. 58: 464-470.

Hultgren, S. J., C. H. Jones, and S. Normark. 1996. Bacterial adhesins and their assembly. *In* Neidhardt, F.C. (ed.), *Escherichia coli and Salmonella*, 2nd Edition, pp. 2730-2756, ASM Press, Washington, DC.

Inoue, T., I. Tanimoto, H. Ohta, K. Kato, Y. Murayama, and K. Fukui. 1998. Molecular characterization of low-molecular-weight component protein, Flp, in *Actinobacillus actinomycetemcomitans* fimbriae. Microbiol. Immunol. 42: 253-258.

Ishihara, K., K. Honma, T. Miura, T. Kato, and K. Okuda. 1997. Cloning and sequence analysis of the fimbriae associated protein (*fap*) gene from *Actinobacillus actinomycetemcomitans*. Microb. Pathog. 23: 63-69.

Israelachvili, J.N., and P.M. McGuiggan. 1988. Forces between surfaces in liquids. Science 241: 795-800.

Izutsu, K. T., C. M. Belton, A. Chan, S. Fatherazi, J. P. Kanter, Y. Park, and R. J. Lamont. 1996. Involvement of calcium in interactions between gingival epithelial cells and *Porphyromonas gingivalis*. FEMS Microbiol. Lett. 144: 145-150.

Jameson, M. W., H. F. Jenkinson, K. Parnell, and P. S. Handley. 1995. Polypeptides associated with tufts of cell-surface fibrils in an oral *Streptococcus*. Microbiol. 141: 2729-2738.

Jenkinson, H. F. 1987. Novobiocin-resistant mutants of *Streptococcus sanguis* with reduced cell hydrophobicity and defective in coaggregation. J. Gen. Microbiol. 133: 1909-1918.

Jenkinson, H. F. 1994. Cell surface protein receptors in oral streptococci. FEMS Microbiol. Lett. 121: 133-140.

Jenkinson, H. F., R. A. Baker, and G. W. Tannock. 1996. A binding-lipoprotein-dependent oligopeptide transport system in *Streptococcus gordonii* essential for uptake of hexa- and heptapeptides. J. Bacteriol. 178: 68-77.

Jenkinson, H. F., and D. R. Demuth. 1997. Structure, function and immunogenicity of streptococcal antigen I/II polypeptides. Mol. Microbiol. 23: 183-190.

Jenkinson, H. F., and R. J. Lamont. 1997. Streptococcal adhesion and colonization. Crit. Rev. Oral Biol. Med. 8: 175-200.

Kataoka, K., A. Amano, M. Kuboniwa, H. Horie, H. Nagata, and S. Shizukuishi. 1997. Active sites of salivary proline-rich protein for binding to *Porphyromonas gingivalis* fimbriae. Infect. Immun. 65: 3159-3164.

Kilian, M., J. Mestecky, and M. W. Russell. 1988. Defense mechanisms involving Fc-dependent functions of immunoglobulin A and their subversion by bacterial immunoglobulin A proteases. Microbiol. Rev. 52: 296-303.

Kinder, S. A., and S. C. Holt. 1989. Characterization of coaggregation between *Bacteroides gingivalis* T22 and *Fusobacterium nucleatum* T18. Infect. Immun. 57: 3425-3433.

Klier, C. M., P. E. Kolenbrander, A. G. Roble, M. L. Marco, S. Cross, and P. S. Handley. 1997. Identification of a 95-kDa putative adhesin from *Actinomyces* serovar WVA963 strain PK1259 that is distinct from type 2 fimbrial subunits. Microbiol. 143: 835-846.

Kolenbrander, P. E., R. N. Andersen, R. A. Baker, and H. F. Jenkinson. 1998. The adhesion-associated *sca* operon in *Streptococcus gordonii* encodes an inducible high-affinity ABC transporter for Mn uptake. J. Bacteriol. 180: 290-295.

Kontani, M., S. Kimura, I. Nakagawa, and S. Hamada. 1997. Adherence of *Porphyromonas gingivalis* to matrix proteins via a fimbrial cryptic receptor exposed by its own arginine-specific protease. Mol. Microbiol. 24: 1179-1187.

Kremer, B. H. A., J. J. E. Bijlsma, J. G. Kusters, J. de Graaff, and T. J. M. van Steenbergen. 1999. Cloning of *fibA*, encoding an immunogenic subunit of the fibril-like surface sturcture of *Peptostreptococcus micros*. J. Bacteriol. 181: 2485-2491.

Kuramitsu, H. K. 1993. Virulence factors of mutans streptococci: role of molecular genetics. Crit. Rev. Oral Biol. Med.4: 159-176.

Lamont, R. J, C. A. Bevan, S. Gil, R. E. Persson, and B. Rosan. 1993. Involvement of *Porphyromonas gingivalis* fimbriae in adherence to *Streptococcus gordonii*. Oral Microbiol. Immunol. 8: 272-276.

Lamont, R. J., A. Chan, C. M. Belton, K. T. Izutsu, D. Vasel, and A. Weinberg. 1995. *Porphyromonas gingivalis* invasion of gingival epithelial cells. Infect. Immun. 63: 3878-3885.

Lamont, R. J., D. R. Demuth, C. A. Davis, D. Malamud, and B. Rosan. 1991. Salivary-agglutinin-mediated adherence of *Streptococcus* mutans to early plaque bacteria. Infect. Immun. 59: 3446-3450.

Lamont, R. J., G. W. Hsiao, and S. Gil. 1994. Identification of a molecule of *Porphyromonas gingivalis* that binds to *Streptococcus gordonii*. Microbial. Pathog. 17: 355-360.

Lamont, R. J. and H. F. Jenkinson. 1998. Life below the gum line: pathogenic mechanisms of *Porphyromonas gingivalis*. Microbiol. Mol. Biol. Rev. 62: 1244-1263.

Lamont, R. J., D. Oda, R. E. Persson, and G. R. Persson. 1992. Interaction of *Porphyromonas gingivalis* with gingival epithelial cells maintained in culture. Oral Microbiol. Immunol. 7: 364-367.

Lancy, P. Jr, J. M. DiRienzo, B. Appelbaum, B. Rosan, and S. C. Holt. 1983. Corncob formation between *Fusobacterium nucleatum* and *Streptococcus sanguis*. Infect.

Immun. 40: 303-309.

Lantz, M. S., R. D. Allen, L. W. Duck, J. L. Blume, L. M. Switalski, and M. Hook. 1991. Identification of *Porphyromonas gingivalis* components that mediate its interactions with fibronectin. J. Bacteriol. 173: 4263-4270.

Lee, S. F., A. Progulske-Fox, G. W. Erdos, D. A. Piacentini, G. Y. Ayakawa, P. J. Crowley, and A. S. Bleiweis. 1989. Construction and characterization of isogenic mutants of *Streptococcus mutans* deficient in major surface protein antigen P1 (antigen I/II). Infect. Immun. 57: 3306-3313.

Lee, J.Y., H. T. Sojar, G. S. Bedi, and R. J. Genco. 1991. *Porphyromonas (Bacteroides) gingivalis* fimbrillin: size, aminoterminal sequence, and antigenic heterogeneity. Infect. Immun. 59: 383-389.

Li, J., R. P. Ellen, C. J. Hoover, and J. R. Felton. 1991. Association of proteases of *Porphyromonas (Bacteroides) gingivalis* with its adhesion to *Actinomyces viscosus*. J. Dent. Res. 70: 82-86.

Li, T., I. Johansson, D. I. Hay, and N. Stromberg. 1999. Strains *of Actinomyces naeslundii* and *Actinomyces viscosus* exhibit structurally variant fimbrial subunit proteins and bind to different peptide motifs in salivary proteins. Infect. Immun. 67: 2053-2059.

Listgarten, M. A. 1976. Structure of the microbial flora associated with periodontal health and disease in man. J. Periodontol. 47: 1-18.

London, J., and J. Allen. 1990. Purification and characterization of a *Bacteroides loescheii* adhesin that interacts with procaryotic and eucaryotic cells. J. Bacteriol. 172: 2527-2534.

Ma, J. K., M. Hunjan, R. Smith, C. Kelly, and T. Lehner. 1990. An investigation into the mechanism of protection by local passive immunization with monoclonal antibodies against *Streptococcus mutans*. Infect. Immun. 58: 3407-3414.

Ma, Y. S., M. O. Lassiter, J. A. Banas, M. Y. Galperin, K. G. Taylor, and R. J. Doyle. 1996. Multiple glucan-binding proteins of *Streptococcus sobrinus*. J. Bacteriol. 178: 1572-1577.

Marshall, K. C. 1980. Bacterial adhesion in natural environments, p. 351-388. *In* R. C. W. Berkeley, J. M. Lynch, J. Melling, P. R. Rutter, and B. Vincent (ed.), Microbial adhesion to surfaces. Ellis Horwood, Chichester, England.

Mathers, D. A., W. K. Leung, J. C. Fenno, Y. Hong, and B. C. McBride. 1996. The major surface protein complex of *Treponema denticola* depolarizes and induces ion channels in HeLa cell membranes. Infect. Immun. 64: 2904-2910.

McNab, R., H. Forbes, P. S. Handley, D. M. Loach, G. W. Tannock, and H. F. Jenkinson. 1999. Cell wall-anchored CshA polypeptide (259 kilodaltons) in *Streptococcus gordonii* forms surface fibrils that confer hydrophobic and adhesive properties. J. Bacteriol. 181: 3087-3095.

McNab, R., A. R. Holmes, J. C. Clarke, G. W. Tannock, and H. F. Jenkinson. 1996. Cell-surface polypeptide CshA mediates binding of *Streptococcus gordonii* to other oral bacteria and to immobilized fibronectin. Infect. Immun. 64: 4204-4210.

McNab, R., and H. F. Jenkinson. 1998. Altered adherence properties of a *Streptococcus gordonii hppA* (oligopeptide permease) mutant result from transcription effects on *cshA* adhesin gene expression. Microbiol. 144: 127-136.

McNab, R., H. F. Jenkinson, D. M. Loach, and G. W. Tannock. 1994. Cell-surface-associated polypeptides CshA and CshB of high molecular mass are colonization determinants in the oral bacterium *Streptococcus gordonii*. Mol. Microbiol. 14: 743-754.

Modun, B., and P. Williams. 1999. The staphylococcal transferrin-binding protein is a cell wall glyceraldehyde-3-phosphate dehydrogenase. Infect. Immun. 67: 1086-1092.

Mouton, C., H. S. Reynolds, and R. J. Genco. 1980. Characterization of tufted streptococci isolated from the "corn cob" configuration of human dental plaque. Infect. Immun. 27: 235-245.

Munro, C. L., S. M. Michalek, and F. L. Macrina. 1995. Sucrose-derived exopolymers have site-dependent roles in *Streptococcus mutans*-promoted dental decay. FEMS Microbiol. Lett. 128: 327-332.

Nakayama, K., F. Yoshimura, T. Kadowaki, and K. Yamamoto. 1996. Involvement of arginine-specific cysteine proteinase (Arg-gingipain) in fimbriation of *Porphyromonas gingivalis*. J. Bacteriol. 178: 2818-2824.

Navarre, W. W., and O. Schneewind. 1994. Proteolytic cleavage and cell wall anchoring at the LPXTG motif of surface proteins in gram-positive bacteria. Mol. Microbiol. 14: 115-121

Novak, R., J. S. Braun, E. Charpentier, and E. Tuomanen. 1998. Penicillin tolerance genes of *Streptococcus penumoniae*: the ABC-type manganese permease complex Psa. Mol. Microbiol. 29: 1285-1296.

Oligino, L., and P. Fives-Taylor. 1993. Overexpression and purification of a fimbria-associated adhesin of *Streptococcus parasanguis*. Infect. Immun. 61: 1016-1022.

Park, Y., and R. J. Lamont. 1998. Contact-dependent protein secretion in *Porphyromonas gingivalis*. Infect. Immun. 66: 4777-4782.

Pavloff, N., P. A. Pemberton, J. Potempa, W. C. Chen, R. N. Pike, V. Prochazka, M. C. Kiefer, J. Travis, and P. J. Barr. 1997. Molecular cloning and characterization of *Porphyromonas gingivalis* lysine-specific gingipain. A new member of an emerging family of pathogenic bacterial cysteine proteinases. J. Biol. Chem. 272: 1595-1600.

Progulske-Fox, A., S. Tumwasorn, and S. C. Holt. 1989. The expression and function of a *Bacteroides gingivalis* hemagglutinin gene in *Escherichia coli*. Oral Microbiol. Immunol. 4: 121-131.

Rogers, J. D., E. M. Haase, A. E. Brown, C. W. I. Douglas, J. P. Gwynn, and F. A. Scannapieco. 1998. Identification and analysis of a gene (*abpA*) encoding a major amylase-binding protein in *Streptococcus gordonii*. Microbiol. 144: 1223-1333.

Rosan, B., J. Slots, R. J. Lamont, G. M. Nelson, and M. Listgarten. 1988. *Actinobacillus actinomycetemcomitans* fimbriae. Oral Microbiol. Immunol. 3: 58-63.

Rosenow, C., P. Ryan, J. N. Weiser, S. Johnson, P. Fontan, A. Ortqvist, and H. R. Masure. 1997. Contribution of novel choline-binding proteins to adherence, colonization, and immunogenicity of *Streptococcus pneumoniae*. Mol. Microbiol. 25: 819-829.

Ruhl, S., A. L. Sandberg, M. F. Cole, and J. O. Cisar. 1996. Recognition of immunoglobulin A1 by oral actinomyces and streptococcal lectins. Infect. Immun. 64: 5421-5424.

Saglie, F. R., A. Marfany, and P. Camargo. 1988. Intragingival occurrence of *Actinobacillus actinomycetemcomitans* and *Bacteroides gingivalis* in active destructive periodontal lesions. J. Periodontol. 59: 259-265.

Sandros, J., P. N. Papapanou, U. Nannmark, and G. Dahlén. 1994. *Porphyromonas gingivalis* invades human pocket epithelium *in vitro*. J. Periodontal Res. 29: 62-69.

Scannapieco, F.A. 1994. Saliva-bacterium interactions in oral microbial ecology. Crit. Rev. Oral Biol. Med. 5: 203-248.

Schilling, K. M., and W. H. Bowen. 1992. Glucans synthesized *in situ* in experimental salivary pellicle function as specific binding sites for *Streptococcus mutans*. Infect. Immun. 60: 284-295.

Shaniztki, B., D. Hurwitz, N. Smorodinsky, N. Ganeshkumar, and E. I. Weiss. 1997. Identification of a *Fusobacterium nucleatum* PK1594 galactose-binding adhesin which mediates coaggregation with periopathogenic bacteria and hemagglutination. Infect. Immun. 65: 5231-5237.

Simpson, C. L., P. M. Giffard, and N. A. Jacques. 1995. *Streptococcus salivarius* ATCC 25975 possesses at least two genes coding for primer-independent glucosyltransferases. Infect. Immun. 63: 609-621.

Skopek, R. J., W. F. Liljemark, C. G. Bloomquist, and J. D. Rudney. 1993. Dental plaque development on defined streptococcal surfaces. Oral Microbiol. Immunol. 8: 16-23.

Slots, J., and R. J. Gibbons. 1978. Attachment of *Bacteroides melanogenicus* subsp. *asaccharolyticus* to oral surfaces and its possible role in the colonization of the mouth and of periodontal pockets. Infect. Immun. 19: 254-264.

Smith, D. J., W. F. King, C. D. Wu, B. I. Shen, M. A. Taubman. 1998. Structural and antigenic characteristics of *Streptococcus sobrinus* glucan binding proteins. Infect. Immun. 66: 5565-5569.

Socransky, S. S., A. D. Haffajee, M. A. Cugini, C. Smith, and R. L. Kent, Jr. 1998. Microbial complexes in subgingival plaque. J. Clin. Periodontol. 25: 134-144.

Spellerberg, B., E. Rozdzinski, S. Martin, J. Weber-Heynemann, N. Schnitzler, R. Lutticken, and A. Podbielski. 1999. Lmb, a protein with similarities to the LraI adhesin family, mediates attachment of *Streptococcus agalactiae* to human laminin. Infect. Immun. 67: 871-878.

Strömberg, N., and T. Borén. 1992. *Actinomyces* tissue specificity may depend on differences in receptor specificity for GalNAcß-containing glycoconjugates. Infect. Immun. 60: 3268-3277.

Sutcliffe, I. C., and R. R. B. Russell. 1995. Lipoproteins of gram-positive bacteria. J. Bacteriol. 177: 1123-1128.

Takahashi, Y., A. L. Sandberg, S. Ruhl, J. Muller, and J. O. Cisar. 1997. A specific cell surface antigen of *Streptococcus gordonii* is associated with bacterial hemagglutination and adhesion to α2-3-linked sialic acid-containing receptors. Infect. Immun. 65: 5042-5051.

Takeshita, A., Y. Murakami, Y. Yamashita, M. Ishida, S. Fujisawa, S. Kitano, and S. Hanazawa. 1998. *Porphyromonas gingivalis* fimbriae use ß$_2$ integrin (CD11/CD18) on mouse peritoneal macrophages as a cellular receptor, and the CD18 beta chain plays a functional role in fimbrial signaling. Infect. Immun. 66: 4056-4060.

Uitto, V. J., Y. M. Pan, W. K. Leung, H. Larjava, R. P. Ellen, B. B. Finlay, and B. C. McBride. 1995. Cytopathic effects of *Treponema denticola* chymotrypsin-like proteinase on migrating and stratified epithelial cells. Infect. Immun. 63: 3401-3410.

Vacca-Smith, A. M., C. A. Jones, M. J. Levine, and M. W. Stinson. 1994. Glucosyltransferase mediates adhesion of *Streptococcus gordonii* to human endothelial cells *in vitro*. Infect. Immun. 62: 2187-2194.

van der Mei, H. C., S. D. Cox, G. I. Geertsema-Doornbusch, R. J. Doyle, and H. J Busscher. 1993. A critical appraisal of positive cooperativity in oral streptococcal adhesion: Scatchard analyses of adhesion data versus analyses of the spatial arrangement of adhering bacteria. J. Gen. Microbiol. 139: 937-948

van Loosdrecht, M. C., J. Lyklema, W. Norde, and A. J. B. Zehnder. 1990. Influence of interfaces on microbial activity. Microbiol. Rev. 54: 75-87.

Vickerman, M. M., and D. B. Clewell. 1997. Regulation of *Streptococcus gordonii* glucosyltransferase. Adv. Exp. Med. Biol.418: 661-664.

Vickerman, M. M., D. B. Clewell, and G. W. Jones. 1991. Ecological implications of glucosyltransferase phase variation in *Streptococcus gordonii*. Appl. Environ. Microbiol. 57: 3648-3651.

von Eichel-Streiber, C., M. Sauerborn, and H. K. Kuramitsu. 1992. Evidence for a modular structure of the homologous repetitive C-terminal carbohydrate-binding sites of *Clostridium difficile* toxins and *Streptococcus mutans* glucosyltransferases. J. Bacteriol. 174: 6707-6710.

Weerkamp, A. H., P. S. Handley, A. Baars, and J. W. Slot. 1986a. Negative staining and immunoelectron microscopy of adhesion-deficient mutants of *Streptococcus salivarius* reveal that the adhesive protein antigens are separate classes of cell surface fibril. J. Bacteriol. 165: 746-755

Weerkamp, A. H., and B. C. McBride. 1981. Identification of a *Streptococcus salivarius* cell wall component mediating coaggregation with *Veillonella alcalescens* V1. Infect. Immun. 32: 723-730.

Weerkamp, A. H., H. C. van der Mei, and R. S. Liem. 1986b. Structural properties of fibrillar proteins isolated from the cell surface and cytoplasm of *Streptococcus salivarius* (K+) cells and nonadhesive mutants. J. Bacteriol. 165: 756-762.

Weinberg, A., C. M. Belton, Y. Park, and R. J. Lamont. 1997. Role of fimbriae in *Porphyromonas gingivalis* invasion of gingival epithelial cells. Infect. Immun. 65: 313-316.

Weiss, E. I., J. London, P. E. Kolenbrander, and R. A. Andersen. 1989. Fimbria-associated adhesin of *Bacteroides loescheii* that recognizes receptors on procaryotic and eucaryotic cells. Infect. Immun. 57: 2912-2913.

Whittaker, C. J., C. M. Klier, and P. E. Kolenbrander. 1996. Mechanisms of adhesion by oral bacteria. Annu. Rev. Microbiol. 50: 513-552.

Williams, R. C., and R. J. Gibbons. 1972. Inhibition of bacterial adherence by secretory immunoglobulin A: a mechanism of antigen disposal. Science 177: 697-699.

Wu, H. and P. Fives-Taylor. 1999. Identification of dipeptide repeats and a cell wall sorting signal in the fimbriae-associated adhesin, Fap1 of *Streptococcus parasanguis*. Mol. Microbiol. 34: 1070-1081.

167

Yeung, M. K., and J. O. Cisar. 1990. Sequence homology between the subunits of two immunologically and functionally distinct types of fimbriae of *Actinomyces* spp. J. Bacteriol. 172: 2462-2468.

Yeung, M. K., and P. A. Ragsdale. 1997. Synthesis and function of *Actinomyces naeslundii* T14V type 1 fimbriae require the expression of additional fimbria-associated genes. Infect. Immun. 65: 2629-2639.

Yeung, M. K., J. A. Donkersloot, J. O. Cisar, and P. A. Ragsdale. 1998. Identification of a gene involved in assembly of *Actinomyces naeslundii* T14V type 2 fimbriae. Infect. Immun. 66: 1482-1491.

Yoshimura, F., Y. Takahashi, E. Hibi, T. Takasawa, H. Kato, and D. P. Dickinson. 1993. Proteins with molecular masses of 50 and 80 kilodaltons encoded by genes downstream from the fimbrillin gene (*fimA*) are components associated with fimbriae in the oral anaerobe *Porphyromonas gingivalis*. Infect. Immun. 61: 5181-5189.

Xie, H., and R. J. Lamont. 1999. Promoter architecture of the *Porphyromonas gingivalis* fimbrillin gene. Infect. Immun. 67: 3227-3235.

From: *Oral Bacterial Ecology: The Molecular Basis*
ISBN 1-898486-22-0 ©2000 Horizon Scientific Press, Wymondham, U.K.

4

ORAL INNATE HOST DEFENSE RESPONSES: INTERACTIONS WITH MICROBIAL COMMUNITIES AND THEIR ROLE IN THE DEVELOPMENT OF DISEASE

Richard P. Darveau

Contents

Introduction

The interactions between the complex oral microbial community and the host and its succession from health to disease might be best understood in the terms by which defense systems and the oral microbial community recognize each other. The host has evolved mechanisms to recognize non-self as opposed to specific microorganisms, and bacteria have evolved mechanisms to sense their environment and evade or modify the host as needed to produce progeny. Bacteria have evolved such that they occupy the ecological niche provided by both the tooth surface and gingival epithelium as well as the surrounding environmental conditions of the oral cavity. However, a highly efficient innate host defense system constantly monitors the bacterial colonization status and prevents bacterial intrusion into local tissues. A dynamic equilibrium exists between dental plaque bacteria and the innate host defense system. This interaction is not a haphazard process, but represents a highly evolved interaction between bacteria and host. The consortium of bacteria that constitute dental plaque represents highly evolved, highly specialized organisms that have adapted and are still adapting to make the best of their local environment. Under conditions of clinical health they may represent a symbiotic relationship with the commensal bacteria providing the host the appropriate inflammatory stimulus to maintain an effective, non-destructive inflammatory barrier against potential pathogens. In polymicrobial infections, typified by periodontitis, the microbial community has established a new dynamic that results in a pathologic state. The innate host response system, particularly the inflammatory response, is central to both the healthy and diseased environments. This chapter will examine how the host sees these different microbial communities and how its response contributes to health and disease. Recent discoveries on the molecular mediators of inflammation including host factors that recognize bacterial components, combined with the molecular characterization of clinically healthy and diseased periodontal tissue preclude a discussion of the role of adaptive immune response (both cellular and antibody mediated) in regulating oral microbial ecology in this chapter. I refer the reader to a recent review by Hajishengallis and Michalek (1999) for more information on this subject.

The Dynamic Between the Microbial Community and the Host in the Periodontium is One of Constant Communication

Dental Plaque is a Microbial Biofilm Community Which Forms on the Tooth and Tooth Root Surface

The oral environment presents the host with a continuous threat of microbial infection. This peril includes both transient species that find temporary favorable environments and those organisms routinely found in the dental plaque microbiota. The primary reason the oral environment poses a continual threat is that dental plaque as a biofilm is recalcitrant to removal and provides attachment and growth sites for numerous microbial species (Bradshaw *et al.*, 1996; Busscher *et al.*, 1995; Le Magrex *et al.*, 1993; Li and Bowden, 1994; Marsh and Bradshaw, 1995; Sissons *et al.*, 1995; Wilson, 1996). In the oral environment, the saliva conditioned tooth serves as the biofilm substratum (Costerton *et al.*, 1994) or one of the major interfaces providing a stable platform for bacterial growth in juxtaposition to normally sterile human oral tissue. There are two types of dental plaque that are defined by their location on the tooth surface. Supragingival plaque forms coronal to the gingival margin and subgingival plaque forms apical to this point.

Bacterial Shedding Informs the Host of the Colonization Status of the Biofilm

Evidence that the host monitors plaque growth comes from both experimental gingivitis studies in humans and a wealth of clinical experience that clearly demonstrates an inflammatory response when supragingival plaque accumulates (Theilade, 1996). Hypothetically, non-specific bacterial shedding represents the major mechanism by which the host is informed of the amount and type of bacterial colonization occurring in the dental plaque biofilm. Indeed, perhaps the most significant ramification of biofilm formation on the tooth surface is that bacteria continuously release cell wall components into the oral cavity and the gingival sulcus (Grenier *et al.*, 1995; Grenier and Mayrand, 1987; Lai *et al.*, 1981; Listgarten and Lai, 1979; Nowotny *et al.*, 1982; Williams and Holt, 1985; Woo *et al.*, 1979). Both Gram-positive and Gram-negative bacteria are known to release large amounts of cell wall material from their cell surface. Gram-positive bacteria secrete lipotechoic acid (LTA), a major cell wall polymer, into culture media (Fischer, 1988). In late stationary phase up to 90% of the total LTA from *Streptococcus mutans* can be found in the culture supernatant (Joseph and Shockman, 1975). Gram-negative bacteria release large amounts of cell surface components as outer membrane vesicles (Fung *et al.*, 1978; Rothfield and Pearlman-Kothencz, 1969; Smalley *et al.*, 1993).

Vesicles contain lipopolysaccharide, lipid, and protein which are found in intact outer membranes (Deslauriers *et al.*, 1990; Russell *et al.*, 1975; Smalley *et al.*, 1993; Zollinger *et al.*, 1972) and are believed to represent a normal mechanism of membrane turnover (Mayrand and Grenier, 1989; Mug-Opstelten and Witholt, 1978). Previous studies have shown that shed bacterial components penetrate deep into periodontal tissue, may concentrate around blood vessels in the lamina propria (Schwartz *et al.*, 1972) and therefore have the capability to come in contact with nearly all cell types present in the periodontium (Hamada *et al.*, 1990; McCoy *et al.*, 1987; Moore *et al.*, 1986; Schwartz *et al.*, 1972; Wilson *et al.*, 1986). It is likely, therefore, that released bacterial material provides a major form of communication between dental plaque and the host. The host continually sorts through this constant microbial exposure and in the clinically healthy periodontium maintains an effective host defense barrier that prevents systemic microbial infection. The effectiveness of the normal innate host response is reflected in the observation that it is extremely rare for the periodontium to be the source of acute bacterial infections (important exceptions in neutropenic individuals are mentioned below).

Innate Host Defense Recognition and Response

The inflammatory response is a fundamental component of the innate host defense system and serves as an excellent example of how the system operates. This response is an immediate recognition and response event, meaning that the host immune system must recognize potential invaders and respond immediately. The response facilitates the movement of serum soluble and cellular components out of the vasculature and into the specific location in tissue where microbial invasion was detected. The goal of the system is to remove or neutralize the microbial invader. A key, unresolved issue in this system is how the host can immediately recognize the

---

Table 1. Characteristics and Examples of Major Bacterial Pattern Recognition Receptor Ligands

Characteristics*

1.	Molecular structures must be present and conserved on large group of microorganisms and represent pattern rather than specific structure.
2.	Structural patterns must be essential, such that mutational alterations in the pattern or structure would be lethal to the microorganism.
3.	The molecular pattern or structure must be sufficiently distinct from self components to be discriminated as non-self by the innate immune system.

Examples

Bacterial ligand	Component	Known host receptors
lipopolysaccharide (LPS)	gram-negative bacteria	CD14, LBP, macrophage scavenger receptor
lipotechoic acid (LTA)	gram-positive bacteria	CD14, macrophage scavenger receptor

*adapted from Medzhitov and Janeway (Medzhitov and Janeway, 1997).

multitude of non-self (microbial) components to which it is exposed and initiate an appropriate response.

There are several known mechanisms by which the host detects potential microbial invaders and initiates an inflammatory response. For example, the complement system, which is a key innate host defense inflammatory mediator pathway, can be activated by both microbial and self components interacting with a highly reactive thioester group buried in C3 (Kinoshita, 1991). When C3 is covalently bound to host cells, a regulatory system down-regulates complement activation, removing the C3 deposition and thereby preventing a response that is detrimental to the host. Most microorganisms do not have the ability to remove C3 or prevent further complement activation and are therefore "marked" for removal by a protective innate host defense response. Alternatively, specific antibody activation of complement induces inflammation, but only after previous exposure to antigen in which a process of distinguishing between self and non-self occurred by clonal deletion (Janeway *et al.*, 1989).

Pattern Recognition Provides a Framework to Understand Innate Host Defense Responses

Pattern recognition is a concept originally proposed by Janeway (1992) to explain a mechanism rationalizing how the host evolved a system to immediately recognize non-self and mount a protective response. The concept of pattern recognition proposes that the host has evolved receptors that recognize common conserved structures found in a variety of different microbes. During evolution, self from non-self discrimination occurred as receptors of the innate host defense system that recognized self were eliminated by natural selection (Janeway, 1992). The "pattern recognition" hypothesis describes the characteristics of microbial pattern recognition receptor ligands as being shared by a large group of microbes and conserving essential structures that are distinct from self-antigens (Medzhitov and Janeway, 1997) (Table 1). Indeed CD14, a member of the pattern recognition family of proteins is able to bind lipoteichoic acid (LTA) (Cleveland *et al.*, 1996; Pugin *et al.*, 1994; Sugawara *et al.*, 1999) and peptidoglycan (Dziarski *et al.*, 1998; Gupta *et al.*, 1999) from gram positive organisms, lipoproteins from spirochetes (Giambartolomei *et al.*, 1999; Sellati *et al.*, 1998; Wooten *et al.*, 1998), and lipopolysaccharide (LPS) from gram negative bacteria (Wright *et al.*, 1990). Numerous other host proteins have been proposed to be members of the pattern recognition family of receptors (Medzhitov and Janeway, 1997). These include LBP (lipopolysaccharide binding protein), collectins, macrophage scavenger receptors, and more recently Toll like receptors (Medzhitov and Janeway, 1997). In the future, almost certainly, additional members of this later family of proteins will be identified. Each of these proteins shares the common characteristic of being able to recognize a wide variety of different microbial structures and are located in positions where they will come in contact with potential microbial threats (Medzhitov and Janeway, 1997).

The LBP/CD14 System

The LBP/CD14 activation pathway (Schumann *et al.*, 1990; Wright *et al.*, 1990) has been described as a pattern recognition system by which the host recognizes and responds to a variety of different microbial components (Pugin *et al.*, 1994). The release of numerous inflammatory mediators including IL-1ß, TNFα, and PGE$_2$ as well as the expression of cell adhesion molecules such as E-selectin and ICAM-1 which are necessary for the cellular infiltration associated with the inflammatory response are all expressed in response to CD14 complexed with microbial components (Pugin *et al.*, 1993; Wright, 1995). CD14 is expressed on the membrane of cells of myeloid origin which includes monocytes and neutrophils, and is found in serum in a soluble form (sCD14) where it participates in the activation of numerous non-myeloid cells such as epithelial, fibroblast, and endothelial cells. With respect to the oral environment both sCD14 and LBP have been found in gingival crevicular fluid (GCF) (R. Darveau, unpublished results). Consistent with its role in the oral environment, most studies have found that CD14 either is required or greatly facilitates host cell activation by oral bacteria or their components (Darveau and Bainbridge, 1999). In addition, it has been reported that secretion of inducible antimicrobial peptides in response to microbial components is also dependent upon CD14 (Diamond *et al.*, 1996). Not surprisingly, an important feature of this system is that the response to non-self components is immediate, occurring in minutes or hours as opposed to days. The importance of CD14 in inflammation has been validated in mice, where CD14 was shown to be required for the development of sepsis, a systemic acute inflammatory event in response to *Escherichia coli* LPS (Haziot *et al.*, 1996). In addition, in a primate model of respiratory infection anti-CD14 therapy was able to reduce *E. coli* LPS-induced inflammation (Leturcq *et al.*, 1996). In the future, it is possible that CD14 will be found to be involved in other cellular responses to bacteria such as differentiation and growth.

Mechanism of Multiple Ligand Binding for Pattern Recognition Receptors
The mechanisms by which proteins are able to bind multiple ligands are currently being elucidated for pattern recognition proteins. Two groups of proteins, which are reported to have a role in innate host defense, are the macrophage scavenger receptor family (Dunne *et al.*, 1994; Krieger, 1997) and a group of molecules termed collectins (Holmskov *et al.*, 1994; Sastry and Ezekowitz, 1993). Truncation mutations of the bovine and murine scavenger receptors have revealed that the collagenous domains are necessary for multiple ligand binding (Acton *et al.*, 1993; Doi *et al.*, 1993). Comparison of the predicted protein sequences of these domains from several different species has revealed conserved regions of homology that contain positively charged lysine residues (Krieger *et al.*, 1993). Substitution mutants replacing specific lysine residues with alanine abolished acetyl-LDL binding under certain conditions suggesting a role for these residues in ligand binding (Doi *et al.*, 1993). Based upon these observations, it has been proposed that the collagenous domain of the macrophage scavenger receptor proteins "provides a selectively sticky surface that functions as a kind of molecular flypaper for the high affinity binding of specific polyanions" (Krieger *et al.*, 1993).

Collectins are a group of molecules that contain a carbohydrate binding region and a collagen tail and which apparently have the ability to discriminate between the patterns of oligosaccharides that are present as the surface components of pathogens and self glycoproteins (Sastry and Ezekowitz, 1993). Examples of this family of proteins include the mannose binding proteins, apoprotein A, and conglutinin (Holmskov *et al.*, 1994; Sastry and Ezekowitz, 1993). The X-ray crystallographic determination of the three-dimensional structure of the carbohydrate-recognition domain of the mannose binding protein (MBP) with bound carbohydrate has been reported (Weis *et al.*, 1992). This structural elucidation coupled with ligand binding studies has revealed that MBP binds equatorial 3- and 4-OH groups, explaining how a variety of diverse carbohydrate ligands may be recognized. These findings also suggest that the structural constraints for recognition may be a means by which only non-self carbohydrate structures may be recognized. Additional studies on the structures of non-self ligands for MBP and self-carbohydrates, which are not recognized, are necessary to validate this potential mechanism.

Little is known about the mechanism by which CD14 is able to bind a wide variety of different ligands. Initial studies with truncation mutants identified the amino terminal region and three leucine repeat domains as sufficient for *E. coli* LPS binding and host cell activation (Juan *et al.*, 1995). Independent studies generated deletion mutations of the most hydrophilic regions of the amino terminus and showed that these mutant proteins could no longer bind *E. coli* LPS (Viriyakosol and Kirkland, 1995). Protease protection experiments identified one of these hydrophilic regions as being necessary for *E. coli* LPS binding (McGinley *et al.*, 1995). A multiple alanine replacement mutation strategy of CD14 has identified another region of the amino terminus of the protein as necessary for *E. coli* LPS binding (Stelter *et al.*, 1997). Initially employing a peptide inhibition approach, single point charge reversal mutations were generated in CD14 which selectively affected the binding of *E. coli* and *P. gingivalis* LPS (Shapiro *et al.*, 1997). Serine replacement mutations did not affect LPS binding, suggesting that the side chain charge group was not directly involved in ligand binding. In contrast, charge reversal mutations in the same residues where serine replacement had no effect significantly reduced LPS binding, suggesting a charge repulsion mechanism. The ability of charge reversal mutations to eliminate LPS binding was highly specific in that mutations of only certain residues would reduce LPS binding (Shapiro *et al.*, 1997). Potential conformational changes induced by the mutations were monitored by the ability of the mutant CD14 proteins to bind different LPS species and to activate human endothelial cells with *E. coli* LPS. These findings indicate that there are several different regions located throughout the amino terminal region of CD14 that could be involved in LPS binding.

Recently, it was proposed that CD14 employs a mechanism of non-self ligand capture by "presenting" a group of both positive and negative charges on one surface of the protein (Darveau, 1998). This tactic is similar to that already reported for the macrophage scavenger protein (Krieger *et al.*, 1993), another pattern recognition receptor (Medzhitov and Janeway, 1997). This hypothesis is consistent with the data generated from several different laboratories that have investigated the LPS ligand properties of this molecule. Accordingly, LPS ligands bind this charged "catch" region in a non-specific fashion, i.e. LPS ligands are capable of interacting with a variety of different charged residues. This is consistent with the observations that numerous single residue serine replacements or multiple alanine replacement mutations had no effect on LPS binding (i.e. no one charge residue is critical for binding). It also provides an explanation for the apparent lack of ligand specific recognition (Delude *et al.*, 1995). The exact mechanism by which CD14 binds multiple non-self ligands must await the precise structural determination afforded by X-ray crystallography. Nevertheless, this hypothesis is consistent with the notion (Wright, 1995) that no one pattern recognition receptor is capable of non-self recognition. Rather the innate host defense system employs multiple proteins with different binding affinities, transfer properties, and locations to arrive at a "consensus" about non-self and mount an appropriate response.

Toll-like Receptors (TLR) Represent a New Family of Pattern Recognition Innate Host Defense Proteins

Recently, a new family of proteins has been identified that may represent a crucial component of the mammalian LBP-CD14 innate host defense microbial recognition pathway. A series of structural (Chaudhary *et al.*, 1998; Medzhitov *et al.*, 1997; Rock *et al.*, 1998) and functional (Chaudhary *et al.*, 1998; Chow *et al.*, 1999; Kirschning *et al.*, 1998; Medzhitov *et al.*, 1997; Yang *et al.*, 1998) mammalian homologues of the *Drosophila melanogaster* Toll protein have been described that appear to act with CD14 as a co-receptor to facilitate activation of host defense cells. Toll was discovered in *Drosophila* and was found to signal transcriptional changes in a manner similar to that employed by the IL-1 receptor (Wright, 1999). Subsequently, it was found that the human homologue of TLR 4 could drive NF-kß activation (Medzhitov *et al.*, 1997) and TLR 2 and 4 could mediate LPS-induced signal transduction in a CD14 dependent manner (Chow *et al.*, 1999; Kirschning *et al.*, 1998; Yang *et al.*, 1998). In addition, it was discovered that the elusive *lps* gene defect in the well-characterized non-responder LPS mice (C3H/HeJ and C57BL10/ ScCr) maps to TLR-4 (Poltorak *et al.*, 1998; Qureshi *et al.*, 1999). These observations suggested that Toll like proteins provide the necessary connection between non-self component binding by CD14 and cellular activation (Wright, 1999).

Where is Non-Self Recognized?

LPS is perhaps the most notable example of a pattern recognition receptor (PRR) ligand. The pattern recognition receptor hypothesis, in fact, helps us to understand the long history of LPS (endotoxin) and it's potent effects on the host immune system. Ever since clinicians recognized in the mid 1800's that bacterial infections were associated with spontaneous remissions of certain neoplasms there has been an intense interest in determining the molecular basis for these effects (Nauts and Fowler, 1953). It was found that LPS was acting as an inducer for a host protein identified as tumor necrosis factor (Carswell *et al.*, 1975). Subsequently, LPS has been shown to be a very potent inducer of a wide variety of innate host defense mediators. Although therapeutic applications of LPS are still considered, the potent toxicity of the molecule has precluded its use as an effective anti-cancer agent to date. Clearly LPS is a key microbial ligand that mammalian hosts have evolved to recognize as non-self (Hoffmann *et al.*, 1999).

However, although LPS molecules obtained from different bacteria are similar in structure, consistent with what is predicted for a pattern recognition receptor (PRR) ligand (Medzhitov and Janeway, 1997; Table 1), they display compositional differences that may significantly alter the innate host response (Table 4) (Darveau,

1998; Golenbock *et al.*, 1991; Loppnow *et al.*, 1989; Somerville *et al.*, 1996; Takada and Kotani, 1992). It is not known how the host recognizes these structural features nor how they may regulate the intensity of the inflammatory response. The best available evidence to date suggests that recognition of specific lipid A structural details most likely occurs after CD14 binding (Delude *et al.*, 1995; Kitchens and Munford, 1995). Kitchens and Munford (1995) by careful titration of LBP and LPS, have shown that deacylated LPS (dLPS) inhibits wild type LPS activation at an uncharacterized site after CD14 binding in the LPS recognition pathway. Further, Delude *et al.* (1995) conclusively demonstrated that inhibition of LPS responses by *R. sphaeroides* lipid A and Lipid IV$_A$ was not due to LPS recognition by CD14. These data are consistent with a proposal by Wright (1995) that CD14 serves as a mechanism to concentrate microbial ligands at the host cell surface. The molecular mechanism of how CD14 complexed with different microbial ligands engages different TLR's after they are presented at the cell surface is not known. Although hypotheses have been proposed to explain this (Wright, 1995), how non-self is recognized by the CD14 TLR pathway remains elusive.

Clinically Healthy Status

Clinically healthy individuals contain a characteristic dental plaque microbiota mostly consisting of gram positive streptococci and actinomyces with about 15 % gram-negative rod species, and display a histological status typical of a low-level inflammatory infiltrate in the adjacent periodontal tissue (Darveau *et al.*, 1997; Tonetti *et al.*, 1998). Are these two observations related? As mentioned above, dental plaque bacteria release microbial components that penetrate host tissue and can be recognized as non-self. Several lines of evidence suggest that the host responds to this material by eliciting a low level inflammatory response that serves to protect clinically healthy tissue from bacterial invasion.

Control of Supragingival Plaque Growth in the Clinically Healthy Individual
The host employs mechanisms to control supragingival plaque accumulation and compositon. Its location in direct contact with the oral tissues renders it subject to intraoral abrasion that serves to restrict its net accumulation (Sissons *et al.*, 1995). In addition, this biofilm is subject to the flow characteristic of saliva and its host defense components. Evidence that saliva flow has a significant effect on supragingival plaque accumulation and composition comes from studies examining the microbial composition of individuals with impaired saliva flow (see below). In addition, those areas of the dentition that are located close to the ducts of the major saliva producing glands where the velocity of saliva is the greatest (lower anterior and upper posterior) are less prone to caries (Edgar, 1992).

Perhaps the best evidence that saliva influences oral microbial ecology comes from studies with patients exhibiting reduced saliva flow (Almstahl *et al.*, 1999; Almstahl and Wikström, 1999; Brown *et al.*, 1975; Celenligil *et al.*, 1998; Fox, 1996; MacFarlane, 1984). Several microbiological studies have been performed with Sjögren's syndrome patients (Almstahl *et al.*, 1999; Celenligil *et al.*, 1998; MacFarlane, 1984). Sjögren's syndrome is an autoimmune disorder that causes a dysfunction of the salivary glands resulting in dryness of the mouth and eyes (Fox, 1996). Microbiological examination of the microbiota of these individuals has revealed a significant increase in the amount of supragingival plaque and alterations in composition. Specifically, a reduction in *Streptococcus salivarius*, *Neisseria pharyngis*, and *Veillonella* sp. was noted (MacFarlane, 1984) and an increase in the amounts of *Candida*, *Staphylococcus aureus* (MacFarlane, 1984), mutans streptococci and lactobacilli (Almstahl *et al.*, 1999) was noted. Similar changes in microbial composition were obtained in a study of patients with xerostomia (dryness of mouth) due to radiation therapy for cancer of the head and neck (Brown *et al.*, 1975). In this study non-cariogenic bacteria were replaced with cariogenic species. Another study that examined individuals with reduced salivary secretion that was not due to either radiation therapy or Sjögren's syndrome also found an increase in cariogenic bacteria and a reduction in α-hemolytic streptococci (Almstahl and Wikström, 1999). These studies describe remarkably consistent changes in supragingival plaque accumulation and composition that occur due to the lack of adequate saliva flow regardless of the physiological basis of the defect. In contrast, the effect of reduced saliva flow on the subgingival plaque community is less clear. Although studies have shown no significant alterations in gram-negative bacteria that normally occupy the subgingival niche (Almstahl *et al.*, 1999; Almstahl and Wikström, 1999), another study has suggested that patients with Sjögren's syndrome may have an increase in periodontal disease and colonization with suspected periodontal pathogens (Celenligil *et al.*, 1998). Further work will be required to determine if localized sites that may be more prone to disease are more susceptible to colonization by periodontopathogens in individuals with reduced saliva flow.

Not only does the mechanical action of saliva secretion and swallowing affect supragingival plaque, saliva contains a plethora of antimicrobial factors (Schenkels *et al.*, 1995; Tenovuo, 1998) that limit both the colonization and spread of the supragingival bacterial biofilm. These antimicrobial factors include sIgA, lactoferrin, lysozyme, and peroxidases, some of which are found in other host secretions. Recently, an excellent review on the role of sIgA on the microbial ecology has been published which accurately describes the dual role that IgA may play in regulating both commensal and pathogenic bacteria. (Marcotte and Lavoie, 1998). Other components of saliva have been shown to work synergistically to limit bacterial growth. For example, lactoferrin and lysozyme display much more potent gram-negative antibacterial activity when combined than when tested individually (Ellison and Giehl, 1991). In addition, IgA (Tenovuo *et al.*, 1982), lactoferrin (Soukka *et al.*, 1991), and lysozyme (Soukka *et al.*, 1991) when combined with lactoperoxidase, have more potent activity against the gram-positive bacterium *S. mutans*. Saliva also contains antimicrobial proteins, most notable of which arc the histatins, which have *in vitro* antifungal (Raj *et al.*, 1990) as well as antibacterial activity (Lamkin and Oppenheim, 1993; Oppenheim *et al.*, 1988). Histatin-derived peptides have been shown to retain their antibacterial activity when examined against *in vitro* generated biofilms and *ex vivo* plaque samples (Helmerhorst *et al.*, 1999). *In vivo*, histatin-5 and a peptide derived from histatin 5, designated P-113, inhibited experimental gingivitis in beagle dogs (Paquette *et al.*, 1997). Human clinical studies attempting to correlate levels of histatins and susceptibility to oral yeast carriage and infection have yielded controversial results. Both higher (Atkinson *et al.*, 1990; Bercier *et al.*, 1999) and decreased (Jainkittivong *et al.*, 1998; Lal *et al.*, 1992; Mandel *et al.*, 1992) levels of histatin have been reported. These differences are more than likely due to the different nature of the study populations employed. However, the significant alterations in histatin levels found in these studies suggest that oral candidiasis may modulate and oral yeast status may be influenced by the levels of salivary histatin. Another antimicrobial peptide which is found in saliva is human ß-defensin 1. This recently discovered antimicrobial peptide is expressed in the parotid gland, gingival tissue and the tongue (Mathews *et al.*, 1999). Due to its cationic nature it is likely that it works synergistically with other components of saliva.

It appears, therefore, that saliva is a major host factor that regulates supragingival plaque accumulation and composition. Recent studies which suggest that the histatin content of saliva may be regulated in response to yeast infection are provocative in that they indicate that feedback systems may exist between colonization status and exocrine gland function. Further work is needed in this area of research to determine if such feedback systems exist.

Control of Subgingival Plaque Growth in the Clinically Healthy Individual

Clinically healthy tissue contains a low-level inflammatory infiltrate. Early histological studies of clinically healthy tissue demonstrated that it contains a cellular infiltrate located in juxtaposition to the colonized tooth surface (Page and Schroeder, 1976). A portion of this cellular infiltrate has been described as forming a "wall" of neutrophils precisely located between bacteria and residing just outside the junctional epithelium, the epithelial cell surface closest to the dental plaque biofilm (Kornman *et al.*, 1997). Other histological observations have described a more widespread distribution of inflammatory cells in clinically normal gingiva (Moskow and Polson, 1991). Consistent with these observations, molecular characterization of healthy periodontal tissue has demonstrated that IL-8 and E-selectin are expressed in clinically healthy tissue (Gemmell *et al.*, 1994; Moughal *et al.*, 1992; Nylander *et al.*, 1993; Tonetti, 1997; Tonetti *et al.*, 1994). These two inflammatory mediators are necessary for leukocyte diapedesis from the vasculature and directed movement through tissue. E-selectin expression on endothelial cells facilitates a tethering interaction between the leukocyte and the endothelial cell wall initiating the "rolling" stage required for leukocyte exit (Springer, 1994). IL-8 is a key neutrophil chemoattractant and evidence demonstrates that its expression is greatest at the gingival epithelial cell surface in contact with the dental plaque biofilm. A recent study has demonstrated that a gradient of IL-8 and ICAM-1 expression exists in clinically healthy tissue (Tonetti *et al.*, 1998). IL-8 expression was greatest at the most superficial junctional epithelial cell layers and the levels of ICAM-1 increased toward areas exposed to bacterial challenges. It has been proposed that highly regulated and ordered expression of these molecular mediators of inflammation facilitate the orderly exit of neutrophils out of the vasculature and into the gingival crevice where they are crucial in protecting the host from bacterial challenge (Tonetti *et al.*, 1998).

As mentioned above, another component of the innate host defense response found in clinically healthy gingival tissue is collectively the antimicrobial peptides (Mathews *et al.*, 1999; Weinberg *et al.*, 1998). At least two antimicrobial peptides with homology to ß-defensins 1 and 2 have been detected in clinically normal gingival tissue and were expressed *in vitro* by gingival epithelial cells (Mathews *et al.*, 1999; Weinberg *et al.*, 1998). RT-PCR analysis revealed constitutive expression of human ß-defensin-1 whereas both RT-PCR (Weinberg *et al.*, 1998) and quantitative RNase protection assays revealed that human ß-defensin 2 was inducible after stimulation with either LPS or IL-1ß (Mathews *et al.*, 1999). Consistent with the *in vitro* results, human ß-defensin 1 was found at the same levels in both normal and inflamed sites from the same patient (Weinberg *et al.*, 1998) and human ß-defensin 2 was more abundant in tissues associated with inflammation (Mathews *et al.*, 1999). Recent studies have shown that ß-defensin antimicrobial peptides display a broad range of antimicrobial activity and, therefore, it is expected that these molecules are an important component of the innate defense response of the periodontium (Krisanaprakornkit *et al.*, 1998; Weinberg *et al.*, 1998).

Relationship Between Microbial Colonization and Innate Host Defense Status
It seems likely that the microbial community that establishes itself in the clinically healthy individual is responsible at least in part for the low-level inflammatory surveillance that has been observed in healthy tissue. Evidence obtained experimentally by histological examination of clinically healthy tissue demonstrating increased expression of both IL-8 and ICAM-1 in tissue closest to bacterial colonization (Tonetti *et al.*, 1998) provides a strong argument for bacterial participation in stimulating expression of molecular mediators. As mentioned above, bacterial components shed from the dental plaque biofilm penetrate gingival tissue (Hamada *et al.*, 1990; McCoy *et al.*, 1987; Moore *et al.*, 1986; Schwartz *et al.*, 1972; Wilson *et al.*, 1986) and may have either direct or indirect effects on host cells (Pugin *et al.*, 1993). Direct effects occur when the bacteria or bacterial extract directly stimulates a cell to respond (for example, by the production of cytokines, chemokines, or cellular adhesion molecules). Indirect effects are defined as those effects which occur when a bacterium activates one cell type and the product from that cell acts upon another cell or cell type. Both types of effects can potentially occur with both myeloid and non-myeloid cells (Gemmell and Seymour, 1993; Heath *et al.*, 1987; Lindemann *et al.*, 1995; Yamazaki *et al.*, 1992). An example of an indirect effect is when bacteria activate myeloid cells (for example, monocytes or neutrophils) to elicit IL-1ß and this cytokine then activates a non-myeloid cell (for example, fibroblasts, endothelial, or epithelial cells) to secrete additional inflammatory mediators.

The majority of the species found in the healthy dental plaque biofilm are gram-positive streptococci (see Chapter 1; Darveau *et al.*, 1997). Consistent with the pattern recognition hypothesis both peptidoglycan and lipoteichoic acids isolated from gram-positive bacteria activate host defense cells *in vitro* (Cleveland *et al.*, 1996; Dziarski *et al.*, 1998; Gupta *et al.*, 1999; Pugin *et al.*, 1994; Sugawara *et al.*, 1999). However, numerous *in vitro* studies have established that, in general, gram-positive bacteria are not potent stimulators of the innate host inflammatory response (Gupta *et al.*,

1996; Kusunoki *et al.*, 1995; Standiford *et al.*, 1994). For example, LTA is consistently 10 to 100 fold less active than *E. coli* LPS in several different *in vivo* and *in vitro* assays of inflammation (Gupta *et al.*, 1996; Kusunoki *et al.*, 1995; Pugin *et al.*, 1994; Standiford *et al.*, 1994). Likewise, crude bacterial extracts from several different streptococcal clinical isolates are much less potent stimulators of IL-8 from either gingival epithelial cells or of E-selectin from human endothelial cells when compared with *E. coli* LPS (R. Darveau, unpublished results). A reduced inflammatory inducing potential from the majority of species residing in clinically healthy plaque is one mechanism by which inflammatory mediator expression may be maintained at low levels in host tissue. The molecular basis for the weaker activation of the host innate inflammatory response by gram-positive bacteria is not known. However, recently it has been shown that Toll like receptor 2 is linked to cell activation with gram-positive bacteria, whereas Toll like receptor 4 is not (Yoshimura *et al.*, 1999). Apparently different pattern receptor ligands activate different Toll like receptors both in humans (Yoshimura *et al.*, 1999) and insects (Lemaitre *et al.*, 1996; Williams *et al.*, 1997). Differential Toll like receptor activation represents one mechanism by which the intensity of the inflammatory response could be regulated with different pattern receptor ligands.

In addition, the maintenance of highly controlled and preferentially expressed molecular mediator expression in clinically healthy tissue may also be due to the limited repertoire of cytokines secreted by epithelial cells in response to bacteria (Agace *et al.*, 1993). Although work in this area is not complete, the evidence suggests that epithelial cells may secrete certain inflammatory mediators such as IL-6 and IL-8 in response to bacterial components but not TNFα or IL-1ß (Agace *et al.*, 1993). In addition, although much more work is needed in this area, it appears that the bacterial requirements for activation of epithelial cells may be different than those for fibroblasts or endothelial cells. For example, it has been shown that uroepithelial cells do not respond to *E. coli* LPS (Hedlund *et al.*, 1999). In addition, it has been shown that whilst endothelial cells respond to isolated *E. coli* LPS by the production of IL-8 many primary gingival epithelial cell lines do not (R. Darveau, unpublished observations). However, potent IL-8 secretion is observed from these same cell lines when *E. coli* or other bacterial cell wall preparations are employed (Darveau *et al.*, 1998). These data would indicate that epithelial cells require more than just LPS to "see" bacterial components. Consistent with these observations is a report that an intact *Heliobacter pylori* membrane is necessary to induce IL-8 secretion from epithelial cell lines (Rieder *et al.*, 1997). In summary, it is likely that gingival epithelial cells constantly exposed to bacterial components are more selective in what they respond to and the responses are more limited when compared with other cell types. These characteristics of epithelial cells may facilitate an ordered inflammatory response that minimizes host tissue damage.

Finally, it is possible that host cells do not see many of the bacterial components that are shed from the dental plaque biofilm into the gingival crevice. Gingival crevicular fluid is a serum exudate and as such it is highly likely that it contains plasma proteins such as collectins and other pattern recognition receptors. These molecules serve as receptors that facilitate the opsonization of non-self components and in that way would remove bacterial components from periodontal tissue reducing their ability to activate other periodontal cells. In addition, the generation of specific antibody and macrophage scavenger receptors (Dunne *et al.*, 1994; Krieger, 1997) is also likely to participate in preventing bacterial components from interacting with host cells.

Is the Oral Ecosystem Associated with Clinical Health a Symbiosis with the Host?

The contribution that bacteria make to the status of innate host defense in the clinically healthy environment is difficult to determine. In the future, gnotobiotic animals could be employed to examine the role of individual or combinations of bacteria on the expression of inflammatory mediators in the oral environment. It is possible that, similar to the intestinal environment, microbial colonization is required for the proper immunological and innate defense functions of oral tissue (Falk *et al.*, 1998). Recently, an interesting and novel hypothesis describing a molecular mechanism by which gram-positive colonization of supragingival plaque may influence the innate host response has been described (Sugawara *et al.*, 1999). It was demonstrated that LTAs obtained from gram-positive bacteria could block the activation of fibroblasts by LPS through competition for binding to CD14. The authors propose that LTA might be a useful agent to suppress periodontal disease due to excessive inflammatory responses caused by gram-negative LPS. Regardless if LTAs ever become the basis for effective therapies, these observations provide a plausible molecular mechanism of symbiosis between the host and the microbial community associated with health. Further studies examining the contribution of the microbial community to clinical health may reveal that the oral ecosystem in health is more symbiotic than previously considered.

Influence of Impaired Innate Host Response on the Development of Disease

Considering the highly structured innate host defense environment in the clinically healthy individual, it is relevant to ask how defects in this barrier affect susceptibility to disease and how they may change the microbial ecology. In fact, an intact innate host defense system has long been recognized as a necessary host component to prevent periodontitis. Some of the early ideas and concepts about the nature of periodontitis were developed based upon disease susceptibility of individuals with congenital defects in innate host defense components (Newman and Rule, 1983). Examination of the effects of congenital and induced deficiencies on disease susceptibility and microbial ecology is a valid approach to determine the role of the innate host defense system in disease prevention. Therefore, in Table 2 and in the following section several congenital and induced deficiencies that have a strong association with the development of periodontitis are discussed. In addition, congenital deficiencies in the innate host defense system that apparently do not render an individual more susceptible to periodontitis are mentioned.

Immunosuppressive Therapy
Perhaps the most profound effects on the innate host defense system are observed in cancer patients undergoing immunosuppressive therapy. It is well known that

Table 2. Induced or Congenital Deficiencies

Condition	Phenotype	Periodontitis	Significant Shift in Periodontopathogenic Microbiota
Immunosuppressive therapy	Significant neutropenia and loss of mucosal barrier	Acute, generalized and severe	Yes, commensal bacteria are found
Leukocyte adhesion deficiency	Loss of all leukocyte movement from vascular compartment to tissue, molecularly defined as failure to express integrin adhesions or selectin receptors	Acute, generalized and severe	Yes, commensal bacteria are found
Chediak-Higashi Syndrome	Non-functioning granulocytes due to defective lysosomal trafficking regulator gene	Acute, generalized and severe	Unknown
Chronic neutropenia	Significantly decreased levels of neutrophils brought about by unknown congenital disorder	Variable, acute and chronic types, generalized, severe	Questionable, appears to have same periodontopathogens associated with adult type disease
Papillon-Lefevre Syndrome	Hyperkeratosis of palms and soles, probable defect in neutrophil chemotaxis, defect localized to chromosome 11q14, unknown gene function	Variable, acute and chronic types, generalized, severe	Questionable, appears to have some periodontopathogens associated with adult type disease
Diabetes	Numerous complications associated with vasculature	Increased incidence and severity	No, appears to contain similar periodontopathogenic species found in non-diabetic individuals
Cigarette smoking	Not clear, may have decreased vascular response in periodontium	Increased incidence and severity	No
Human Immunodeficiency Virus	Impaired cell mediated immunity	Unusual forms which can be acute, severe, and generalized	Yes, Candida and Borrelia can be found

myelosuppressive agents severely reduce the number of circulating neutrophils rendering the patients susceptible to infectious disease. In addition, however, the direct toxicity of chemotherapeutic agents causes a decrease in the proliferation of the basal epithelial cells resulting in a thinning of the oral mucosa (Toth *et al.*, 1990), rendering these patients also susceptible to a variety of different infectious oral complications. One complication is the development of an acute form of periodontitis.

Several studies have been performed examining the microbial etiology of oral infectious complications associated with immunosuppressive therapy (Dreizen *et al.*, 1974; Dreizen *et al.*, 1977; Galili *et al.*, 1992; Overholser *et al.*, 1982; Peterson *et al.*, 1987; Toth *et al.*, 1990). Most of these studies involve culturing the more obvious oral ulcerations with emphasis on detecting microbial agents that are associated with septicemia. This is understandable considering that infectious complication is the primary reason for mortality in these patients. Nevertheless these studies have demonstrated that nosocomial pathogens such as *Candida spp*, *Staphylococcus*, *Pseudomonas*, and gram-negative enteric rods can colonize the gingiva in these patients and may temporarily replace the normal microbiota (Dreizen *et al.*, 1974). A comprehensive study of the effects of myelosuppressive therapy on periodontitis has revealed that both nosocomial pathogens and indigenous subgingival microbiota are associated with disease (Peterson *et al.*, 1987). This study found that out of 27 episodes of acute periodontitis 17 were associated with nosocomial pathogens of which *Staphylococcus epidermidis*, *Candida albicans*, *S. aureus*, and *Pseudomonas aeruginosa* predominated. The remaining 10 episodes of periodontitis were associated with indigenous subgingival microbiota including black-pigmented *Bacteriodes*, *S. mutans* and *Veillonella spp*. It was pointed out, however, that the proportions of these bacteria were different from those characteristically observed in comparable periodontally diseased sites in non-cancer patients. Based upon the data presented it appears that the myelosuppressed cancer patient has a disproportionate increase in *Veillonella spp*. Although, additional studies will be required to more fully understand the effect of suppressive therapy on subgingival oral ecology and its possible role in the development of acute periodontitis, it appears that numerous different bacteria are capable of inducing disease in these patients.

Leukocyte Adhesion Deficiency
Leukocyte adhesion deficiency (LAD) is one of the best characterized defects in neutrophil function that results in some of the most severe forms of periodontitis (Anderson and Springer, 1987; Waldrop *et al.*, 1987). This congenital disorder results in either the defective biosynthesis of the B chain of CD18 intregrin adhesion molecules (LAD I) or deficient expression of sialyl-Lewis X (LAD II), a constitutively expressed carbohydrate ligand for selectins (Etzioni, 1997). The resultant phenotype for both of these disorders is a failure of leukocytes to move from the vasculature to the tissue. In LAD I, leukocytes fail to negotiate the high affinity intregrin interaction with their ICAM receptors whereas in LAD II there is a selectin interaction failure which results in the lack of the typical "rolling" phenotype, a key initial step in leukocyte extravasation. In the absence of a bone marrow transplant, the most severe forms of the LAD I are fatal due to infection

(Majorana *et al.*, 1999). However, patients with moderate forms of the disease (expression of CD18 3 to 10 % of normal) survive past infancy (Majorana *et al.*, 1999). These patients invariably display severe general periodontitis that may involve both the primary and permanent dentitions. Often the complete loss of dentition occurs. Consistent with the LAD I adhesion defect being directly responsible for the development of periodontitis is the observation that greater CD18 expression is associated with less severe periodontitis (Anderson *et al.*, 1985).

There has been no comprehensive study examining the microbiology associated with this form of periodontitis, certainly not the degree that has been afforded other forms of early onset and adult type diseases. One of the first case reports describing LAD I deficiency (prior to its classification) described the consistent isolation of *Capnocytophaga spp.* from periodontal pockets in two patients (Bowen *et al.*, 1982). However, the amount or diversity of the subgingival microbial population was not discussed. The lack of a comprehensive microbiological analysis in LAD patients is not surprising considering the serious nature of this disorder and the need for the clinician to rapidly initiate and focus on successful treatment regimes. For example, a recent study described the management of periodontitis in an LAD I patient who also manifested a large ulcer on the tongue (Majorana *et al.*, 1999). An oral culture, taken presumably to address the ulcer, revealed *Pseudomonas aeruginosa* and *Klebsiella pneumoniae* and was treated successfully with an antibiotic. The microbial content of the periodontal pockets was not addressed. In contrast, another study (Roberts and Atkinson, 1990) described specimens taken from periodontal pockets in a LAD I patient as containing *Proteus mirabilis*, *Pseudomonas aeruginosa*, and alpha hemolytic *Streptococcus* as well as other normal aerobic intraoral microorganisms. Anaerobes *Bacterioides melaninogenicus* and *Peptostreptococcus micros* were also present in small quantities. *A. actinomycetemcomitans* was not isolated nor was previous exposure indicated due to the lack of specific serum IgG directed against this bacterium. The authors concluded that at least in this patient periodontal disease resulted from numerous invasive and/or opportunistic organisms as opposed to specific periodontopathogens. This is consistent with the observations stated above in drug induced neutropenic cancer patients indicating that nosocomial bacteria can replace the normal oral microbiota of the gingival crevice (Dreizen *et al.*, 1974).

Chediak-Higashi Syndrome
Chediak-Higashi syndrome (CHS) is a congenital disorder that is often referenced as rendering individuals susceptible to gingivitis and periodontitis (Blume and Wolff, 1972; Charon *et al.*, 1985; Hamilton and Giansanti, 1974). However, similar to LAD, very little is known concerning the microbiology of these oral manifestations. Affected individuals have frequent pyogenic infections with intermittent febrile episodes (Hamilton and Giansanti, 1974). As explained above for LAD, it is likely that priorities in case management have precluded an analysis of the microbial etiology of this form of periodontitis. One case report describes the isolation of streptococci from pus found at the gingival margin in a case of severe periodontitis (Hamilton and Giansanti, 1974). Recently, the defective gene in Chediak-Higashi syndrome was identified as the lysosomal trafficking regulator (LYST) (Barrat *et*

al., 1999). The precise function of LYST is not known; however, it is believed to be involved in regulating the secretory processes of intracellular lysosomal vesicles. This function is consistent with histological studies that reveal abnormally large granules in all granule-containing cells (Blume and Wolff, 1972). Ineffective lysosomal trafficking that reduces bacterial killing in neutrophils is a likely explanation for the observed relationship with gingivitis and periodontitis.

Chronic Neutropenia

Another congenital neutrophil disorder that results in a severe generalized periodontitis is chronic idiopathic neutropenia (Deasy *et al.*, 1980; Kalkwarf and Gutz, 1981). This disorder, and the less common cyclic neutropenia (Cohen and Morris, 1961), is similar to LAD and CHS disorders in their association with periodontitis and other infectious complications. They are very different, however, in that whilst LAD patients usually display an increase in the number of circulating neutrophils (they just cannot exit the vasculature normally) chronic and cyclic neutropenic patients display severely reduced numbers of circulating neutrophils. The molecular basis for the absence of neutrophils is not known. In addition, while there is a paucity of information concerning the microbiology of periodontitis in LAD patients, the microbiological composition of chronic and cyclic neutropenia patient associated periodontitis has been reported more frequently (Baehni *et al.*, 1983; Carrassi *et al.*, 1989; Kamma *et al.*, 1998; Vaughan *et al.*, 1990).

The most comprehensive studies examining the microbiology of chronic neutropenia associated periodontitis have revealed a typical gram-negative predominant microbiota. In addition, periodontopathic species associated with disease in normal individuals were usually present (Baehni *et al.*, 1983; Kamma *et al.*, 1998). In one microscopic study there was a preponderance of spirochetes found among numerous bacterial morphotypes, leading the authors to conclude that perhaps spirochetes play an important role in this form of periodontitis (Carrassi *et al.*, 1989). In the same study no invasion of the gingival tissue was observed, prompting the authors to conclude that the host reaction to dental plaque was sufficient to prevent invasion of host tissue in spite of the neutropenic status of the patient. Another microscopic study (Vaughan *et al.*, 1990) also noted little or no bacterial invasion into host tissues and described the presence of bacterial morphotypes typically associated with rapidly progressive forms of periodontitis. In perhaps the most comprehensive study to date (Kamma *et al.*, 1998), microbiological analysis of the subgingival microbiota of a chronic neutropenic patient resulted in the isolation of 27 different microbial species. Predominant among these were *P. gingivalis, Prevotella intermedia, Capnocytophaga sputigena, Capnocytophaga ochracea,* and *A. actinomycetemcomitans.* Consistent with the identification of periodontopathogens in chronic neutropenic patients, an earlier report on one of the patients described above (Vaughan *et al.*, 1990) revealed serum antibody to *A. actinomycetemcomitans* (Baehni *et al.*, 1983). It is difficult to find published microbiological information about cyclic neutropenic patients. However, in one study, neither *A. actinomycetemcomitans* nor *P. gingivalis* was identified in two different patients (Pernu *et al.*, 1996).

The limited number of case reports precludes the drawing of general conclusions about the microbiota and the influence of the dysfunctional inflammatory response in CHS, LAD, or chronic neutropenia patients. However, clearly these congenital deficiencies result in an altered inflammatory response as evidenced by painful inflamed gingiva. In fact, in cases of chronic neutropenia it has been noted that chronic gingival inflammation may be the first clinical presentation of the disease that warrants professional treatment. The best available evidence to date suggests that the microbial flora of chronic neutropenia periodontitis, unlike LAD, contains specific periodontopathogens and resembles the more common early onset or adult type disease. If, indeed, specific periodontopathogens are required for initiation of periodontitis in chronic neutropenic patients but not LAD, it may be due to the fact that the nature of the dysfunctional inflammatory response is different in these two patient populations. In chronic neutropenia, there is only a deficiency in neutrophils, whereas in LAD a general exit deficiency of all leukocytes including monocytes and lymphocytes exists. Evidently, the host response in chronic neutropenic patients is sufficient to normally prevent bacterial invasion of gingival tissues (Carrassi *et al.*, 1989; Vaughan *et al.*, 1990). Therefore, the role of specific periodontopathogens may be to augment bacterial penetration of host tissue resulting in a more destructive and/or chronic host inflammatory response.

Papillon-Lefevre Syndrome

Another congenital deficiency that results in periodontitis is Papillon-Lefevre syndrome (Clerehugh *et al.*, 1996; D'Angelo *et al.*, 1992; Dreizen *et al.*, 1977; Eronat *et al.*, 1993; Hart and Shapira, 1994). This condition is a form of palmoplantar keratoderma, a general term for a heterogeneous group of skin diseases characterized by hyperkeratosis of the palms and soles. Papillon-Lefevre syndrome is defined and differentiated from other forms of palmoplantar keratoderma by the development of periodontitis. Recently, two different homozygosity mapping studies have localized the gene responsible for Papillon-Lefevre syndrome to chromosome 11q14 (Fischer *et al.*, 1997; Laass *et al.*, 1997). Both studies examined several different families from different geographical areas and concluded that a single gene was responsible for the syndrome. Further analysis will be necessary to identify the gene affected. The characteristic phenotype of patients with Papillon-Lefevre syndrome is hyperkeratosis of the palms and soles as well as periodontitis. Few additional infectious complications are associated although several have been recorded (Eronat *et al.*, 1993; Tinanoff *et al.*, 1986). Molecular explanations for the development of periodontitis are lacking. However, two predominant manifestations of the disease are believed to be responsible. One, the hyperkeratoderma is suspected to contribute to a breach in the gingival epithelial cell barrier, perhaps by changing gingival epithelial cell function. This then facilitates pathogenic bacterial colonization and resultant disease. However, it has been argued that such a change is not responsible for the development of disease. The rationale for this is that hyperkeratosis palmoplantaris should affect the epithelial barrier in both the primary and permanent dentition. However, since some case reports documented periodontitis only in the primary dentition an epithelial cell barrier explanation was viewed as not tenable (D'Angelo *et al.*, 1992). Although this is possible, this explanation does not take into account that the Papillon-Lefevre syndrome may manifest itself in different degrees of severity depending upon the genetic background in which it is expressed. Less severe forms may only manifest disease of the primary dentition.

Another, not mutually exclusive explanation, is a defect in neutrophil chemotaxis (Firatli *et al.*, 1996). Several reports have documented a defect in chemotaxis of neutrophils obtained from Papillon-Lefevre syndrome patients (D'Angelo *et al.*, 1992; Tinanoff *et al.*, 1986). Although there are reports to the contrary (Lyberg, 1982; Rateitschak-Pluss and Schroeder, 1984; Schroeder *et al.*, 1983), the data are strong that such a defect exists (Firatli *et al.*, 1996). Indeed, chemotaxis data suggest that the molecular defect of the syndrome involves components of cell movement distal to the chemotactic signal (Van Dyke *et al.*, 1984). An intriguing possibility that has been presented (D'Angelo *et al.*, 1992; Tinanoff *et al.*, 1986) is that bacterial components induce the chemotactic defect in neutrophils obtained from Papillon-Lefevre syndrome patients. A bacteria-induced chemotactic defect has been proposed in these studies since during infection a neutrophil chemotaxis defect was manifested that disappeared in the same patients years later when periodontitis was not evident. Consistent with a bacteria-induced chemotaxis defect is an early report that describes bacterially induced neutrophil chemotaxis defects in two patients with recurrent bacterial infections (Shurin *et al.*, 1979). It was proposed that *Capnocytophaga spp.* were responsible for inducing the neutrophil defect. Subsequent to these studies, it has been clearly demonstrated that certain bacterial components, in particular the lipopolysaccharide portion of the cell wall, inhibits neutrophil chemotaxis (Bignold *et al.*, 1991). The role that these effectors play in the development of periodontitis in Papillon-Lefevre syndrome patients remains to be determined.

Several excellent comprehensive studies have been performed that have examined the microbial composition of subgingival plaque in Papillon-Lefevre syndrome patients (Clerehugh *et al.*, 1996; D'Angelo *et al.*, 1992; Lundgren *et al.*, 1998; Tinanoff *et al.*, 1986). Early work suggested that the subgingival microbiota was similar to other forms of early onset disease which have as a prominent member *A. actinomycetemcomitans* (Eronat *et al.*, 1993). Subsequent studies employing serological (Clerehugh *et al.*, 1996) or DNA probe (Lundgren *et al.*, 1998) analysis as well as microbiology have found a wider range of periodontopathogens present. In one study it was concluded that there was probably not a Papillon-Lefevre syndrome specific microbiota (Lundgren *et al.*, 1998). It appears that although there is clearly a host defect that results in periodontitis, the initiation of disease is still associated with the presence of subgingival periodontopathogens.

Diabetes
Diabetes is a systemic disorder that is associated with an increase in the severity of periodontitis. There has been a long historical association between diabetes and the risk and incidence of periodontitis. Excellent reviews on this subject have been published recently (reviewed in Grant-Theule, 1996; Soskolne, 1998; Yalda *et al.*, 1994). Diabetes is associated with many complications some of which are major and life threatening, including accelerated atherosclerosis, cerebrovascular, and peripheral vascular diseases (Yalda *et al.*, 1994). It is the major cause of blindness and lower limb amputations in industrialized countries. Diabetes has also been reported to alter the gingival vasculature and this provides one possible explanation

for the increase in the severity of periodontitis in diabetic patients (Frantzis *et al.*, 1971; McMullen *et al.*, 1967; Zatz and Brenner, 1986). Other possible explanations which are not mutually exclusive include impaired neutrophil function, an excessive inflammatory response, and impaired wound healing (Yalda *et al.*, 1994). All of these effects are believed to be the result of either the generation of sorbitol, a tissue toxin, or the production of advanced glycation end products due to the non-enzymatic addition of hexoses to host proteins (Lalla *et al.*, 1998).

Several comprehensive studies have been performed to determine the composition of the subgingival microbiota in diabetic patients with periodontitis (Mandell *et al.*, 1992; Mashimo *et al.*, 1983; Shordone *et al.*, 1995; Zambon *et al.*, 1988). One study suggested that a unique constellation of organisms containing *Capnocytophaga spp.* and *A. actinomycetemcomitans* was associated with insulin-dependent diabetes (Mashimo *et al.*, 1983). This idea is consistent with the specific plaque hypothesis that contends that certain forms of periodontitis have unique microbiological components. Another study has found these bacteria as well as *P. gingivalis* and *P. intermedia* (Sastrowijoto *et al.*, 1989), leading these authors to conclude that the same kinds of microorganisms arc associated with periodontitis in both diabetic and non-diabetic individuals. A study by Mandell *et al.* (1992) found increased levels of *P. intermedia*, *Bacteroides gracilis*, *Eikenella corrodens*, *Fusobacterium nucleatum*, and *Campylobacter rectus* in the subgingival microbiota of insulin-dependent diabetic patients. The authors postulated that the increased levels of these five organisms may have been the result of the impaired host defense function of the diabetic patients. A study of noninsulin-dependent diabetes implicated *Bacteriodes intermedus*, *P. gingivalis*, and *Wolinella recta* as important in the etiology in this form of periodontitis (Zambon *et al.*, 1988). Regardless of whether there are specific microbial compositions associated with different types of diabetes it is clear that periodontopathogens are present in high numbers in all forms of the disease. This in itself is remarkable considering the major effects that diabetes can have on neutrophil function, the vasculature, and the role of the highly vascularized periodontal tissue and neutrophils in protecting the host from infection (see Clinically Healthy Status on page 177). It may have been expected that these major effects on vascularity would render individuals susceptible to infection by numerous other bacteria similar to that found in severely neutropenic patients (see above).

Cigarette Smoking

Another clear risk factor associated with the incidence and severity of periodontitis is cigarette smoking (Tonetti, 1998). An excellent microbiological analysis has demonstrated that cigarette smoking increases the risk for subgingival colonization with periodontal pathogens (Zambon *et al.*, 1996). Specifically, the risk for *Bacteriodes forsythus* was significantly increased and there was a trend for more colonization with *P. gingivalis* in the smoking group. Although the reasons for the increased risk for disease and colonization are not clear, a decrease in the vascular response in smokers during experimental gingivitis has been reported (Bergström *et al.*, 1988). Considering the important role that the highly vascularized periodontium contributes to the delivery of neutrophils and innate defense components it is possible that smokers are compromised in their ability to contain subgingival plaque growth.

189

Human Immunodeficiency Virus

The role of cell mediated immunity in the development of gingivitis and periodontitis is perhaps best understood in the study of individuals infected with human immunodeficiency virus (HIV). Three forms of periodontal disease (linear gingival erythema (LGE), necrotizing (ulcerative) gingivitis (NUG), and necrotizing (ulcerative) periodontitis (NUP)) have been described in these patients (Holmstrup and Westergaard, 1998; Lamster *et al.*, 1998). Patients with LGE often contain many of the same bacteria found in conventional gingivitis; however, this form of gingivitis is unusual in that it is associated with *Candida* infection (Holmstrup and Westergaard, 1998; Lamster *et al.*, 1998). The microbiology associated with NUG is limited but appears to be unusual in that *Borrelia* and gram positive cocci are often isolated from infected individuals. NUP is characterized by a more rapid and severe tissue attachment loss than that observed in conventional periodontitis (Holmstrup and Westergaard, 1998; Lamster *et al.*, 1998). Individuals with NUP often contain many of the same periodontopathogens associated with periodontitis in non-HIV infected individuals; however, they also have an increase in enteric species and *Candida*. Although correlations of CD4 + T cell counts and disease susceptibility and severity for all forms of HIV associated periodontal disease are still controversial, defective cell mediated immunity clearly alters the gingival and periodontal microbiota. The contribution of these microorganisms not normally isolated from periodontal lesions to the unusual clinical manifestations of HIV associated disease remains to be determined.

Congenital or Induced Immune Host Defects Generally Not Associated with the Development of Periodontitis

In contrast to the studies presented above several immune host defense defects fail to demonstrate an association with the development of periodontitis. These are noteworthy due to their lack of effect.

Chronic Granulomatous Disease

Perhaps the congenital disorder that stands in most stark contrast is chronic granulomatous disease (CGD). The fundamental cellular defect in patients who have inherited this disorder is the inability to mount a respiratory burst during phagocytosis (Mills and Quie, 1980). This results in a significant impairment of the neutrophil to kill bacteria after ingestion. However, cell motility, phagocytosis, granule enzyme content, and degranulation are usually normal (Mills and Quie, 1980). The net result of this defect is the inability to kill catalase-positive organisms such as *S. aureus* and *E. coli*. Accordingly, patents suffer from deep, recurrent, serious bacterial infections involving the bones and vital organs. Despite this impairment, several studies have demonstrated that CGD patients are no more susceptible and do not have any more severe periodontitis than normal individuals (Cohen *et al.*, 1985). The reasons for this apparent contradiction have been addressed (Cohen *et al.*, 1985) and include explanations such as 1) non-oxidative mechanisms of bacterial killing are sufficient to protect against periodontal pathogens; and / or 2) the oxidative mechanism of neutrophils is important for promoting the tissue damage normally associated with periodontitis. Regardless of the explanations, considering the role of the neutrophil in the clinically healthy periodontium, and the impact that other neutrophil disorders have on periodontitis, it is remarkable that such a severe neutrophil bactericidal defect such as seen in CGD apparently has no clinical effect. Perhaps due to the lack of periodontitis in these patients very little has been published concerning their microbiota. Additional studies on the microbial composition of CGD patients may clarify the role of non-oxidative neutrophil mechanisms in dental plaque biofilm composition in both healthy and diseased CGD patients.

Complement
Although numerous congenital defects in nearly all of the complement cascade proteins, including accessory factors, have been identified (Lokki and Colten, 1995; Morgan and Walport, 1991), apparently there are no cases of gingivitis or periodontitis being reported as an associated clinical manifestation. It is clear that complement proteins are present in gingival crevicular fluid (Embery and Waddington, 1994). However, based upon the apparent lack of a gingivitis or periodontitis phenotype in complement-deficient patients their role in the control of subgingival plaque must be questioned. It is possible that the subgingival microbiota in patients with complement deficiencies is altered in composition but there are no clearly recognizable clinical manifestations. Analysis of the subgingival microbiota of complement deficient patients is needed and should provide some insight on the role of complement in subgingival ecology.

IgA Deficiency and Agammaglobulinemia
It is surprising also that in patients with IgA deficiency and agammaglobulinemia no significant differences in gingivitis or periodontitis were observed when compared to normal age matched controls (Robertson *et al.*, 1980). However, as pointed out by Tolo (1991), these study populations were young and hence were not susceptible to periodontitis. This would suggest that IgA deficiency and agammaglobulinemia are not sufficient to induce periodontal disease in normally resistant individuals. However, it does not address the role of Ig in modulating the oral microbial flora in individuals that have had disease or may be more susceptible for a variety of reasons. It is likely that adaptive immunity, specifically antibody responses which occur due to sufficient exposure to antigen presenting cells in the course of an individual's lifetime, is important in modifying the microbiota.

Summary of the Effects of Induced and Congenital Deficiencies on the Oral Microbiota
In most cases of innate or acquired host deficiencies there is a clear increase in the incidence and severity of gingivitis and/or periodontitis. A general pattern emerges underscoring the importance of a proper functioning vasculature and regulated movement of neutrophils and other host defense cells into surrounding tissue. This provides strong evidence that the low-level expression of inflammatory mediators in clinically healthy tissue is important for the proper functioning of the periodontium. Therefore, it would be difficult to underestimate the importance of the innate host defense/microbial dynamic that exists in clinically healthy tissue. Cell mediated immunity also contributes to regulating microbial composition as evidenced in HIV

191

positive individuals. These individuals are subject to periodontal disease that may be associated with different microorganisms. It is also likely that IgA and complement regulate microbial composition but this has not been formally examined since there has not been a clear association with disease.

Another pattern that emerges is that in most cases periodontitis is associated with the same periodontopathogens (see below) that are found in non-immunocompromised individuals. The notable exceptions are chemotherapy-induced neutropenia and LAD. The innate host defense system is severely compromised in these individuals, which may explain in part the ability of nosocomial and commensal microorganisms to be associated with disease. These observations demonstrate that the gingival environment in the absence of the innate host defense barrier can provide a suitable habitat for numerous and diverse microorganisms. Therefore, the continued association of known periodontopathogens with the development of periodontitis in most immunocompromised individuals is all the more consistent with their proposed pathogenic contribution to the etiology of the disease. The mechanisms employed by individual periodontopathogens resulting in a dysfunctional inflammatory response in periodontitis are becoming elucidated (Madianos *et al.*, 1997; Darveau *et al.*, 1998; and Huang *et al.*, 1998). Considering the important role that innate host defense contributes to containing the dental plaque biofilm it is not surprising that additional impairments of the innate host defense barrier are involved (see below).

Gingivitis and Periodontitis Represent Two Clinical Conditions Where the Microbial Compositions are Linked to an Altered Innate Host Response

Gingivitis and periodontitis are two major clinical manifestations that are associated with unique microbial communities. Compared to clinically healthy individuals, gingivitis is associated with an increased microbial load (10^4-10^5 organisms/ml) of which 15-50% are Gram-negative bacteria. Periodontitis is associated with a further increase in microbial load (10^5-10^8 organisms/ml) and a clear association with specific gram-negative bacteria often referred to as periodontopathogens (Darveau *et al.*, 1997). The bacterial species most commonly associated with these two clinical conditions are listed in Table 3. It is clear that in both of these clinical conditions the microbial biofilm bacteria increase in numbers and diversity. Gingivitis is characterized by an increase in the number and types of inflammatory cells in the gingiva surrounding the tooth root surface. Periodontitis is associated with the expression of more and different inflammatory mediators compared with gingivitis, and results in the loss of fibrous connective tissue and the alveolar bone that surrounds and supports the tooth root. In contrast to the clinically healthy situation where few studies have been conducted to correlate bacterial composition to inflammatory response (see above), there is an abundance of evidence that bacterial levels and/or composition is directly responsible for the clinical manifestations of gingivitis and periodontitis. Experimental gingivitis studies in humans have demonstrated that increased numbers and a shift in the types of bacteria are invariably associated with increased inflammation. In addition, removal of dental plaque remains the most effective mechanism of restoring the clinical response to normal in both gingivitis and periodontitis patients.

Table 3. Subgingival Species Associated with Health, Gingivitis and Disease

Health	Gingivitis	Periodontitis
Streptococcus oralis[g]	*Streptococcus oralis*	*Porphyromonas gingivalis*[g]
Streptococcus sanguis[g]	*Streptococcus sanguis*	*Actinobacillus actinomycetemcomitans*
Streptococcus mitis[g]	*Streptococcus mitis*	serotype b
Streptococcus gordonii	*Streptococcus intermedius*	*Bacteroides forsythus*[g]
Streptococcus mutans	*Capnocytophaga ochracea*	PRO spirochete
Streptococcus anginosus	*Capnocytophaga gingivalis*	*Treponema denticola*
Streptococcus intermedius	*Capnocytophaga gracilis*	*Prevotella intermedia*[g]
Gemella morbillorum	*Prevotella loescheii*	*Prevotella nigrescens*[g]
Rothia dentocariosa	*Peptostreptococcus micros*	*Campylobacter rectus*[g]
Actinomyces naeslundii	*Eubacterium nodatum*	*Peptostreptococcus micros*
Actinomyces gerencseriae	*Actinomyces naeslundii*	*Fusobacterium nucleatum*
Actinomyces odontolyticus	*Actinomyces israelii*	subsp. *vincentii*
Peptostreptococcus micros	*Campylobacter concisus*	*Fusobacterium nucleatum*
Eubacterium nodatum	*Actinomyces odontolyticus*	subsp. *nucleatum*
Capnocytophaga ochracea	*Fusobacterium nucleatum*	*Selenomonas noxia*
Capnocytophaga gingivalis	subsp. *nucleatum*	*Selenomonas flueggeii*[g]
Campylobacter gracilis	*Eubacterium brachy*	Enteric species
Fusobacterium nucleatum	*Eikenella corrodens*	*Fusobacterium alocis*
subsp.*polymorphum*	*Actinobacillus*	*Lactobacillu uli*[g]
	actinomycetemcomitans	*Veillonella parvula*[g]
	serotype a	

Source: text references and laboratory data. [g] also associated with gingivitis. Any subject can at any time be colonized from other clinical categories, but usually these additional species constitute a minor segment of the subgingival microbiota. Species in the table are ordered as being more likely to be found at the top and less likely at the bottom. Reprinted by permission from Darveau *et al.* 1997.

Although the microbial composition of dental plaque associated with either gingivitis or periodontitis is well characterized (Moore and Moore, 1994; Socransky and Haffajee, 1994), little is known about how these different compositions influence the inflammatory response. It is recognized that the host inflammatory response varies with different bacteria. However, neither the mechanisms the host employs to recognize different bacteria or their released antigens nor how the host regulates the intensity of the inflammatory response are understood.

Gingivitis
The Increased Microbial Load Associated with Gingivitis Contributes to the Increase in the Host Inflammatory Response
The inflammation characteristic of gingivitis is due to both increases in the number of host cells responding to dental plaque and the amount of inflammatory mediators secreted by these cells. This would be expected since an increase in the microbial burden would lead to more bacterial material being released from both the supra- and subgingival plaque biofilms. The best available evidence to date suggests that the bacterial components recognized by the host are pattern recognition receptor ligands (Medzhitov and Janeway, 1997). As mentioned above, these are components of the bacterial cell wall that are essential for bacterial survival, are structurally similar for a wide variety of different bacteria, and do not resemble self components

(Table 4). Examples of bacterial components that bind pattern recognition receptors to date include LTA (Cleveland *et al.*, 1996; Pugin *et al.*, 1994; Sugawara *et al.*, 1999), LPS (Wright *et al.*, 1990), and peptidoglycan (Dziarski *et al.*, 1998; Gupta *et al.*, 1999) (Table 4), however, other bacterial components such as flagella and pilli may need to be considered as pattern recognition receptor ligands in the future. In the healthy situation due to the prevalence of gram-positive bacteria the host most likely sees a preponderance of LTA and peptidoglycan. In gingivitis the host tissue experiences an increase in LTA and peptidoglycan along with a significant increase in new components such as lipoproteins from spirochetes and LPS from gram-negative bacteria.

The increase in pattern recognition receptor ligands likely overwhelms the resident neutrophil protective defense and results in excessive neutrophil activation and/or bacterial component penetration of host tissue. Neutrophils and epithelial cells are the first innate host defense cells to encounter bacteria and their released components. Although little is known about their cooperative functions, normally their combined efforts of cellular recruitment and controlled amplification (Agace *et al.*, 1993; Hedlund *et al.*, 1999; Rieder *et al.*, 1997) of the inflammatory response prevent excessive cytokine release and bacterial component penetration of host tissue. Increased bacterial component penetration of periodontal tissue is likely in gingivitis due to both a limited capacity of neutrophils to inactivate bacteria and saturation of soluble innate host defense components of the gingival crevicular fluid.

Although not formally examined, serum components such as collectins, C reactive protein, LBP, sCD14, and others (Medzhitov and Janeway, 1997) are presumably present in GCF and are available to bind and neutralize bacterial components. These factors can effectively neutralize non-self components either before they enter the periodontal tissue or prevent recognition by host epithelial, fibroblast, and endothelial cells when complexed after they enter tissue. Neutralization occurs either by opsonization via pattern recognition receptors such as the macrophage mannose or scavenger receptor and detoxification within the monocyte. An example of a detoxifying enzyme within the monocyte is acyloxyacyl hydrolase, which removes the acyloxyacyl-linked (secondary) fatty acids (Munford and Hall, 1986). Alternatively, and not mutually exclusive, detoxification may occur in the liver or spleen or by transfer to serum proteins which form complexes with the bacterial component. These interactions help to maintain the low-level inflammatory status observed in clinically healthy tissue and represent a balance between neutralization and inflammatory activation. For example, studies have shown that high levels of sCD14 may potentiate activation of host cells by LPS (Ferrero *et al.*, 1993), but sCD14 may also serve to neutralize LPS and protect the host from inflammatory damage (Goyert and Haziot, 1995; Leturcq *et al.*, 1996). Similar results have been observed with LBP (Lamping *et al.*, 1998). In addition, as mentioned above (Clinically Healthy Status on page 177), LTAs from gram positive bacteria can compete with LPS for binding to CD14, and since LTAs are much less active this essentially prevents excessive host cell activation (Sugawara *et al.*, 1999). The result of these transfer reactions is a balance between neutralization and host cell activation. In gingivitis, although more serum components are present due to increased gingival crevicular fluid flow, bacterial components both directly and

Table 4. Lipid A and Core Composition and Select Biological Activity of LPS from Various Oral Bacteria

Organism	Heptose	KDO	#P/LA	3-Hydroxy fatty acids				Shwartzman	E-selectin	Ref.
				Short	Medium	Long	Other			
E. coli[a]	+	+	2	–	H14	–	–	+	+	(Morrison and Ryan, 1992; Ogawa, 1994)
A. actinomycetemcomitans	+	+	2	–	H14	DD-Hep	–	+	+	(Mashimo et al., 1985; Masoud et al., 1991; Perry et al., 1996;
C. rectus	+	+	2	–	H14,H16	–	–	+		Perry et al., 1996;
F. nucleatum	+	+	2	–	H14,H16	–	–	+		(Kumada et al., 1989) (Mashimo et al., 1985; Onoue et al., 1996)
H. parainfluenzae	+		2	–	H14	–	–	+		(Tuyau and Sims, 1983)
L. buccalis	+			–	–	–	–	+		(Knox and Parker, 1973)
B. fragilis	+	+	1	–	–	H7i	–	–		(Sveen, 1977; Weintraub et al., 1989)
P. gingivalis	+/–	+	1	–	–	H7i	P-KDO	–		(Bramanti et al., 1989; Fujiwara et al., 1990; Johne et al., 1988; Kumada et al., 1995; Kumada et al., 1997; Mashimo et al., 1985;
B. forsythus				–	–	H7i	–	–		Ogawa, 1993; Schifferle et al., 1989)
B. loescheii				–	–	H7i,H16	–			(Gersdorf et al., 1993)
P. intermedia	+/–	+		–	–	H7i	P-KDO			(Johne et al., 1988; Mashimo et al., 1985) (Johne and Bryn, 1986; Johne et al., 1988; Mashimo et al., 1985;
C. sputigena				–	–	H17i,H15i	–			(Dees et al., 1982; Poirier et al., 1983)
S. sputigena	+	+		H13	–	–	–			(Kumada et al., 1989)
E. corrodens	+	+		H12	–	–	GalN in LA			(Mashimo et al., 1985)
C. periodonti	+			H13	–	–	–	+		(Kokeguchi et al., 1990)
Veillonella spp.	+	+		H13	–	–	–	+	+	(Bishop et al., 1971)
T. denticola?				H12,H13	–	–	–			(Dahle et al., 1996)

[a]Nonoral species; DD-Hep = D-glycero-D-mannoheptose; P-KDO = phosphorylated KDO; (+) = Component has been reported; (–) = absence of component has been reported; #P/LA = number of phosphates/lipid A; H = hydroxy fatty acid; P-KDO indicates the component or activity has not been reported. Reprinted by permission from D...

195

indirectly shift this balance toward activation. For example, increased release of cytokines from host cells located in normally sterile periodontal tissue has been observed (Matsuki *et al.*, 1992).

The increased species complexity of bacterial biofilms associated with gingivitis presents the host with a multitude of structurally similar pattern recognition receptor ligands. The increase in the number of different bacterial species found in dental plaque associated with gingivitis results in a variety of different types of bacterial components being released into the gingival space and periodontal tissue. For example, there is an increase in the number of gram-negative bacteria releasing LPS in gingivitis-associated plaque. The LPS of oral bacterial species have diverse chemical compositions (Table 4) indicating a great diversity in structure (Darveau and Bainbridge, 1999). The host is therefore presented with a plethora of different LPS structures that may interact with pattern recognition receptors. A convenient classification of the different types of oral bacterial LPS is based upon the chain length of the fatty acids found attached to lipid A (Darveau and Bainbridge, 1999). The bacterial LPS consist of three groups: medium chain fatty acid (14C) LPS similar to *E. coli*, long chain fatty acid (17C) LPS similar to *Bacteroides fragilis* and short chain fatty acid (12-13C) LPS similar to *Pseudomonas aeruginosa*. Species of gram-negative bacteria that produce LPS containing fatty acids of short (*Eikenella spp.*), medium (*Campylobacter spp.* and *Fusobacterium spp.*), and long (*Prevotella spp.* and *Capnocytophaga spp.*) chain lengths are all found in gingivitis associated plaque (Tables 3 and 4). In addition the oral spirochete *Treponema denticola*, which is found in some gingivitis associated plaque samples, secretes lipoproteins, which in other spirochetes has been shown to activate host defense cells by a CD14 dependent mechanism (Sellati *et al.*, 1998).

Different LPS Structures Elicit Different Types and Intensities of Inflammatory Responses from Host Cells

Studies with chemically synthesized and partially degraded *E. coli* lipid A analogues provide a basis for understanding how different LPS structures elicit different types and intensities of inflammatory responses. For example, previous studies have revealed that the amount and position of fatty acid acylation and phosphorylation are key components in host cell responses to *E. coli* LPS (Golenbock *et al.*, 1991; Kirikae *et al.*, 1994; Loppnow *et al.*, 1989; Pohlman *et al.*, 1987; Takada and Kotani, 1992). *In vivo*, *E. coli* lipid A analogues lacking one or more fatty acids or phosphates demonstrate significantly reduced pyrogenicity and do not elicit a Shwartzman reaction, a standard *in vivo* test of LPS tissue toxicity (Takada and Kotani, 1992). *In vitro*, an *E. coli* lipid A analogue lacking both acyloxyacyl-linked (secondary) fatty acids but containing both lipid A phosphates failed to elicit the production of inflammatory mediators from either myeloid or non-myeloid cells (Pohlman *et al.*, 1987; Takada and Kotani, 1992). This is consistent with the detoxifying function of the leukocyte enzyme acyloxyacyl hydrolase which removes the acyloxyacyl-linked (secondary) fatty acids (Munford and Hall, 1986). These findings demonstrate that relatively minor changes in fatty acid composition can have profound effects on the intensity of the inflammatory response.

196

The structurally distinct LPSs of oral bacteria also display different biological activities. To obtain a positive Schwartzman reaction, for example, it is known that a lipid A structure containing at least one acyloxyacyl group in addition to a biphosphorylated glucosamine disaccharide is required (Takada and Kotani, 1992). Many oral LPSs such as those from *A. actinomycetemcomitans, C. rectus, E. corrodens*, and *F. nucleatum*, are capable of inducing a positive Schwartzman reaction and meet or appear to meet these structural requirements (Table 4). In contrast, LPS from *P. gingivalis, Prevotella spp.*, and other oral species (Table 4) yield a negative reaction consistent with their underacylated lipid A structure. Furthermore, since it is known that adhesion molecule expression is required for a positive reaction (Argenbright and Barton, 1992), one would predict that Schwartzman-positive LPS would also have the ability to stimulate up-regulation of adhesion molecules *in vitro*. Consistent with this, it has been demonstrated that in contrast to *E. coli* LPS, a potent stimulator of E-selectin, the Schwartzman-negative LPS of *P. gingivalis* and *B. forsythus* do not up-regulate the expression of this adhesion molecule (Darveau *et al.*, 1995).

The periodontal tissue is therefore subjected to an assortment of structurally similar LPS types, yet the innate host defense system should recognize them all in the context of pattern recognition receptors. Although structurally similar it is clear that LPS obtained from some species of bacteria are much more potent than others (Table 3). In gingivitis the net result of these interactions is an increase in the inflammatory response. The intensity and duration of the inflammatory response will depend upon how long it takes the host to neutralize the bacterial components. In a manner similar to that observed in clinically healthy tissue, this will depend upon binding affinities and transfer rates to different innate host defense components. For example, LPS obtained from different bacteria have significantly different binding affinities for LBP (Cunningham *et al.*, 1996). LPS from both *H. pylori* and *P. gingivalis* bind LPB 10 to 100-fold less efficiently than *E. coli* (Cunningham *et al.*, 1996). It has been proposed that poor LBP binding accounts for the lower transfer rates to CD14 and the decreased ability of these LPS species to activate human monocytes (Cunningham *et al.*, 1996). Less efficient LBP binding also provides one mechanism by which different LPS species may persist or be removed from tissue more effectively.

Toll-Like Receptors May Modulate the Intensity and Type of Inflammatory Response to Different Bacterial Components
Recently, as mentioned above, the discovery of Toll-like receptors has provided a missing link in our knowledge of how the host recognizes and responds to non-self components (Wright, 1999). Initial data strongly indicated that different Toll proteins were responsible for responses to different microbial components. For example, in *Drosophila* Toll regulates anti-fungal responses (Lemaitre *et al.*, 1996) while a Toll homologue designated "18-wheeler" is responsible for responses to gram-negative bacteria (Williams *et al.*, 1997). In a similar fashion mammalian TLR-2, when transfected into CHO cells, responds to both gram-positive and gram-negative cell wall components whereas TLR-4 in the same system does not respond to gram-positive components (Yoshimura *et al.*, 1999). This latter finding is consistent with

the observation that LPS non-responder mice are defective in TLR 4 and yet have normal responses to gram positive bacteria. In other words, since gram positive bacteria do not engage TLR 4, a defect in this gene has no effect on the innate response to these bacteria. Since the innate host response to gram-positive and gram-negative bacteria is very different with respect to its intensity both *in vivo* and *in vitro* it also suggests that engagement of different TLRs results in different types of inflammatory responses. Further work will be needed to clarify the role of different Toll-like receptor proteins in responses to different bacterial components. However, this family of proteins offers an exciting possible mechanism by which the host may regulate the nature and type of inflammatory response to different bacterial components.

The goal of the host is to remove the inflammatory stimulus from the normally sterile tissue and return to the low level, highly ordered, inflammatory surveillance status observed in clinically healthy individuals. It is easy to understand, therefore, that the increase in inflammation associated with gingivitis is meant to effectively neutralize and remove potential non-self components from normally sterile tissue. The host does not "see" bacteria, rather the amount of bacterial components present in tissue and the relative binding affinities of these components to host cell recognition proteins determines the intensity of the inflammatory response.

Periodontitis
Periodontitis is significantly different from gingivitis both in its clinical manifestations and correlations with microbial communities. It is a disease that rarely goes spontaneously into complete remission, but rather is cyclical in nature, vacillating between periods of quiescence and over active inflammatory stimulation (Becker *et al.*, 1979; Goodson *et al.*, 1981). A new microbial community is associated with periodontitis that almost invariably involves members of the black pigmented, anaerobic, gram negative rod families of *Porphyromonas* or *Prevotella* (Chapter 6; Socransky *et al.*, 1998). The correlation with the presence of these periodonto-pathogens is strong (1996). Furthermore, as discussed above, these periodonto-pathogens remain associated with disease in most instances of induced or congenital deficiencies of innate host defense. These observations strongly support a specific role for these bacteria in either maintaining or creating the dysfunctional inflammatory response characteristic of periodontitis.

Dysfunctional Inflammatory Response
The nature of the inflammatory response in periodontitis is different than in gingivitis. Several studies have shown increased levels of the same inflammatory mediators present in gingivitis as well as new mediators in tissue and gingival cervical fluid in subjects with periodontitis (Lamster, 1997; Moskow and Polson, 1991). For example, in situ analysis has shown that although monocytes are the predominant cell type activated in gingivitis, T and B lymphocytes, fibroblasts, and endothelial cells all express cytokine mRNA in periodontitis (Matsuki *et al.*, 1992). In addition, elevated expression of IL-1β and IL-1β receptor antagonist in the connective tissue of periodontitis versus healthy subjects was observed (Roberts *et al.*, 1997). More recently, examination of the relationship between tissue concentrations of three

198

inflammatory mediators and adjacent sulcular depth revealed an interesting change in the ratios of IL-8, IL-6 and IL-1β (McGee *et al.*, 1998). It was found that as the sulcular depth increased, the level of IL-8 decreased and the levels of IL-6 and IL-1β increased. Therefore, it is likely that both increases in inflammatory mediators and changes in local concentrations of individual mediators occur in response to the periodontitis bacterial challenge. These types of analysis underscore the fact that in periodontitis activation of cell types not normally exposed to bacterial components is occurring with increased frequency. As mentioned above, as the host attempts to eliminate these bacterial components it increases the inflammatory response. In periodontitis, however, this increase is dysfunctional and results in tissue and bone destruction.

There are several possible mechanisms for a dysfunctional inflammatory response in periodontitis. Perhaps the most significant is that cells located deep in the periodontium, in contrast to epithelial cells, respond to bacteria or their components by increased production of numerous, different inflammatory mediators. Instead of a more limited response characteristic of epithelial cells and proposed to serve a host protective function due to their location in contact with the environment (Agace *et al.*, 1993; Hedlund *et al.*, 1999), fibroblasts and endothelial cells secrete a plethora of inflammatory mediators in response to bacterial components or cytokine stimulation (for review see Darveau and Bainbridge, 1999; Darveau *et al.*, 1997). In addition, it has been demonstrated that gingival fibroblasts in healthy tissue are CD14-negative, whereas in inflamed tissue they are CD14-positive (Hiraoka *et al.*, 1998). Along with the change in phenotype of the fibroblasts is an increased IL-6 response to LPS. The production of inflammatory mediators deep within periodontal tissue can have devastating effects on the highly organized innate host defense molecular architecture necessary for an effective inflammatory response. For example, numerous cell types in the periodontium are capable of producing PGE_2 (Offenbacher *et al.*, 1993), and this particular mediator has been closely correlated with disease progression as measured by bone loss (Offenbacher *et al.*, 1993). It has been demonstrated *in vitro* (Armstrong, 1995) and by at least one clinical correlation *in vivo* (Garzetti *et al.*, 1998) that PGE_2 blocks the ability of leukocytes to respond to chemotactic stimuli. It is becoming increasing clear that chemotaxis is one of the most important mechanisms by which the host maintains a non-destructive, highly organized, inflammatory response in clinically healthy tissue. The disorientation of this neutrophil property alone severely hinders their ability to remove the source of the bacterial component penetration into host tissue and may contribute to premature degranulation contributing to host tissue destruction.

Bacterial components obtained from a wide variety of different bacteria can all elicit the production of these inflammatory mediators, which makes it difficult to determine the specific contribution of each organism. It is possible that certain bacterial components are directly responsible for the production of specific more destructive types of inflammatory mediators. Therefore, engaging different Toll-like receptors with distinct bacterial components provides a mechanism for producing graded inflammatory responses. However, it is more likely that the specific contribution of each organism may not be relevant at this point in the inflammatory response. Rather, members of the subgingival microbial community may all participate in disorienting the host as part of a pathologic process either initiated or maintained by specific periodontopathogens.

Role of Periodontopathogens in the Innate Host Response

It is not clear if periodontopathogens initiate or facilitate bacterial component penetration of the epithelial barrier generating a destructive inflammatory response. Strong arguments for either a passive or an active role can be made. It is possible that during episodes of gingivitis changes in epithelial cell permeability occur such that new environments are created and periodontopathogens "find" themselves passively becoming predominant members of the community due to enriched physiological conditions. Alternatively, the ability of *P. gingivalis*, perhaps the best studied periodontopathogen, to invade epithelial cells is consistent with a pathogen-initiated mechanism to breach the epithelial cell barrier, gain access to deeper tissue, and disrupt the normal inflammatory response. Regardless of their initial role, the high numbers and deep penetration of periodontopathogens insure their participation in the host inflammatory response.

Examination of the innate host inflammatory response to periodontopathogens has been mostly performed with *A. actinomycecomitans* and *P. gingivalis*. Interestingly, these two bacteria evoke markedly different host responses (Darveau and Bainbridge, 1999; Huang *et al.*, 1998). *A. actinomycecomitans* whole cells (or LPS) appear to be similar to *E. coli* in that they are potent activators of the cytokine responses in epithelial cells, monocytes, or neutrophils (for review see Darveau and Bainbridge, 1999). This is consistent with *A. actinomycecomitans'* LPS's structural similarities to *E. coli* lipid A (Table 4). In contrast, *P. gingivalis* is either less potent or differentially activates these cells such that some cytokines are made at levels similar to *E. coli* whilst others are produced in much lower amounts (for review see Darveau and Bainbridge, 1999). Direct comparisons of *A. actinomycecomitans* and *P. gingivalis* interactions with gingival epithelial cells have revealed that *A. actinomycecomitans* is a potent activator of IL-8 and ICAM-1 whereas *P. gingivalis* decreases constitutive expression of these inflammatory mediators (Huang *et al.*, 1998). Consistent with this, it has been demonstrated that *P. gingivalis* can repress IL-8 accumulation by epithelial cells and inhibit neutrophil transwell migration through epithelial cell monolayers grown *in vitro* (Madianos *et al.*, 1997). These events can occur through both protease dependent and independent mechanisms (Darveau *et al.*, 1998). Studies with neutrophils and monocytes (Darveau and Bainbridge, 1999) are consistent with less potent and differential activation of cytokines when *P. gingivalis* LPS or lipid A are employed compared to *E. coli* or *A. actinomycecomitans*. Finally, it has been demonstrated that *P. gingivalis* LPS is a natural antagonist of endothelial cells, inhibiting their ability to respond to other bacteria by the secretion of IL-8 or expression of E selectin (Darveau *et al.*, 1995). *P. gingivalis* LPS did not block cytokine-induced responses from these cells (Darveau *et al.*, 1995). It is clear from these studies that *P. gingivalis* LPS is recognized by the innate host defense system but fails to elicit the type of responses normally attributed to the classic endotoxin LPS from *E. coli*.

The microbial community changes to a pathologic state in periodontitis. Periods of active subgingival dental plaque growth, appropriately termed bacterial blooms

(Sissons *et al.*, 1995) are followed by apparently no or little microbial activity. The host response fluctuates accordingly with periods of greater bone resorption during active disease. A new host/microbial dynamic ensues which is strongly associated with periodontopathogens, especially gram-negative black pigmented rods. One possible role of these periodontopathogens is to maintain the chronic inflammatory state. For example, forms of periodontitis that do not contain these organisms (in LAD and neutropenic patients for example) have an acute as opposed to a chronic manifestation of the disease. Localized juvenile periodontitis which has a strong association with *A. actinomycetemcomitans* is also a more rapid, rather than chronic form of the disease. Consistent with this, the innate host response to *A. actinomycetemcomitans* is much more potent than for other periodontopathogens (for review see Darveau and Bainbridge, 1999). By possessing an LPS that is markedly lower in its activation potential it is possible that black pigmented periodontopathogenic members of the subgingival community maintain longevity by dampening or altering the host response. When viewed in terms of pattern recognition, it is likely that these periodontopathogens exploit the nature of host recognition to maintain a pathologic state. It is possible that co-evolution of a symbiotic community in the gut containing *Bacteroides spp.* where innate defense is quiescent in the healthy host, provided the essential low activation structure of the closely related *P. gingivalis* LPS. However, in the oral environment, where highly regulated activation of the innate host defense system is key to clinical health, these same structural properties of the LPS are detrimental and aid in the survival of the whole community. In the intestine, pattern recognition of this type of LPS results in a commensal relationship, whereas in the oral environment the same pattern recognition events are pathologic. It is not clear if this would augment initial growth of the subgingival community, aid in the generation of bacterial blooms, or have no effect on microbial growth but rather simply alter the host response to a destructive phase. In addition, a bacteria-induced defect in innate host defense would not be necessary in localized juvenile periodontitis since individuals with this disease have been reported to already have neutrophil defects (Dennison and Van Dyke, 1997) rendering them more susceptible to infection.

Conclusions

The host constantly monitors and responds to the colonization status of the oral cavity. It is clear that subgingival bacterial numbers and compositions have profound effects on the host innate response, whereas provocative data suggest that saliva composition may be altered in response to *Candida* infection. Studies of individuals with impaired innate host responses clearly demonstrate that a proper innate inflammatory response is key to periodontal health. Examination of the innate host response status in clinically healthy individuals has revealed a low level "inflammatory surveillance" state where the host maintains an effective barrier against bacterial infection. It is possible that the normal oral microflora is not merely a series of non-pathogenic commensal communities, but rather these communities participate in establishing this protective state, elevating them to symbiotic partners with the host. Mechanisms of bacterial recognition are emerging that may help explain how different members of the microbiotia contribute to maintenance of an effective host defense barrier. Likewise, information on the host activation potential of periodontopathogens suggests that dysfunctional host responses may be created. Additional studies examining the bacterial/host dynamic in the clinically normal and diseased host are needed to more fully understand how these different microbial communities are created.

References

Anonymous. 1996. Position paper: epidemiology of periodontal diseases. American Academy of Periodontology [see comments]. J Periodontol. 67: 935-945.

Acton, S., D. Resnick, M. Freeman, Y. Ekkel, J. Ashkenas, and M. Krieger. 1993. The collagenous domains of macrophage scavenger receptors and complement component C1q mediate their similar, but not identical, binding specificities for polyanionic ligands. J Biol Chem. 268: 3530-3537.

Agace, W., S. Hedges, U. Andersson, J. Andersson, M. Ceska, and C. Svanborg. 1993. Selective cytokine production by epithelial cells following exposure to *Escherichia coli*. Infect Immun. 61: 602-609.

Almstahl, A., U. Kroneld, A. Tarkowski, and M. Wikström. 1999. Oral microbial flora in Sjogren's syndrome. J Rheumatol. 26: 110-114.

Almstahl, A., and M. Wikström. 1999. Oral microflora in subjects with reduced salivary secretion. J Dent Res. 78: 1410-1416.

Anderson, D. C., F. C. Schmalsteig, M. J. Finegold, B. J. Hughes, R. Rothlein, L. J. Miller, S. Kohl, M. F. Tosi, R. L. Jacobs, T. C. Waldrop, and *et al*. 1985. The severe and moderate phenotypes of heritable Mac-1, LFA-1 deficiency: their quantitative definition and relation to leukocyte dysfunction and clinical features. J Infect Dis. 152: 668-689.

Anderson, D. C., and T. A. Springer. 1987. Leukocyte adhesion deficiency: an inherited defect in the Mac-1, LFA-1, and p150,95 glycoproteins. Annu Rev Med. 38: 175-194.

Argenbright, L. W., and R. W. Barton. 1992. Interactions of leukocyte integrins with intercellular adhesion molecule 1 in the production of inflammatory vascular injury *in vivo*. The Shwartzman reaction revisited. J Clin Invest. 89: 259-272.

Armstrong, R. A. 1995. Investigation of the inhibitory effects of PGE_2 and selective EP agonists on chemotaxis of human neutrophils. Br J Pharmacol. 116: 2903-2908.

Atkinson, J. C., C. Yeh, F. G. Oppenheim, D. Bermudez, B. J. Baum, and P. C. Fox. 1990. Elevation of salivary antimicrobial proteins following HIV-1 infection. J Acquir Immune Defic Syndr. 3: 41-48.

Baehni, P. C., P. Payot, C. C. Tsai, and G. Cimasoni. 1983. Periodontal status associated with chronic neutropenia. J Clin Periodontol. 10: 222-230.

Barrat, F. J., F. Le Deist, M. Benkerrou, P. Bousso, J. Feldmann, A. Fischer, and G. de Saint Basile. 1999. Defective CTLA-4 cycling pathway in Chediak-Higashi syndrome: a possible mechanism for deregulation of T lymphocyte activation. Proc Natl Acad Sci U S A. 96: 8645-8650.

Becker, W., L. Berg, and B. E. Becker. 1979. Untreated periodontal disease: a longitudinal study. J Periodontol. 50: 234-244.

Bercier, J. G., I. Al-Hashimi, N. Haghighat, T. D. Rees, and F. G. Oppenheim. 1999. Salivary histatins in patients with recurrent oral candidiasis. J Oral Pathol Med. 28: 26-29.

Bergström, J., L. Persson, and H. Preber. 1988. Influence of cigarette smoking on vascular reaction during experimental gingivitis. Scand J Dent Res. 96: 34-39.

Bignold, L. P., S. D. Rogers, T. M. Siaw, and J. Bahnisch. 1991. Inhibition of chemotaxis of neutrophil leukocytes to interleukin-8 by endotoxins of various bacteria. Infect Immun. 59: 4255-4258.

Bishop, D. G., M. J. Hewett, and K. W. Knox. 1971. Biochemical studies on lipopolysaccharides of *Veillonella*. Eur J Biochem. 19: 169-175.

Blume, R. S., and S. M. Wolff. 1972. The Chediak-Higashi syndrome: studies in four patients and a review of the literature. Medicine (Baltimore). 51: 247-280.

Bowen, T. J., H. D. Ochs, L. C. Altman, T. H. Price, D. E. Van Epps, D. L. Brautigan, R. E. Rosin, W. D. Perkins, B. M. Babior, S. J. Klebanoff, and R. J. Wedgwood. 1982. Severe recurrent bacterial infections associated with defective adherence and chemotaxis in two patients with neutrophils deficient in a cell-associated glycoprotein. J Pediatr. 101: 932-940.

Bradshaw, D. J., P. D. Marsh, K. M. Schilling, and D. Cummins. 1996. A modified chemostat system to study the ecology of oral biofilms. J Appl Bacteriol. 80: 124-130.

Bramanti, T. E., G. G. Wong, S. T. Weintraub, and S. C. Holt. 1989. Chemical characterization and biologic properties of lipopolysaccharide from *Bacteroides gingivalis* strains W50, W83, and ATCC 33277. Oral Microbiol Immunol. 4: 183-192.

Brown, L. R., S. Dreizen, S. Handler, and D. A. Johnston. 1975. Effect of radiation-induced xerostomia on human oral microflora. J Dent Res. 54: 740-750.

Busscher, H. J., R. Bos, and H. C. van der Mei. 1995. Initial microbial adhesion is a determinant for the strength of biofilm adhesion. FEMS Microbiol Lett. 128: 229-234.

Carrassi, A., S. Abati, G. Santarelli, and G. Vogel. 1989. Periodontitis in a patient with chronic neutropenia. J Periodontol. 60: 352-357.

Carswell, E. A., L. J. Old, R. L. Kassel, S. Green, N. Fiore, and B. Williamson. 1975. An endotoxin-induced serum factor that causes necrosis of tumors. Proc Natl Acad Sci U S A. 72: 3666-3670.

Celenligil, H., K. Eratalay, E. Kansu, and J. L. Ebersole. 1998. Periodontal status and serum antibody responses to oral microorganisms in Sjogren's syndrome. J Periodontol. 69: 571-577.

Charon, J. A., S. E. Mergenhagen, and J. I. Gallin. 1985. Gingivitis and oral ulceration in patients with neutrophil dysfunction. J Oral Pathol. 14: 150-155.

Chaudhary, P. M., C. Ferguson, V. Nguyen, O. Nguyen, H. F. Massa, M. Eby, A. Jasmin, B. J. Trask, L. Hood, and P. S. Nelson. 1998. Cloning and characterization of two Toll/Interleukin-1 receptor-like genes TIL3 and TIL4: evidence for a multi-gene receptor family in humans. Blood. 91: 4020-4027.

Chow, J. C., D. W. Young, D. T. Golenbock, W. J. Christ, and F. Gusovsky. 1999. Toll-like receptor-4 mediates lipopolysaccharide-induced signal transduction. J

Biol Chem. 274: 10689-10692.

Clerehugh, V., D. B. Drucker, G. J. Seymour, and P. S. Bird. 1996. Microbiological and serological investigations of oral lesions in Papillon-Lefevre syndrome. J Clin Pathol. 49: 255-257.

Cleveland, M. G., J. D. Gorham, T. L. Murphy, E. Tuomanen, and K. M. Murphy. 1996. Lipoteichoic acid preparations of gram-positive bacteria induce interleukin-12 through a CD14-dependent pathway. Infect Immun. 64: 1906-1912.

Cohen, D. W., and A. L. Morris. 1961. Periodontal Manifestations of Cyclic Neutropenia. J Periodontol. 32: 159-168.

Cohen, M. S., P. A. Leong, and D. M. Simpson. 1985. Phagocytic cells in periodontal defense. Periodontal status of patients with chronic granulomatous disease of childhood. J Periodontol. 56: 611-617.

Costerton, J. W., Z. Lewandowski, D. DeBeer, D. Caldwell, D. Korber, and G. James. 1994. Biofilms, the customized microniche. J Bacteriol. 176: 2137-2142.

Cunningham, M. D., C. Seachord, K. Ratcliffe, B. Bainbridge, A. Aruffo, and R. P. Darveau. 1996. *Helicobacter pylori* and *Porphyromonas gingivalis* lipopolysaccharides are poorly transferred to recombinant soluble CD14. Infect Immun. 64: 3601-3608.

Dahle, U. R., L. Tronstad, and I. Olsen. 1996. 3-hydroxy fatty acids in a lipopolysaccharide-like material from *Treponema denticola* strain FM. Endod Dent Traumatol. 12: 202-205.

D'Angelo, M., V. Margiotta, P. Ammatuna, and F. Sammartano. 1992. Treatment of prepubertal periodontitis. A case report and discussion [see comments]. J Clin Periodontol. 19: 214-219.

Darveau, R. P. 1998. Lipid A diversity and the innate host response to bacterial infection. Curr Opin Microbiol. 1: 36-42.

Darveau, R. P., and B. Bainbridge. 1999. Lipopolysaccharide from oral bacteria: Role in innate host defense and chronic inflammatory disease, p. 2216. *In* Brade, Morrison, Opal, and Vogel (ed.), Endotoxin in Health and Disease. Marcel Dekker Incorporated.

Darveau, R. P., C. M. Belton, R. A. Reife, and R. J. Lamont. 1998. Local chemokine paralysis, a novel pathogenic mechanism for *Porphyromonas gingivalis*. Infect Immun. 66: 1660-1665.

Darveau, R. P., M. D. Cunningham, T. Bailey, C. Seachord, K. Ratcliffe, B. Bainbridge, M. Dietsch, R. C. Page, and A. Aruffo. 1995. Ability of bacteria associated with chronic inflammatory disease to stimulate E-selectin expression and promote neutrophil adhesion. Infect Immun. 63: 1311-1317.

Darveau, R. P., M. D. Cunningham, C. L. Seachord, R. C. Page, and A. Aruffo. 1995. The ability of bacteria associated with chronic inflammatory disease to stimulate E-selectin expression and neutrophil adhesion. Prog Clin Biol Res. 392: 69-78.

Darveau, R. P., A. Tanner, and R. C. Page. 1997. The microbial challenge in periodontitis. Periodontol 2000. 14: 12-32.

Deasy, M. J., R. I. Vogel, B. Macedo-Sobrinho, G. Gertzman, and B. Simon. 1980. Familial benign chronic neutropenia associated with periodontal disease. A case report. J Periodontol. 51: 206-210.

Dees, S. B., D. E. Karr, D. Hollis, and C. W. Moss. 1982. Cellular fatty acids of *Capnocytophaga* species. J Clin Microbiol. 16: 779-783.

Delude, R. L., R. Savedra, Jr., H. Zhao, R. Thieringer, S. Yamamoto, M. J. Fenton, and D. T. Golenbock. 1995. CD14 enhances cellular responses to endotoxin without imparting ligand-specific recognition. Proc Natl Acad Sci U S A. 92: 9288-9292.

Dennison, D. K., and T. E. Van Dyke. 1997. The acute inflammatory response and the role of phagocytic cells in periodontal health and disease. Periodontol 2000. 14: 54-78.

Deslauriers, M., D. ni Eidhin, L. Lamonde, and C. Mouton. 1990. SDS-PAGE analysis of protein and lipopolysaccharide of extracellular vesicles and Sarkosyl-insoluble membranes from *Bacteroides gingivalis*. Oral Microbiol Immunol. 5: 1-7.

Diamond, G., J. P. Russell, and C. L. Bevins. 1996. Inducible expression of an antibiotic peptide gene in lipopolysaccharide-challenged tracheal epithelial cells. Proc Natl Acad Sci U S A. 93: 5156-5160.

Doi, T., K. Higashino, Y. Kurihara, Y. Wada, T. Miyazaki, H. Nakamura, S. Uesugi, T. Imanishi, Y. Kawabe, H. Itakura, and et al. 1993. Charged collagen structure mediates the recognition of negatively charged macromolecules by macrophage scavenger receptors. J Biol Chem. 268: 2126-2133.

Dreizen, S., G. P. Bodey, and L. R. Brown. 1974. Opportunistic gram-negative bacillary infections in leukemia. Oral manifestations during myelosuppression. Postgrad Med. 55: 133-139.

Dreizen, S., T. E. Daly, J. B. Drane, and L. R. Brown. 1977. Oral complications of cancer radiotherapy. Postgrad Med. 61: 85-92.

Dunne, D. W., D. Resnick, J. Greenberg, M. Krieger, and K. A. Joiner. 1994. The type I macrophage scavenger receptor binds to gram-positive bacteria and recognizes lipoteichoic acid. Proc Natl Acad Sci U S A. 91: 1863-1867.

Dziarski, R., R. I. Tapping, and P. S. Tobias. 1998. Binding of bacterial peptidoglycan to CD14. J Biol Chem. 273: 8680-8690.

Edgar, W. M. 1992. Saliva: its secretion, composition and functions. Br Dent J. 172: 305-312.

Ellison, R. T. 3rd., and T. J. Giehl. 1991. Killing of gram-negative bacteria by lactoferrin and lysozyme. J Clin Invest. 88: 1080-1091.

Embery, G., and R. Waddington. 1994. Gingival crevicular fluid: biomarkers of periodontal tissue activity. Adv Dent Res. 8: 329-336.

Eronat, N., F. Ucar, and G. Kilinc. 1993. Papillon Lefevre syndrome: treatment of two cases with a clinical microbiological and histopathological investigation. J Clin Pediatr Dent. 17: 99-104.

Etzioni, A. 1997. Leukocyte adhesion defects. Arch Immunol Ther Exp. 45: 31-36.

Falk, P. G., L. V. Hooper, T. Midtvedt, and J. I. Gordon. 1998. Creating and maintaining the gastrointestinal ecosystem: what we know and need to know from gnotobiology. Microbiol Mol Biol Rev. 62: 1157-1170.

Ferrero, E., D. Jiao, B. Z. Tsuberi, L. Tesio, G. W. Rong, A. Haziot, and S. M. Goyert. 1993. Transgenic mice expressing human CD14 are hypersensitive to lipopolysaccharide. Proc Natl Acad Sci U S A. 90: 2380-2384.

Firatli, E., B. Tuzun, and A. Efeoglu. 1996. Papillon-Lefevre syndrome. Analysis of neutrophil chemotaxis. J Periodontol. 67: 617-620.

Fischer, J., C. Blanchet-Bardon, J. F. Prud'homme, S. Pavek, P. M. Steijlen, L. Dubertret, and J. Weissenbach. 1997. Mapping of Papillon-Lefevre syndrome to the chromosome 11q14 region. Eur J Hum Genet. 5: 156-160.

Fischer, W. 1988. Physiology of lipoteichoic acids in bacteria. Adv Microb Physiol. 29: 233-302.

Fox, R. I. 1996. Sjogren's syndrome: immunobiology of exocrine gland dysfunction. Adv Dent Res. 10: 35-40.

Frantzis, T. G., C. M. Reeve, and A. L. Brown, Jr. 1971. The ultrastructure of capillary basement membranes in the attached gingiva of diabetic and nondiabetic patients with periodontal disease. J Periodontol. 42: 406-411.

Fujiwara, T., T. Ogawa, S. Sobue, and S. Hamada. 1990. Chemical, immunobiological and antigenic characterizations of lipopolysaccharides from *Bacteroides gingivalis* strains. J Gen Microbiol. 136: 319-326.

Fung, J., T. J. MacAlister, and L. I. Rothfield. 1978. Role of murein lipoprotein in morphogenesis of the bacterial division septum: phenotypic similarity of lkyD and lpo mutants. J Bacteriol. 133: 1467-1471.

Galili, D., A. Donitza, A. Garfunkel, and M. N. Sela. 1992. Gram-negative enteric bacteria in the oral cavity of leukemia patients. Oral Surg Oral Med Oral Pathol. 74: 459-462.

Garzetti, G. G., A. Ciavattini, M. Provinciali, M. Amati, M. Muzzioli, and M. Governa. 1998. Decrease in peripheral blood polymorphonuclear leukocyte chemotactic index in endometriosis: role of prostaglandin E2 release. Obstet Gynecol. 91: 25-29.

Gemmell, E., and G. J. Seymour. 1993. Interleukin 1, interleukin 6 and transforming growth factor-beta production by human gingival mononuclear cells following stimulation with *Porphyromonas gingivalis* and *Fusobacterium nucleatum*. J Periodontal Res. 28: 122-129.

Gemmell, E., L. J. Walsh, N. W. Savage, and G. J. Seymour. 1994. Adhesion molecule expression in chronic inflammatory periodontal disease tissue. J Periodontal Res. 29: 46-53.

Gersdorf, H., A. Meissner, K. Pelz, G. Krekeler, and U. B. Gobel. 1993. Identification of *Bacteroides forsythus* in subgingival plaque from patients with advanced periodontitis. J Clin Microbiol. 31: 941-946.

Giambartolomei, G. H., V. A. Dennis, B. L. Lasater, and M. T. Philipp. 1999. Induction of pro- and anti-inflammatory cytokines by *Borrelia burgdorferi* lipoproteins in monocytes is mediated by CD14. Infect Immun. 67: 140-147.

Golenbock, D. T., R. Y. Hampton, N. Qureshi, K. Takayama, and C. R. Raetz. 1991. Lipid A-like molecules that antagonize the effects of endotoxins on human monocytes. J Biol Chem. 266: 19490-19498.

Goodson, J. M., A. C. R. Tanner, A. D. Haffajee, and S. S. Socransky. 1981. Evidence for episodic periodontal diseases activity. J Dent Res. 60: Abst 305.

Goyert, S. M., and A. Haziot. 1995. Recombinant soluble CD14 inhibits LPS-induced mortality in a murine model. Prog Clin Biol Res. 392: 479-483.

Grant-Theule, D. A. 1996. Periodontal disease, diabetes, and immune response: a review of current concepts. J West Soc Periodontal Periodontal Abstr. 44: 69-77.

Grenier, D., J. Bertrand, and D. Mayrand. 1995. *Porphyromonas gingivalis* outer membrane vesicles promote bacterial resistance to chlorhexidine. Oral Microbiol Immunol. 10: 319-320.

Grenier, D., and D. Mayrand. 1987. Functional characterization of extracellular vesicles produced by *Bacteroides gingivalis*. Infect Immun. 55: 111-117.

Gupta, D., T. N. Kirkland, S. Viriyakosol, and R. Dziarski. 1996. CD14 is a cell-activating receptor for bacterial peptidoglycan. J Biol Chem. 271: 23310-23316.

Gupta, D., Q. Wang, C. Vinson, and R. Dziarski. 1999. Bacterial peptidoglycan induces CD14-dependent activation of transcription factors CREB/ATF and AP-1. J Biol Chem. 274: 14012-14020.

Hajishengallis, G., and S. M. Michalek. 1999. Current status of a mucosal vaccine against dental caries. Oral Microbiol Immunol. 14: 1-20.

Hamada, S., H. Takada, T. Ogawa, T. Fujiwara, and J. Mihara. 1990. Lipopolysaccharides of oral anaerobes associated with chronic inflammation: chemical and immunomodulating properties. Int Rev Immunol. 6: 247-261.

Hamilton, R. E., Jr., and J. S. Giansanti. 1974. The Chediak-Higashi syndrome. Report of a case and review of the literature. Oral Surg Oral Med Oral Pathol. 37: 754-761.

Hart, T. C., and L. Shapira. 1994. Papillon-Lefevre syndrome. Periodontol 2000. 6: 88-100.

Haziot, A., E. Ferrero, F. Kontgen, N. Hijiya, S. Yamamoto, J. Silver, C. L. Stewart, and S. M. Goyert. 1996. Resistance to endotoxin shock and reduced dissemination of gram- negative bacteria in CD14-deficient mice. Immunity. 4: 407-414.

Heath, J. K., S. J. Atkinson, R. M. Hembry, J. J. Reynolds, and M. C. Meikle. 1987. Bacterial antigens induce collagenase and prostaglandin E_2 synthesis in human gingival fibroblasts through a primary effect on circulating mononuclear cells. Infect Immun. 55: 2148-2154.

Hedlund, M., C. Wachtler, E. Johansson, L. Hang, J. E. Somerville, R. P. Darveau, and C. Svanborg. 1999. P fimbriae-dependent, lipopolysaccharide-independent activation of epithelial cytokine responses. Mol Microbiol. 33: 693-703.

Helmerhorst, E. J., R. Hodgson, W. van 't Hof, E. C. Veerman, C. Allison, and A. V. Nieuw Amerongen. 1999. The effects of histatin-derived basic antimicrobial peptides on oral biofilms. J Dent Res. 78: 1245-1250.

Hiraoka, T., Y. Izumi, and T. Sueda. 1998. Immunochemical detection of CD14 on human gingival fibroblasts *in vitro*. Oral Microbiol Immunol. 13: 246-252.

Hoffmann, J. A., F. C. Kafatos, C. A. Janeway, and R. A. Ezekowitz. 1999. Phylogenetic perspectives in innate immunity. Science. 284: 1313-1318.

Holmskov, U., R. Malhotra, R. B. Sim, and J. C. Jensenius. 1994. Collectins: collagenous C-type lectins of the innate immune defense system. Immunol Today. 15: 67-74.

Holmstrup, P., and J. Westergaard. 1998. HIV infection and periodontal diseases. Periodontol 2000. 18: 37-46.

Huang, G. T., S. K. Haake, J. W. Kim, and N. H. Park. 1998. Differential expression of interleukin-8 and intercellular adhesion molecule-1 by human gingival epithelial cells in response to *Actinobacillus actinomycetemcomitans* or *Porphyromonas gingivalis* infection. Oral Microbiol Immunol. 13: 301-309.

Jainkittivong, A., D. A. Johnson, and C. K. Yeh. 1998. The relationship between salivary histatin levels and oral yeast carriage. Oral Microbiol Immunol. 13: 181-187.

Janeway, C. A., Jr. 1992. The immune system evolved to discriminate infectious nonself from noninfectious self. Immunol Today. 13: 11-16.

Janeway, C. A., Jr., U. Dianzani, P. Portoles, S. Rath, E. P. Reich, J. Rojo, J. Yagi, and D. B. Murphy. 1989. Cross-linking and conformational change in T-cell receptors: role in activation and in repertoire selection. Cold Spring Harb Symp Quant Biol. 54: 657-666.

Johne, B., and K. Bryn. 1986. Chemical composition and biological properties of a lipopolysaccharide from *Bacteroides intermedius*. Acta Pathol Microbiol Immunol Scand [B]. 94: 265-271.

Johne, B., I. Olsen, and K. Bryn. 1988. Fatty acids and sugars in lipoplysaccharides from *Bacteroides intermedius*, *Bacteroides gingivalis* and *Bacteroides loescheii*. Oral Microbiol Immunol. 3: 22-27.

Joseph, R., and G. D. Shockman. 1975. Synthesis and excretion of glycerol teichoic acid during growth of two streptococcal species. Infect Immun. 12: 333-338.

Juan, T. S., M. J. Kelley, D. A. Johnson, L. A. Busse, E. Hailman, S. D. Wright, and H. S. Lichenstein. 1995. Soluble CD14 truncated at amino acid 152 binds lipopolysaccharide (LPS) and enables cellular response to LPS. J Biol Chem. 270: 1382-1387.

Kalkwarf, K. L., and D. P. Gutz. 1981. Periodontal changes associated with chronic idiopathic neutropenia. Pediatr Dent. 3: 189-195.

Kamma, J. J., N. A. Lygidakis, and M. Nakou. 1998. Subgingival microflora and treatment in prepubertal periodontitis associated with chronic idiopathic neutropenia. J Clin Periodontol. 25: 759-765.

Kinoshita, T. 1991. Biology of complement: the overture. Immunol Today. 12: 291-295.

Kirikae, T., F. U. Schade, U. Zahringer, F. Kirikae, H. Brade, S. Kusumoto, T. Kusama, and E. T. Rietschel. 1994. The significance of the hydrophilic backbone and the hydrophobic fatty acid regions of lipid A for macrophage binding and cytokine induction. FEMS Immunol Med Microbiol. 8: 13-26.

Kirschning, C. J., H. Wesche, T. Merrill Ayres, and M. Rothe. 1998. Human toll-like receptor 2 confers responsiveness to bacterial lipopolysaccharide. J Exp Med. 188: 2091-2097.

Kitchens, R. L., and R. S. Munford. 1995. Enzymatically deacylated lipopolysaccharide (LPS) can antagonize LPS at multiple sites in the LPS recognition pathway. J Biol Chem. 270: 9904-9910.

Knox, K. W., and R. B. Parker. 1973. Isolation of a phenol-soluble endotoxin from *Leptotrichia buccalis*. Arch Oral Biol. 18: 85-93.

Kokeguchi, S., O. Tsutsui, K. Kato, and T. Matsumura. 1990. Isolation and characterization of lipopolysaccharide from *Centipeda periodontii* ATCC 35019. Oral Microbiol Immunol. 5: 108-112.

Kornman, K. S., R. C. Page, and M. S. Tonetti. 1997. The host response to the microbial challenge in periodontitis: assembling the players. Periodontol 2000. 14: 33-53.

Krieger, M. 1997. The other side of scavenger receptors: pattern recognition for host defense. Curr Opin Lipidol. 8: 275-280.

Krieger, M., S. Acton, J. Ashkenas, A. Pearson, M. Penman, and D. Resnick. 1993. Molecular flypaper, host defense, and atherosclerosis. Structure, binding properties, and functions of macrophage scavenger receptors. J Biol Chem. 268: 4569-4572.

Krisanaprakornkit, S., A. Weinberg, C. N. Perez, and B. A. Dale. 1998. Expression of the peptide antibiotic human beta-defensin 1 in cultured gingival epithelial cells and gingival tissue. Infect Immun. 66: 4222-4228.

Kumada, H., Y. Haishima, T. Umemoto, and K. Tanamoto. 1995. Structural study on the free lipid A isolated from lipopolysaccharide of *Porphyromonas gingivalis*. J Bacteriol. 177: 2098-2106.

Kumada, H., K. Watanabe, A. Nakamu, Y. Haishima, S. Kondo, K. Hisatsune, and T. Umemoto. 1997. Chemical and biological properties of lipopolysaccharide from *Selenomonas sputigena* ATCC 33150. Oral Microbiol Immunol. 12: 162-167.

Kumada, H., K. Watanabe, T. Umemoto, K. Kato, S. Kondo, and K. Hisatsune. 1989. Chemical and biological properties of lipopolysaccharide, lipid A and degraded polysaccharide from *Wolinella recta* ATCC 33238. J Gen Microbiol. 135: 1017-1025.

Kusunoki, T., E. Hailman, T. S. Juan, H. S. Lichenstein, and S. D. Wright. 1995. Molecules from *Staphylococcus aureus* that bind CD14 and stimulate innate immune responses. J Exp Med. 182: 1673-1682.

Laass, M. W., H. C. Hennies, S. Preis, H. P. Stevens, M. Jung, I. M. Leigh, T. F. Wienker, and A. Reis. 1997. Localisation of a gene for Papillon-Lefevre syndrome to chromosome 11q14-q21 by homozygosity mapping. Hum Genet. 101: 376-382.

Lai, C. H., M. A. Listgarten, and B. F. Hammond. 1981. Comparative ultrastructure of leukotoxic and non-leukotoxic strains of *Actinobacillus actinomycetemcomitans*. J Periodontal Res. 16: 379-389.

Lal, K., J. J. Pollock, R. P. d. Santarpia, H. M. Heller, H. W. Kaufman, J. Fuhrer, and R. T. Steigbigel. 1992. Pilot study comparing the salivary cationic protein concentrations in healthy adults and AIDS patients: correlation with antifungal activity. J Acquir Immune Defic Syndr. 5: 904-914.

Lalla, E., I. B. Lamster, and A. M. Schmidt. 1998. Enhanced interaction of advanced glycation end products with their cellular receptor RAGE: implications for the pathogenesis of accelerated periodontal disease in diabetes. Ann Periodontol. 3: 13-19.

Lamkin, M. S., and F. G. Oppenheim. 1993. Structural features of salivary function. Crit Rev Oral Biol Med. 4: 251-259.

Lamping, N., R. Dettmer, N. W. Schroder, D. Pfeil, W. Hallatschek, R. Burger, and R. R. Schumann. 1998. LPS-binding protein protects mice from septic shock caused by LPS or gram-negative bacteria. J Clin Invest. 101: 2065-2071.

Lamster, I. B. 1997. Evaluation of components of gingival crevicular fluid as diagnostic tests. Ann Periodontol. 2: 123-137.

Lamster, I. B., J. T. Grbic, D. A. Mitchell-Lewis, M. D. Begg, and A. Mitchell. 1998. New concepts regarding the pathogenesis of periodontal disease in HIV infection. Ann Periodontol. 3: 62-75.

Le Magrex, E., L. F. Jacquelin, J. Carquin, L. Brisset, and C. Choisy. 1993. Antiseptic activity of some antidental plaque chemicals on *Streptococcus mutans* biofilms. Pathol Biol (Paris). 41: 364-368.

Lemaitre, B., E. Nicolas, L. Michaut, J. M. Reichhart, and J. A. Hoffmann. 1996. The dorsoventral regulatory gene cassette spatzle/Toll/cactus controls the potent antifungal response in Drosophila adults. Cell. 86: 973-983.

Leturcq, D. J., A. M. Moriarty, G. Talbott, R. K. Winn, T. R. Martin, and R. J. Ulevitch. 1996. Antibodies against CD14 protect primates from endotoxin-induced shock. J Clin Invest. 98: 1533-1538.

Li, Y. H., and G. H. Bowden. 1994. The effect of environmental pH and fluoride from the substratum on the development of biofilms of selected oral bacteria. J Dent Res. 73: 1615-1626.

Lindemann, R. A., M. Kjeldsen, and M. Cabret. 1995. Effect of whole oral bacteria and extracted lipopolysaccharides on peripheral blood leukocyte interleukin-2 receptor expression. J Periodontal Res. 30: 264-271.

Listgarten, M. A. 1994. The structure of dental plaque. Periodontol 2000. 5: 52-65.

Listgarten, M. A., and C. H. Lai. 1979. Comparative ultrastructure of *Bacteroides melaninogenicus* subspecies. J Periodontal Res. 14: 332-340.

Lokki, M. L., and H. R. Colten. 1995. Genetic deficiencies of complement. Ann Med. 27: 451-459.

Loppnow, H., H. Brade, I. Durrbaum, C. A. Dinarello, S. Kusumoto, E. T. Rietschel, and H. D. Flad. 1989. IL-1 induction-capacity of defined lipopolysaccharide partial structures. J Immunol. 142: 3229-3238.

Lundgren, T., S. Renvert, P. N. Papapanou, and G. Dahlen. 1998. Subgingival microbial profile of Papillon-Lefevre patients assessed by DNA-probes. J Clin Periodontol. 25: 624-629.

Lyberg, T. 1982. Immunological and metabolical studies in two siblings with Papillon- Lefevre syndrome. J Periodontal Res. 17: 563-568.

MacFarlane, T. W. 1984. The oral ecology of patients with severe Sjogren's syndrome. Microbios. 41: 99-106.

Madianos, P. N., P. N. Papapanou, and J. Sandros. 1997. *Porphyromonas gingivalis* infection of oral epithelium inhibits neutrophil transepithelial migration. Infect Immun. 65: 3983-3990.

Majorana, A., L. D. Notarangelo, E. Savoldi, G. Gastaldi, and F. Lozada-Nur. 1999. Leukocyte adhesion deficiency in a child with severe oral involvement. Oral Surg Oral Med Oral Pathol Oral Radiol Endod. 87: 691-694.

Mandel, I. D., C. E. Barr, and L. Turgeon. 1992. Longitudinal study of parotid saliva in HIV-1 infection. J Oral Pathol Med. 21: 209-213.

Mandell, R. L., J. Dirienzo, R. Kent, K. Joshipura, and J. Haber. 1992. Microbiology of healthy and diseased periodontal sites in poorly controlled insulin dependent diabetics. J Periodontol. 63: 274-279.

Marcotte, H., and M. C. Lavoie. 1998. Oral microbial ecology and the role of salivary immunoglobulin A. Microbiol Mol Biol Rev. 62: 71-109.

Marsh, P. D., and D. J. Bradshaw. 1995. Dental plaque as a biofilm. J Ind Microbiol. 15: 169-175.

Mashimo, J., M. Yoshida, K. Ikeuchi, S. Hata, S. Arata, N. Kasai, K. Okuda, and I. Takazoe. 1985. Fatty acid composition and Shwartzman activity of lipopolysaccharides from oral bacteria. Microbiol Immunol. 29: 395-403.

Mashimo, P. A., Y. Yamamoto, J. Slots, B. H. Park, and R. J. Genco. 1983. The periodontal microflora of juvenile diabetics. Culture, immunofluorescence, and serum antibody studies. J Periodontol. 54: 420-430.

Masoud, H., S. T. Weintraub, R. Wang, R. Cotter, and S. C. Holt. 1991. Investigation of the structure of lipid A from *Actinobacillus actinomycetemcomitans* strain Y4 and human clinical isolate PO 1021-7. Eur J Biochem. 200: 775-781.

Mathews, M., H. P. Jia, J. M. Guthmiller, G. Losh, S. Graham, G. K. Johnson, B. F. Tack, and P. B. McCray, Jr. 1999. Production of beta-defensin antimicrobial peptides by the oral mucosa and salivary glands. Infect Immun. 67: 2740-2745.

Matsuki, Y., T. Yamamoto, and K. Hara. 1992. Detection of inflammatory cytokine messenger RNA (mRNA)-expressing cells in human inflamed gingiva by combined in situ hybridization and immunohistochemistry. Immunology. 76: 42-47.

Mayrand, D., and D. Grenier. 1989. Biological activities of outer membrane vesicles. Can J Microbiol. 35: 607-613.

McCoy, S. A., H. R. Creamer, M. Kawanami, and D. F. Adams. 1987. The concentration of lipopolysaccharide on individual root surfaces at varying times following *in vivo* root planing. J Periodontol. 58: 393-399.

McGee, J. M., M. A. Tucci, T. P. Edmundson, C. L. Serio, and R. B. Johnson. 1998. The relationship between concentrations of proinflammatory cytokines within gingiva and the adjacent sulcular depth. J Periodontol. 69: 865-871.

McGinley, M. D., L. O. Narhi, M. J. Kelley, E. Davy, J. Robinson, M. F. Rohde, S. D. Wright, and H. S. Lichenstein. 1995. CD14: physical properties and identification of an exposed site that is protected by lipopolysaccharide. J Biol Chem. 270: 5213-5218.

McMullen, J. A., M. Legg, R. Gottsegen, and R. Camerini-Davalos. 1967. Microangiopathy within the gingival tissues of diabetic subjects with special reference to the prediabetic state. Periodontics. 5: 61-69.

Medzhitov, R., and C. A. Janeway, Jr. 1997. Innate immunity: impact on the adaptive immune response. Curr Opin Immunol. 9: 4-9.

Medzhitov, R., P. Preston-Hurlburt, and C. A. Janeway, Jr. 1997. A human homologue of the *Drosophila* Toll protein signals activation of adaptive immunity [see comments]. Nature. 388: 394-397.

Mills, E. L., and P. G. Quie. 1980. Congenital disorders of the function of polymorphonuclear neutrophils. Rev Infect Dis. 2: 505-517.

Moore, J., M. Wilson, and J. B. Kieser. 1986. The distribution of bacterial lipopolysaccharide (endotoxin) in relation to periodontally involved root surfaces. J Clin Periodontol. 13: 748-751.

Moore, W. E., and L. V. Moore. 1994. The bacteria of periodontal diseases. Periodontol 2000. 5: 66-77.

Morgan, B. P., and M. J. Walport. 1991. Complement deficiency and disease. Immunol Today. 12: 301-306.

Morrison, D. C., and J. L. Ryan. 1992. Bacterial Endotoxic Lipopolysaccharides. CRC Press, Boca Raton, FL.

Moskow, B. S., and A. M. Polson. 1991. Histologic studies on the extension of the inflammatory infiltrate in human periodontitis. J Clin Periodontol. 18: 534-542.

211

Moughal, N. A., E. Adonogianaki, M. H. Thornhill, and D. F. Kinane. 1992. Endothelial cell leukocyte adhesion molecule-1 (ELAM-1) and intercellular adhesion molecule-1 (ICAM-1) expression in gingival tissue during health and experimentally-induced gingivitis. J Periodontal Res. 27: 623-630.

Mug-Opstelten, D., and B. Witholt. 1978. Preferential release of new outer membrane fragments by exponentially growing *Escherichia coli*. Biochim Biophys Acta. 508: 287-295.

Munford, R. S., and C. L. Hall. 1986. Detoxification of bacterial lipopolysaccharides (endotoxins) by a human neutrophil enzyme. Science. 234: 203-205.

Nauts, H. C., and G. A. Fowler. 1953. A review of the influence of bacterial infections and of bacterial products (Coley's toxin) on malignant tumors in man. Acta Med. Scand. 145: 1-103.

Newman, H. N., and D. C. Rule. 1983. Plaque-host imbalance in severe periodontitis. A discussion based on two cases. J Clin Periodontol. 10: 137-147.

Newman, M. G., V. Grinenco, M. Weiner, I. Angel, H. Karge, and R. Nisengard. 1978. Predominant microbiota associated with periodontal health in the aged. J Periodontol. 49: 553-559.

Nowotny, A., U. H. Behling, B. Hammond, C. H. Lai, M. Listgarten, P. H. Pham, and F. Sanavi. 1982. Release of toxic microvesicles by *Actinobacillus actinomycetemcomitans*. Infect Immun. 37: 151-154.

Nylander, K., B. Danielsen, O. Fejerskov, and E. Dabelsteen. 1993. Expression of the endothelial leukocyte adhesion molecule-1 (ELAM-1) on endothelial cells in experimental gingivitis in humans. J Periodontol. 64: 355-357.

Offenbacher, S., P. A. Heasman, and J. G. Collins. 1993. Modulation of host PGE_2 secretion as a determinant of periodontal disease expression. J Periodontol. 64: 432-444.

Ogawa, T. 1993. Chemical structure of lipid A from *Porphyromonas* (*Bacteroides*) *gingivalis* lipopolysaccharide. FEBS Lett. 332: 197-201.

Ogawa, T. 1994. Immunobiological properties of chemically defined lipid A from lipopolysaccharide of *Porphyromonas* (*Bacteroides*) *gingivalis*. Eur J Biochem. 219: 737-742.

Onoue, S., M. Niwa, Y. Isshiki, and K. Kawahara. 1996. Extraction and characterization of the smooth-type lipopolysaccharide from *Fusobacterium nucleatum* JCM 8532 and its biological activities. Microbiol Immunol. 40: 323-331.

Oppenheim, F. G., T. Xu, F. M. McMillian, S. M. Levitz, R. D. Diamond, G. D. Offner, and R. F. Troxler. 1988. Histatins, a novel family of histidine-rich proteins in human parotid secretion. Isolation, characterization, primary structure, and fungistatic effects on *Candida albicans*. J Biol Chem. 263: 7472-7477.

Overholser, C. D., D. E. Peterson, L. T. Williams, and S. C. Schimpff. 1982. Periodontal infection in patients with acute nonlymphocyte leukemia. Prevalence of acute exacerbations. Arch Intern Med. 142: 551-554.

Page, R. C., and H. E. Schroeder. 1976. Pathogenesis of inflammatory periodontal disease. A summary of current work. Lab Invest. 34: 235-249.

Paquette, D. W., G. S. Waters, V. L. Stefanidou, H. P. Lawrence, P. M. Friden, S. M. O'Connor, J. D. Sperati, F. G. Oppenheim, L. H. Hutchens, and R. C. Williams.

1997. Inhibition of experimental gingivitis in beagle dogs with topical salivary histatins. J Clin Periodontol. 24: 216-222.

Pernu, H. E., U. H. Pajari, and M. Lanning. 1996. The importance of regular dental treatment in patients with cyclic neutropenia. Follow-up of 2 cases. J Periodontol. 67: 454-459.

Perry, M. B., L. L. MacLean, R. Gmur, and M. E. Wilson. 1996. Characterization of the O-polysaccharide structure of lipopolysaccharide from *Actinobacillus actinomycetemcomitans* serotype b. Infect Immun. 64: 1215-1219.

Perry, M. B., L. M. MacLean, J. R. Brisson, and M. E. Wilson. 1996. Structures of the antigenic O-polysaccharides of lipopolysaccharides produced by *Actinobacillus actinomycetemcomitans* serotypes a, c, d and e. Eur J Biochem. 242: 682-688.

Peterson, D. E., G. E. Minah, C. D. Overholser, J. B. Suzuki, L. G. DePaola, D. M. Stansbury, L. T. Williams, and S. C. Schimpff. 1987. Microbiology of acute periodontal infection in myelosuppressed cancer patients. J Clin Oncol. 5: 1461-1468.

Pohlman, T. H., R. S. Munford, and J. M. Harlan. 1987. Deacylated lipopolysaccharide inhibits neutrophil adherence to endothelium induced by lipopolysaccharide *in vitro*. J Exp Med. 165: 1393-1402.

Poirier, T. P., R. Mishell, C. L. Trummel, and S. C. Holt. 1983. Biological and chemical comparison of butanol- and phenol-water extracted lipopolysaccharide from *Capnocytophaga sputigena*. J Periodontal Res. 18: 541-557.

Poltorak, A., X. He, I. Smirnova, M. Y. Liu, C. V. Huffel, X. Du, D. Birdwell, E. Alejos, M. Silva, C. Galanos, M. Freudenberg, P. Ricciardi-Castagnoli, B. Layton, and B. Beutler. 1998. Defective LPS signaling in C3H/HeJ and C57BL/10ScCr mice: mutations in Tlr4 gene. Science. 282: 2085-2088.

Pugin, J., I. D. Heumann, A. Tomasz, V. V. Kravchenko, Y. Akamatsu, M. Nishijima, M. P. Glauser, P. S. Tobias, and R. J. Ulevitch. 1994. CD14 is a pattern recognition receptor. Immunity. 1: 509-516.

Pugin, J., C. C. Schurer-Maly, D. Leturcq, A. Moriarty, R. J. Ulevitch, and P. S. Tobias. 1993. Lipopolysaccharide activation of human endothelial and epithelial cells is mediated by lipopolysaccharide-binding protein and soluble CD14. Proc Natl Acad Sci U S A. 90: 2744-2748.

Qureshi, S. T., L. Lariviere, G. Leveque, S. Clermont, K. J. Moore, P. Gros, and D. Malo. 1999. Endotoxin-tolerant mice have mutations in Toll-like receptor 4 (Tlr4) [see comments] [published erratum appears in J Exp Med 1999 May 3;189(9): following 1518]. J Exp Med. 189: 615-625.

Raj, P. A., M. Edgerton, and M. J. Levine. 1990. Salivary histatin 5: dependence of sequence, chain length, and helical conformation for candidacidal activity. J Biol Chem. 265: 3898-3905.

Rateitschak-Pluss, E. M., and H. E. Schroeder. 1984. History of periodontitis in a child with Papillon-Lefevre syndrome. A case report. J Periodontol. 55: 35-46.

Rieder, G., R. A. Hatz, A. P. Moran, A. Walz, M. Stolte, and G. Enders. 1997. Role of adherence in interleukin-8 induction in *Helicobacter pylori*-associated gastritis. Infect Immun. 65: 3622-3630.

Roberts, F. A., R. D. Hockett, Jr., R. P. Bucy, and S. M. Michalek. 1997. Quantitative assessment of inflammatory cytokine gene expression in chronic adult periodontitis.

Oral Microbiol Immunol. 12: 336-344.

Roberts, M. W., and J. C. Atkinson. 1990. Oral manifestations associated with leukocyte adhesion deficiency: a five-year case study. Pediatr Dent. 12: 107-111.

Robertson, P. B., B. F. Mackler, T. E. Wright, and B. M. Levy. 1980. Periodontal status of patients with abnormalities of the immune system. II. Observations over a 2-year period. J Periodontol. 51: 70-73.

Rock, F. L., G. Hardiman, J. C. Timans, R. A. Kastelein, and J. F. Bazan. 1998. A family of human receptors structurally related to *Drosophila* Toll. Proc Natl Acad Sci U S A. 95: 588-593.

Rothfield, L., and M. Pearlman-Kothencz. 1969. Synthesis and assembly of bacterial membrane components. A lipopolysaccharide-phospholipid-protein complex excreted by living bacteria. J Mol Biol. 44: 477-492.

Russell, R. R., K. G. Johnson, and I. J. McDonald. 1975. Envelope proteins in Neisseria. Can J Microbiol. 21: 1519-1534.

Sastrowijoto, S. H., P. Hillemans, T. J. van Steenbergen, L. Abraham-Inpijn, and J. de Graaff. 1989. Periodontal condition and microbiology of healthy and diseased periodontal pockets in type 1 diabetes mellitus patients. J Clin Periodontol. 16: 316-322.

Sastry, K., and R. A. Ezekowitz. 1993. Collectins: pattern recognition molecules involved in first line host defense [published erratum appears in *Curr Opin Immunol* 1993 Aug;5(4): 566]. Curr Opin Immunol. 5: 59-66.

Schenkels, L. C., E. C. Veerman, and A. V. Nieuw Amerongen. 1995. Biochemical composition of human saliva in relation to other mucosal fluids. Crit Rev Oral Biol Med. 6: 161-175.

Schifferle, R. E., M. S. Reddy, J. J. Zambon, R. J. Genco, and M. J. Levine. 1989. Characterization of a polysaccharide antigen from *Bacteroides gingivalis*. J Immunol. 143: 3035-3042.

Schroeder, H. E., R. A. Seger, H. U. Keller, and E. M. Rateitschak-Pluss. 1983. Behavior of neutrophilic granulocytes in a case of Papillon-Lefevre syndrome. J Clin Periodontol. 10: 618-635.

Schumann, R. R., S. R. Leong, G. W. Flaggs, P. W. Gray, S. D. Wright, J. C. Mathison, P. S. Tobias, and R. J. Ulevitch. 1990. Structure and function of lipopolysaccharide binding protein. Science. 249: 1429-1431.

Schwartz, J., F. L. Stinson, and R. B. Parker. 1972. The passage of tritiated bacterial endotoxin across intact gingival crevicular epithelium. J Periodontol. 43: 270-276.

Sellati, T. J., D. A. Bouis, R. L. Kitchens, R. P. Darveau, J. Pugin, R. J. Ulevitch, S. C. Gangloff, S. M. Goyert, M. V. Norgard, and J. D. Radolf. 1998. *Treponema pallidum* and *Borrelia burgdorferi* lipoproteins and synthetic lipopeptides activate monocytic cells via a CD14-dependent pathway distinct from that used by lipopolysaccharide. J Immunol. 160: 5455-5464.

Shapiro, R. A., M. D. Cunningham, K. Ratcliffe, C. Seachord, J. Blake, J. Bajorath, A. Aruffo, and R. P. Darveau. 1997. Identification of CD14 residues involved in specific lipopolysaccharide recognition. Infect Immun. 65: 293-297.

Shordone, L., L. Ramaglia, A. Barone, R. N. Ciaglia, A. Tenore, and V. J. Iacono. 1995. Periodontal status and selected cultivable anaerobic microflora of insulin-dependent juvenile diabetics. J Periodontol. 66: 452-461.

Shurin, S. B., S. S. Socransky, E. Sweeney, and T. P. Stossel. 1979. A neutrophil disorder induced by *Capnocytophaga*, a dental micro-organism. N Engl J Med. 301: 849-854.

Sissons, C. H., L. Wong, and T. W. Cutress. 1995. Patterns and rates of growth of microcosm dental plaque biofilms. Oral Microbiol Immunol. 10: 160-167.

Slots, J. 1977. Microflora in the healthy gingival sulcus in man. Scand J Dent Res. 85: 247-254.

Smalley, J. W., D. Mayrand, and D. Grenier. 1993. Vesicles, p. 259-292. *In* H. N. Shah and D. Myrand (ed.), Biology of the species *Porphyromonas gingivalis*. CRC Press, Boca Raton, FL.

Socransky, S. S., and A. D. Haffajee. 1994. Evidence of bacterial etiology: a historical perspective. Periodontol 2000. 5: 7-25.

Socransky, S. S., A. D. Haffajee, M. A. Cugini, C. Smith, and R. L. Kent, Jr. 1998. Microbial complexes in subgingival plaque. J Clin Periodontol. 25: 134-144.

Somerville, J. E., Jr., L. Cassiano, B. Bainbridge, M. D. Cunningham, and R. P. Darveau. 1996. A novel *Escherichia coli* lipid A mutant that produces an antiinflammatory lipopolysaccharide. J Clin Invest. 97: 359-365.

Soskolne, W. A. 1998. Epidemiological and clinical aspects of periodontal diseases in diabetics. Ann Periodontol. 3: 3-12.

Soukka, T., M. Lumikari, and J. Tenovuo. 1991. Combined bactericidal effect of human lactoferrin and lysozyme against *Streptococcus mutans* serotype c. Microb. Ecol. Health Dis. 4: 259-264.

Soukka, T., M. Lumikari, and J. Tenovuo. 1991. Combined inhibitory effect of lactoferrin and lactoperoxidase system on the viability of Streptococcus mutans, serotype c. Scand J Dent Res. 99: 390-396.

Springer, T. A. 1994. Traffic signals for lymphocyte recirculation and leukocyte emigration: the multistep paradigm. Cell. 76: 301-314.

Standiford, T. J., D. A. Arenberg, J. M. Danforth, S. L. Kunkel, G. M. VanOtteren, and R. M. Strieter. 1994. Lipoteichoic acid induces secretion of interleukin-8 from human blood monocytes: a cellular and molecular analysis. Infect Immun. 62: 119-125.

Stelter, F., M. Bernheiden, R. Menzel, R. S. Jack, S. Witt, X. Fan, M. Pfister, and C. Schutt. 1997. Mutation of amino acids 39-44 of human CD14 abrogates binding of lipopolysaccharide and *Escherichia coli*. Eur J Biochem. 243: 100-109.

Sugawara, S., R. Arakaki, H. Rikiishi, and H. Takada. 1999. Lipoteichoic acid acts as an antagonist and an agonist of lipopolysaccharide on human gingival fibroblasts and monocytes in a CD14-dependent manner. Infect Immun. 67: 1623-1632.

Sveen, K. 1977. The capacity of lipopolysaccharides from bacteroides, fusobacterium and *Veillonella* to produce skin inflammation and the local and generalized Shwartzman reaction in rabbits. J Periodontal Res. 12: 340-350.

Takada, H., and S. Kotani. 1992. Bacterial Endotoxic Lipopolysaccharides, p. 107-130. *In* D. C. Morrison and J. L. Ryan (ed.), Molecular Biochemistry and Cellular Biology, vol. 1. CRC Press, Boca Raton.

Tenovuo, J. 1998. Antimicrobial function of human saliva—how important is it for oral health? Acta Odontol Scand. 56: 250-256.

Tenovuo, J., Z. Moldoveanu, J. Mestecky, K. M. Pruitt, and B. M. Rahemtulla. 1982. Interaction of specific and innate factors of immunity: IgA enhances the antimicrobial effect of the lactoperoxidase system against Streptococcus mutans. J Immunol. 128: 726-731.

Theilade, E. 1996. The experimental gingivitis studies: the microbiological perspective. J Dent Res. 75: 1434-1438.

Tinanoff, N., J. M. Tanzer, K. S. Kornman, and E. G. Maderazo. 1986. Treatment of the periodontal component of Papillon-Lefevre syndrome. J Clin Periodontol. 13: 6-10.

Tolo, K. 1991. Periodontal disease mechanisms in immunocompromised patients. J Clin Periodontol. 18: 431-435.

Tonetti, M. S. 1998. Cigarette smoking and periodontal diseases: etiology and management of disease. Ann Periodontol. 3: 88-101.

Tonetti, M. S. 1997. Molecular factors associated with compartmentalization of gingival immune responses and transepithelial neutrophil migration. J Periodontal Res. 32: 104-109.

Tonetti, M. S., M. A. Imboden, L. Gerber, N. P. Lang, J. Laissue, and C. Mueller. 1994. Localized expression of mRNA for phagocyte-specific chemotactic cytokines in human periodontal infections. Infect Immun. 62: 4005-4014.

Tonetti, M. S., M. A. Imboden, and N. P. Lang. 1998. Neutrophil migration into the gingival sulcus is associated with transepithelial gradients of interleukin-8 and ICAM-1. J Periodontol. 69: 1139-1147.

Toth, B. B., J. W. Martin, and T. J. Fleming. 1990. Oral complications associated with cancer therapy. An M. D. Anderson Cancer Center experience. J Clin Periodontol. 17: 508-515.

Tuyau, J. E., and W. Sims. 1983. Aspects of the pathogenicity of some oral and other haemophili. J Med Microbiol. 16: 467-475.

Van Dyke, T. E., M. A. Taubman, J. L. Ebersole, A. D. Haffajee, S. S. Socransky, D. J. Smith, and R. J. Genco. 1984. The Papillon-Lefevre syndrome: neutrophil dysfunction with severe periodontal disease. Clin Immunol Immunopathol. 31: 419-429.

Vaughan, A. G., T. P. Vrahopoulos, F. Joachim, K. Sati, P. Barber, and H. N. Newman. 1990. A case report of chronic neutropenia: clinical and ultrastructural findings. J Clin Periodontol. 17: 435-445.

Viriyakosol, S., and T. N. Kirkland. 1995. A region of human CD14 required for lipopolysaccharide binding. J Biol Chem. 270: 361-368.

Waldrop, T. C., D. C. Anderson, W. W. Hallmon, F. C. Schmalstieg, and R. L. Jacobs. 1987. Periodontal manifestations of the heritable Mac-1, LFA-1, deficiency syndrome. Clinical, histopathologic and molecular characteristics. J Periodontol. 58: 400-416.

Weinberg, A., S. Krisanaprakornkit, and B. A. Dale. 1998. Epithelial antimicrobial peptides: review and significance for oral applications. Crit Rev Oral Biol Med. 9: 399-414.

Weintraub, A., U. Zahringer, H. W. Wollenweber, U. Seydel, and E. T. Rietschel. 1989. Structural characterization of the lipid A component of *Bacteroides fragilis* strain NCTC 9343 lipopolysaccharide. Eur J Biochem. 183: 425-431.

Weis, W. I., K. Drickamer, and W. A. Hendrickson. 1992. Structure of a C-type mannose-binding protein complexed with an oligosaccharide. Nature. 360: 127-134.

Williams, G. D., and S. C. Holt. 1985. Characteristics of the outer membrane of selected oral Bacteroides species. Can J Microbiol. 31: 238-250.

Williams, M. J., A. Rodriguez, D. A. Kimbrell, and E. D. Eldon. 1997. The 18-wheeler mutation reveals complex antibacterial gene regulation in Drosophila host defense. EMBO J. 16: 6120-6130.

Wilson, M. 1996. Susceptibility of oral bacterial biofilms to antimicrobial agents. J Med Microbiol. 44: 79-87.

Wilson, M., J. Moore, and J. B. Kieser. 1986. Identity of limulus amoebocyte lysate-active root surface materials from periodontally involved teeth. J Clin Periodontol. 13: 743-747.

Woo, D. D., S. C. Holt, and E. R. Leadbetter. 1979. Ultrastructure of Bacteroides species: *Bacteroides asaccharolyticus, Bacteroides fragilis, Bacteroides melaninogenicus* subspecies *melaninogenicus*, and *B. melaninogenicus* subspecies *intermedius*. J Infect Dis. 139: 534-546.

Wooten, R. M., T. B. Morrison, J. H. Weis, S. D. Wright, R. Thieringer, and J. J. Weis. 1998. The role of CD14 in signaling mediated by outer membrane lipoproteins of *Borrelia burgdorferi*. J Immunol. 160: 5485-5492.

Wright, S. D. 1995. CD14 and innate recognition of bacteria. J Immunol. 155: 6-8.

Wright, S. D. 1999. Toll, a new piece in the puzzle of innate immunity [comment]. J Exp Med. 189: 605-609.

Wright, S. D., R. A. Ramos, P. S. Tobias, R. J. Ulevitch, and J. C. Mathison. 1990. CD14, a receptor for complexes of lipopolysaccharide (LPS) and LPS binding protein [see comments]. Science. 249: 1431-1433.

Yalda, B., S. Offenbacher, and J. G. Collins. 1994. Diabetes as a modifier of periodontal disease expression. Periodontol 2000. 6: 37-49.

Yamazaki, K., F. Ikarashi, T. Aoyagi, K. Takahashi, T. Nakajima, K. Hara, and G. J. Seymour. 1992. Direct and indirect effects of *Porphyromonas gingivalis* lipopolysaccharide on interleukin-6 production by human gingival fibroblasts. Oral Microbiol Immunol. 7: 218-224.

Yang, R. B., M. R. Mark, A. Gray, A. Huang, M. H. Xie, M. Zhang, A. Goddard, W. I. Wood, A. L. Gurney, and P. J. Godowski. 1998. Toll-like receptor-2 mediates lipopolysaccharide-induced cellular signalling [see comments]. Nature. 395: 284-288.

Yoshimura, A., E. Lien, R. R. Ingalls, E. Tuomanen, R. Dziarski, and D. Golenbock. 1999. Cutting edge: recognition of Gram-positive bacterial cell wall components by the innate immune system occurs via Toll-like receptor 2. J Immunol. 163: 1-5.

Zambon, J. J., S. G. Grossi, E. E. Machtei, A. W. Ho, R. Dunford, and R. J. Genco. 1996. Cigarette smoking increases the risk for subgingival infection with periodontal pathogens. J Periodontol. 67: 1050-1054.

Zambon, J. J., H. Reynolds, J. G. Fisher, M. Shlossman, R. Dunford, and R. J. Genco. 1988. Microbiological and immunological studies of adult periodontitis in patients with noninsulin-dependent diabetes mellitus. J Periodontol. 59: 23-31.

Zatz, R., and B. M. Brenner. 1986. Pathogenesis of diabetic microangiopathy. The hemodynamic view. Am J Med. 80: 443-453.

Zollinger, W. D., D. L. Kasper, B. J. Veltri, and M. S. Artenstein. 1972. Isolation and characterization of a native cell wall complex from *Neisseria meningitidis*. Infect Immun. 6: 835-851.

From: *Oral Bacterial Ecology: The Molecular Basis*
ISBN 1-898486-22-0 ©2000 Horizon Scientific Press, Wymondham, U.K.

5

ECOLOGICAL BASIS FOR DENTAL CARIES

Ian R. Hamilton

Contents

Introduction

In recognition of W. D. Miller's historic book "The microorganisms of the human mouth" published in 1890, an appropriate sub-title for this Chapter might be "From Miller to the Millenium". This book resulted from Miller's work in Robert Koch's laboratory in Berlin in which he brought together the idea of acid and microorganisms in his "chemicoparasitic theory" of dental caries. While earlier theories had implicated "parasites" in caries etiology, Miller was the first to establish the role of oral bacteria in dental plaque in the demineralizing tooth enamel. He showed that the degradation of carbohydrate-containing foods resulted in acid formation and was able to demonstrate this process *in vitro* with isolated oral bacteria and extracted teeth. He also visualized that the final destruction of enamel and dentin was accompanied by proteolysis of the organic matrix of enamel. This led to Miller's major conclusion that dental caries was caused by multiple species of oral bacteria, in what Loesche (1982) later described as the "non-specific plaque hypothesis". In fact, it can be argued that the practice of toothbrushing, flossing and professional toothcleaning as a means of preventing dental diseases have their basis in Miller's original 1890 experiments.

As one would expect, however, a vast amount of newer information is available on the caries process, particularly in relation to the bacteria associated with this disease, including the manner in which they adhere to the tooth surface, the metabolic processes that result in acid formation in the presence of carbohydrate substrates, and the resulting physical-chemical reactions with tooth enamel. Even though the recent studies are providing information on many events associated with caries development in finer detail, some of it at the molecular level, there is still a need on occasion to stand back and view the basic elements of the caries process in light of this new information. This discussion can be enhanced by examining situations where caries has occurred and where it has not occurred. Good examples of this are approximal sites on adjacent teeth where caries has occurred in one surface, but not on an adjacent tooth only a millimeter away. For example, in Figure 1 [A], a lesion is evident on tooth #35 of an 11 year old male patient, while the adjacent site on tooth #36 is free of caries. Another example (Figure 1[B]) is that of two approximal lesions on tooth #45 of a 28 year male patient with no caries on the adjacent teeth (#44 and #46). As we will see later, these two apparently similar situations probably occurred for different reasons, but in order to explain these lesions, we need to understand the nature of the site involved, the bacteria that inhabit that site, and the local environment. In other words, we need to understand the ecology of the specific site in question. Before a discussion of the microbial ecology of dental plaque, it is perhaps useful to review some of the salient features of the caries process.

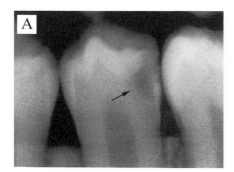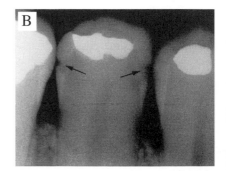

Figure 1. Approximal caries lesions on one tooth and not on the adjacent approximal site. [A] Lesion on tooth #35 of an eleven year old male patient, [B] Mesial and distal lesions on tooth #45 of a 28 year old male patient. (Photographs courtesy of B. Cleghorn, University of Manitoba)

A considerable literature exists on microbial ecology, dental caries, the oral microflora associated with dental plaque and caries lesions, and the pertinent metabolism associated with these bacteria. Consequently, it will not be possible in this chapter to cite all of the relevant references; however, attempts have been made to cite key papers and reviews, particularly, the more current references which will provide the reader with an avenue to trace the important studies that relate to the ecology of dental caries.

Dental Caries

Dental caries is a complex and dynamic process that involves the progressive destruction of tooth enamel, dentine and cementum by the action of bacteria in dental plaque. The major form of tooth 'destruction' is demineralization of tooth hydroxyapatite by acids generated during carbohydrate metabolism by the acidogenic oral microflora, although the disease process inevitably involves the degradation of the organic matrix as well. Acid formation drives the dissolution of calcium and phosphate in the hydroxyapatite (HA) crystal structure (Reaction 1, Equation 1), provided that the pH remains below the 'critical pH' of ~5.5 - 5.3, which is defined as the pH at which saliva and the fluid phase of plaque is saturated with respect to calcium and phosphate (Larsen and Bruun, 1994). Aqueous plaque fluid in the interstitial spaces between bacteria in the plaque biofilm is very important because of its intimate association with the acid-generating bacterial cells in the film and proximity to the enamel surface.

$$Ca_{10}(PO_4)_6(OH)_2 \quad \underset{base\ (2)}{\overset{acid\ (1)}{\rightleftharpoons}} \quad 10\ Ca^{2+} + 6PO_4^{3-} + 2OH^- \qquad \text{(Equation 1)}$$

Detailed examination of the progression of the caries process in enamel reveals the incremental microscopic erosion of the periphery of surface HA crystals with small increases in enamel porosity (Thylstrup and Fejerskov, 1994). With a consistent acid challenge over time, a larger number of crystals are eroded; however, as saliva is supersaturated with calcium and phosphate, the surface mineral becomes less porous than the subsurface and the lesion develops beneath an apparently intact enamel surface. Histological examination of enamel caries over time reveals that the subsurface demineralized zone will reach the underlying dentin before the lesion is visible on a bite-wing radiograph, and this zone will continue to enlarge before the visual appearance of the white spot, or incipient lesion (Figure 2, Stages 2-4) (Chapter 10, Nikiforuk, 1985). The critical and irreversible step in this process is the progression of the incipient lesion to an overt lesion in which cavitation of the surface mineral layer has occurred (Stage 5). Extensive clinical research (Loesche, 1986) has indicated a differential rate of caries formation on various tooth surfaces with the rate for occlusal fissures > approximal surfaces of molars and pre-molars > approximal surfaces of maxillary incisors > approximal surfaces of mandibular incisors.

Decalcification by acid from oral bacteria is dependent on a variety of factors, including the concentration of the carbon source, the amount and activity of the plaque microflora, the flow rate of saliva, and the nature of the site on the tooth

221

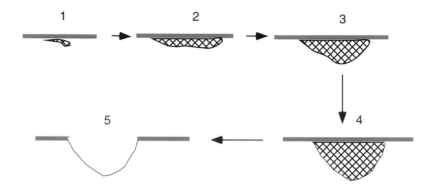

Figure 2. Diagrammatic representation of the progression of a caries lesion to an overt lesion (Stage 5).

surface. The relationship of some of these factors was examined over 50 years ago in a classic study by R. M. Stephan, who gave a 10% glucose rinse to subjects with increasing degrees of caries activity and then monitored the pH on the enamel surface with a 'touch' antimony electrode (Stephan, 1944). The pH vs time 'Stephan Curves' of 3-4 day old plaque on the labial surface of the maxillary anterior teeth (Figure 3) revealed five useful pieces of information in relation to caries and acid formation by human dental plaque: (1) The 'resting' plaque pH, obtained prior to the addition of the glucose rinse, was inversely related to caries activity, with lower values as the activity increased such that, even without added sugar, those individuals in this study with extreme caries activity had resting pH values at or near the critical pH. (2) The rate of acid formation immediately following sugar addition (Phase I) was not significantly different between the five groups, indicating similar metabolic capacity by the plaque microflora (see later). (3) The pH minimum (Phase II), like the resting plaque pH, was inversely related to caries activity, with all groups except the 'caries-free' and 'caries inactive' exhibiting pH minima below the critical pH. (4) In all groups, Phase III was initiated by an increase in pH, presumably because the glucose was either depleted, or the plaque bacteria were inhibited by the low pH, and glucose and acid were cleared through the action of saliva. (5) The final observation was that the amount of time the enamel was exposed to acid at pH values below the critical pH increased with increasing caries activity.

Fortunately, as seen in Figure 3, the consumption of readily-degradable carbohydrate substrates by plaque bacteria in subjects with less severe caries activity is followed by pH increases above the critical pH. Thus, provided calcium and phosphate are available at the sites of the decay, remineralization of the enamel can occur as the pH becomes more basic (Reaction 2, Equation 1). One can, therefore, envisage continual cycles of demineralization and remineralization in individuals with moderate to low caries activity following the consumption of carbohydrate-containing foods. Even when the equilibrium in Equation 1 is to the right, resulting in extensive subsurface demineralization (Stage 4, Figure 2), this process can be reversed. For example, ultrastructural studies of extracted non-carious teeth have

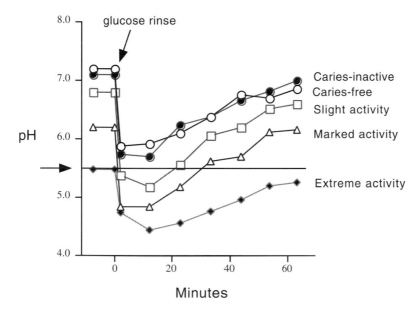

Figure 3. The pH response of *in vivo* dental plaque following a 10% glucose rinse in caries-free subjects and those with increasing caries activity rinse as measured by a touch pH electrode. Arrow indicates 'critical pH'. Data derived from Stephan (1944).

revealed HA crystals that are larger and more electron dense than those of normal enamel, indicating remineralization of previous caries lesions (Silverstone *et al.,* 1981).

Another important factor in discussing the process of dental caries is the matter of time. Keeping in mind the basic concepts of Equation 1, active demineralization of enamel can be interrupted by quiescent periods characterized by no loss of mineral or even some remineralization of lesions. As one might expect, the overall process is highly variable between individuals and even at specific sites in the same mouth (Figure 1). In spite of these differences, certain sites on the dentition are more susceptible to lesion development and certain situations can lead to predictions of high caries activity. For example, individuals afflicted with xerostomia because of aplasia, irradiation of the salivary glands, or Sjögren's syndrome, are very prone to dental caries due to the reduction in the saliva output of 90% or more. Without an effective preventive program, such subjects can develop rampant caries within three months (Dreizen and Brown, 1976). Nursing caries (Ripa, 1978), particularly in some populations of pre-school children, can also occur rapidly resulting in massive destruction of the dentition (Milnes, 1987; Milnes and Bowden, 1985). Enamel dissolution can be initiated very rapidly under conditions promoting acid retention as observed in an *in situ* study of children having teeth extracted during orthodontic treatment (Boyar *et al.,* 1989). Banded teeth were extracted over a period of 14 days and histological examination revealed that, in some cases, enamel dissolution could be observed within two days of band application.

The Plaque Ecosystems

An ecosystem is comprised of biotic (biological) and abiotic (environment) components located at a specific site, or habitat. Thus, reference to 'plaque ecosystems' in this chapter indicates a group of sites, on or associated with the tooth surface, each of which possesses a specific collection or community of oral bacteria under the influence of the environment of the specific site. Conventionally, dental plaque is divided into three types depending on whether it is above (supragingival), at (gingival), or below (subgingival) the gingival margin. Most of the discussion on caries ecology will focus on the microbial activity at the various sites that comprise supragingival plaque, although it should be recognized that the recession of the gingival margin can expose the root surface to caries activity. In addition, the architecture of a single tooth provides a considerable diversity with smooth surfaces, interproximal areas, occusal pits and fissures, each with a specific disposition to caries activity. Nevertheless, the key factor in a discussion of the microbial ecology of caries is the recognition that not all sites are equally susceptible to the disease, and each small individual site on the tooth surface should be considered a unique ecological entity. Consequently, the development of dental caries is a 'site-specific' process (see Figure 1). One should add that normal dental plaque in ecological homeostasis is a protective entity suppressing the overgrowth by opportunistic pathogens.

At first glance, the environment at the tooth surface would appear to be fairly simple with plenty of moisture, air and the periodic input of nutrients during food consumption. The environment is; however, a much more complex factor that must be discussed in relation to the location of the site. For example, food sources for supragingival plaque include saliva and gingival crevicular fluid, as well as dietary components, while subgingival plaque depends predominately on host-derived components of crevicular fluid, which has a composition similar to serum. Even within the supragingival group of sites, there is considerable variability in access to nutrients, with significant differences between the microflora in fissures, approximal sites or those on the smooth surface near the gingival margin. Even at the microscopic level, those 'pioneer' organisms present on the tooth surface during the initial stages of plaque formation become increasingly deprived of nutrients as the plaque develops above and around them, and the competition for utilizable nutrients becomes more intense. Within a large occusal fissure, the access to any nutrient source for those organisms in the deeper regions is even more severe, with starvation and cell death predictable.

The mouth is an ideal habitat for the growth of bacteria because of the consistent temperature, a variety of soft and hard tissue surfaces to grow on, and a variable supply of nutrients in aqueous solution. Nevertheless, within this overall environment, some factors have a significant impact on the ecology of dental caries, including the presence or absence of oxygen, pH, mineral salts (e.g., calcium and phosphate), anti-microbial factors, key acid- or base-producing substrates, and microbial interactions. Microbial interactions can have a significant effect on the environment of plaque ecosystems, promoting both positive and negative effects with respect to growth and survival. One positive relationship is the emergence of anaerobic bacteria in plaque preceded by a reduction in the redox potential generated through the metabolic action of the facultative populations. Other positive interactions include, the coaggregation of different species resulting in the colonization of less adherent organisms that bind to those with a higher affinity for the tooth surface (Scannapieco, 1994). A complex series of multigeneric cell-cell interactions are now known to exist between oral bacteria that are associated with the formation of plaque (Kolenbrander and London, 1992). These interactions are important not only for adherence, but for the exchange of essential nutrients and for the cooperative degradation of complex macromolecules. Unlike the brief periods of exposure to dietary constituents, most bacteria in the plaque are exposed to a thin film of saliva for prolonged periods, although the rate of flow and distribution will vary in different regions of the mouth (Dawes and Macpherson, 1993). Saliva contains low concentrations of free sugar (i.e., glucose < 50 μM), amino acids (< 200 μM) and other components (Carlsson, 1986); thus, oral bacteria must rely on the degradation of complex macromolecules, such as mucin and glycoproteins, for energy. Plaque bacteria produce a variety of extracellular glycosidases, peptidases and esterases that permit them to degrade these large macromolecules in saliva, a process that requires the concerted action of several species with overlapping enzyme activities (Beighton *et al.,* 1986; De Jong and van der Hoeven, 1987; Bradshaw *et al.,* 1994). Furthermore, saliva from different glands has been shown to support different oral bacteria suggesting that saliva itself will apply a certain selective pressure on the ecosystems at various sites in the mouth (van der Hoeven *et al.,* 1989).

Relative to other factors, the microflora that exists at a site is a particularly unpredictable component of the ecosystem. This is somewhat understandable when one considers that each supragingival plaque community can be comprised of more than 30 different bacterial genera with 300-400 taxa with total cell numbers as high as 10^{11} organisms per gram wet weight (Hamilton and Bowden, 2000). Historically, even though Miller espoused the 'non-specific plaque' hypothesis (according to Loesche), oral microbiologists in the 20th century were driven by Koch's 'Postulates' in order to find the single organism in the plaque community that was the causal agent for caries. Consequently, the pre-radiograph period in dentistry was characterized by the 'lactobacillus era' in caries etiology, largely because species of this genus could be isolated consistently from overt caries lesions. This was followed by the 'streptococcus era', which made significant advances in the 1960's with the use of animal models to dissect the plaque microflora resulting in the identification of *Streptococcus mutans* as a major etiological agent in animal caries (Gibbons, 1964). Even though significant information is now available to indicate that various bacterial agents can be associated with dental caries (Bowden, 1991), a large scientific enterprise has evolved with *S. mutans* as the focal point. This research has generated a massive amount of information on the activity and properties of this organism and these will be discussed later in this chapter.

At this stage, however, some attention should be paid to the concept of the individual plaque community since it is often assumed that only a few 'specific' organisms within these communities possess the requisite 'virulence' properties that are important in the caries process. In recent years, however, more attention has been paid to the activity of other resident populations that may modulate the activity of such agents. Microbial interactions in plaque communities have been recognized for many years, particularly in relation to food chains, but important advances in our understanding of community control or modulation of individual oral bacteria has come from continuous culture studies with mixtures of oral bacteria in which elements of the environment can be varied under controlled conditions. A number of these studies have involved mixtures of oral bacteria in liquid (planktonic) phase, and such cultures have provided useful information on the basic biochemical and metabolic properties of the constituent oral bacteria. In recent years, however, an increasing interest has been directed at the characteristics and properties of oral bacteria when associated with surfaces in microbial biofilms as they are in dental plaque. An increasing body of knowledge indicates that bacteria in biofilms have unique properties not apparent when they are grown in liquid or planktonic phase.

The Dental Plaque Biofilm

In plaque formation, an organism must first adhere to the tooth surface and then grow in the prevailing environment. Adherence is now known to involve a complex array of both specific and non-specific factors between components of saliva and bacterial surfaces (Gibbons, 1989; Scannapieco, 1994). An organism in a biofilm ecosystem on the tooth surface has a number of advantages compared with life in a liquid culture: (1) physical attachment to a surface enhances survival, (2) the surface provides sequestered nutrients, (3) promotes the formation of consortia for the cooperative degradation of complex molecules, and (4) provides increased protection from anti-microbial agents. A variety of model systems have been employed to study biofilms of oral bacteria ranging from *in vitro* systems, such as, 'artificial mouths', submerged surfaces in static media, various chemostat systems, constant-depth film fermenters, flow cells, as well as *in vivo* models employing implanted surfaces on teeth of human subjects (Bowden, 1995; Marsh, 1995). The information derived from these various systems is highly diverse ranging from data on cellular survival under various conditions to events at the molecular level within an individual species. Since no single system can provide this range of information with the multitude of bacteria comprising 'natural' dental plaque, it is necessary to synthesize the fundamental information in these studies into a developing picture of the 'real world' analogous to placing pieces in an evolving 'jig-saw puzzle'.

Biofilm development generally follows similar stages in many environments and involves the adherence of bacteria to a thin organic surface or conditioning film that is equivalent to the acquired pellicle of dental plaque. Surface attachment is followed by a period of rapid surface growth (accumulation), and then slower accumulation resulting in relatively stable cell numbers. Human and animal studies with oral bacteria have demonstrated that *in vivo* the generation or doubling time (g) of the attached cells during the initial stages (12-24 hours) of biofilm formation is short, reminiscent of exponential growth in batch cultures (Nyvad and Kilian, 1987; Beckers and van der Hoeven, 1982, 1984). The rate of growth is dependent on the nutrients available, the competing organisms in the community, and the environmental conditions in the biofilm at the time. For example, *Streptococcus mutans* and *Actinomyces viscosus* growing together on the teeth of germ-free rats attained initial doubling times only slightly higher than the maximum rates when grown with glucose or sucrose in the same rats in monoculture (1.1 and 2.8 hours, respectively). Within two days of accumulation in the biofilm, the doubling times become longer with g values commonly exceeding 24 hours with cells equivalent to, or approaching the status of cells in the stationary-phase of growth in batch cultures. Furthermore, growth will not be consistent throughout the biofilm and will usually be limited at various levels in the film by low levels of one or more essential nutrients, or the formation of high concentrations of metabolic acid or other toxic products.

Thus, for the majority of the time, the bacteria in dental plaque will be in a status equivalent to cells in the stationary phase of growth. The important factor with respect to the ecology of dental plaque is the physiological status of the cells during the various stages of biofilm development, as well as throughout the various sections of the biofilm. Cells at the base of the film may be dead or lysing, while those near the surface may be actively growing, however, even with such physiological diversity, it can be argued that the 'community' may be a form of 'steady state' (Wimpenny, 1995). A variety of fluorogenic compounds are being used in conjunction with epifluorescence microscopy to assess 'live' and 'dead' cells within biofilms usually by a combination of cell permeant and impermeant stains. Intact cells take up the permeant dye, but not the impermeant stain, whereas the cells with compromised membranes will take up both dyes and produce a colour different from that of the permeant dye alone. Thus, cells with compromised membranes are considered 'dead' cells. From the perspective of the acidogenic flora in plaque and caries; however, one might imagine that such 'dead' cells might maintain sufficient glycolytic activity to promote enamel demineralization. Thus, from the metabolic perspective, they will not be 'dead' and would, theoretically, be able to contribute to demineralization of tooth enamel.

Many physiological and biochemical differences in the characteristics of biofilm versus planktonic cells have been reported suggesting alterations in growth rate and yield, substrate utilization, fermentation patterns, pH optimum, susceptibility to toxic conditions and other parameters. In an early study, increased lactate production by *S. mutans* was observed during growth on hydroxyapatite surfaces compared to planktonic cells, which was not attributed to buffering by the hydroxyapatite (Berry and Henry, 1977). More recently, the application of molecular biology to oral streptococci has made it possible to demonstrate in a more definitive fashion the relationship of bacterial metabolism and growth on surfaces. For example, Hudson and Curtiss (1990) constructed fusion strains of *S. mutans* using the promoterless chloramphenicol acetyltransferase reporter gene (*cat*) between the regulatory regions for the genes for fructosyltransferase gene (*ftf*) and the glucosyltransferase B/C (*gtfB/C*) operon to examine their expression in a salivary-coated hydroxyapatite model system. The expression of these genes was induced by sucrose and the expression of the *gtf* operon, but not the fructosyltransferase gene, was increased in cells adherent to saliva-coated hydroxylapatite. Of interest from the point of view of diversity is that one of the explanations advanced by the authors for this effect was that the population tested was heterologous, with those cells producing more glucan exhibiting enhanced binding to the surface (i.e., selective adherence).

Wexler *et al.,* (1993) used the same *S. mutans* strains to examine the impact of environmental perturbation on transcription of the *gtf*BC and *ftf* gene promoters and confirmed the effect of sucrose. Continuous culture showed that the environmental pH, growth rate and nutrient levels influenced the expression of *gtf* and *ftf* genes by *S. mutans* and that this species responded rapidly to changes in the environment. Importantly, the re-establishment of steady-state conditions following environmental change resulted in a return of enzyme activities to previous steady-state levels indicating that regulation of expression, and not selection of specific populations within the chemostat, was responsible for the variation in expression. Burne *et al.,* (1997) have also used *cat* fusions to the *gtfB/C* and *ftf* promoters, as well as *lacZ* gene fusions to the *scrA* gene (Enzyme II of the sucrose-P-enolpyruvate phosphotransferase sugar transport system [PTS][Figure 8]), to examine gene expression in *S. mutans* in a continuous culture biofilm reactor following growth for 2 or 7 days. The results indicated that reporter gene expression at 2 days was comparable with that in planktonic cells; however, the 7 day cells expressed significantly higher levels of the activity from the *gtfB/C* promoter, while that from the *ftf* promoter was lower. In addition, sucrose increased reporter gene expresssion from these two promoters, but not the *scrA* promoter. The authors concluded that the observed differences in gene expression were not due solely to growth in the biofilm, but were the result of a combination of environmental factors, some unique to the biofilm micro-environment. An increasing body of evidence indicates that the survival of oral bacteria to changes in the environment involves signficant alterations in gene expression (Bowden and Hamilton, 1998). The ability of an organism to induce such 'global' changes in protein synthesis to counter adverse conditions in dental plaque may be the key to why some sites on the tooth develop dental caries, while other similar sites do not. This topic will be discussed in relation to the response of oral streptococci to acid in a later section of this chapter.

The Bacteria

Mutans Streptococci

Even the casual reader of the dental literature will recognize the name *Streptococcus mutans* or the term 'mutans streptococci' (MS) in relation to dental caries. *S. mutans*, first isolated by Clark in 1924, was 'rediscovered' in the 1960's and brought to prominence in animal model experiments and human clinical studies examining the etiology of caries (Hamada and Slade, 1980; Loesche, 1986). Subsequent taxonomic assessment of various human and animal strains of *S. mutans* revealed considerable serological and genetic heterogeneity, and from this group emerged eight new species. The major species in humans are *S. mutans* and *S. sobrinus* with the occasional isolation of *S. rattus* and *S. cricetus*, and all of these strains are cariogenic in animal models (Table 1). The 'mutans streptococci' designation generally refers to studies prior to reclassification when it was not possible to assign the newer name, or where a semi-selective medium was used eliminating organisms not in the mutans streptococci group. As pointed out by Loesche in 1986, studies of plaque samples from various parts of the world indicated that 70-100% of the human MS strains can be classified as *S. mutans*. Consequently, the organism has received a vast amount

Table 1. Bacteria Associated with Dental Caries

Actinomyces
    A. odontolyticus
    A. naeslundii genospecies 2[a]
    A. israelii
    A. gerencseriae
    Other *Actinomyces?*

Lactobacillus
    L. casei
    Other species?

Streptococcus
    S. mutans
    S. sobrinus
    S. rattus[b]
    S. cricetus[b]
    Non-mutans strains
    S. mitis (biovar 1)

Other acid-resistant genera?

[a]Includes strains previously designated *A. viscosus* serotype II.
[b]Strains predominant in only some parts of the world.

of attention with respect to its characteristics and properties, as well as, the frequency of isolation in longitudinal and cross-sectional clinical studies.

Although the methodology is available to differentiate the various species within the mutans streptococci, many of the recent clinical studies compare the numbers of MS on semi-selective medium, such as Mitis Salivarius Bacitracin agar, with respect to the total streptococci or total colony-forming units (CFU) in plaque samples from various tooth sites. Comparisons usually include sound tooth surfaces with either 'white spot' lesions prior to cavitation and/or advanced lesions following cavitation of the intact surface at various locations, such as coronal fissures, approximal and smooth sites, or comparisons between posterior and anterior teeth. Subject selection may also be highly variable, including comparisons of caries-free and caries-active individuals with and without prior caries experience. With a few exceptions, the early clinical data could be characterized as being narrowly focused on the mutans streptococci and this research indicated a major role for the MS in the initiation and progression of caries. For example, the general observations showed that the number of MS isolated from the surface of 'white spot' lesions was higher than that from similar sites on sound enamel (Loesche, 1986; Carlsson *et al.,* 1987), although, in many cases, the numerical status of other plaque organisms was unknown.

The establishment of the mutans streptococci as important etiological agents in dental caries naturally led to studies on the role of parents in the acquisition of these bacteria by children. Early work suggested that mothers were probably the reservoir for the transmission of MS to children, an observation confirmed by Köhler and Bratthall (1978). These workers showed that mothers with high counts of MS in saliva tended to have children with high levels of MS, whereas mothers with low counts tended to have non-infected children. Later work with first-time mothers with high salivary counts of MS ($\geq 10^6$ colony-forming units (CFU)/ml) demonstrated that the introduction of preventive measures to reduce the MS below 3×10^5 CFU/

ml prevented or delayed the transmission of MS to the children for periods up to three years of age (Köhler *et al.*, 1983). More recently, Caulfield and coworkers have shown that the acquisition of mutans streptococci by small children (0-5 years) appears to occur during a discrete period designated as the "window of infectivity" (Caulfield *et al.*, 1993). In their study of 46 infants considered to be at risk because of the high MS counts in the saliva of their mothers, 38 of the children acquired MS between the ages of 19 (25% MS) and 31 (75% MS) months (median age = 26 months) indicating a finite period for maximum susceptibility for the acquisition of MS once the teeth have erupted. A more recent study has shown that intrafamilial transmission of *S. mutans* can occur in children who acquired the organism between the ages of 5 and 11 years (van Loveren *et al.*, 1999).

Other 'Cariogenic' Bacteria
In spite of the strong association of the mutans streptococci with caries, a variety of studies have reported anomalous situations with respect to the numbers of mutans streptococci in human subjects with and without dental caries. For example, Carlsson and co-workers (Carlsson *et al.*, 1987) examined a group of children in rural Sudan with a very low prevalence of caries (DMFT=0.17). Microbiological examination revealed that 96% of the children had mutans streptococci with 45% of these subjects possessing high cell numbers. Both *S. mutans* and *S. sobrinus* were identified, indicating that high numbers of mutans streptococci can be present without high caries activity. Interestingly, the strains isolated from this latter study were later shown to cause caries in hamsters maintained on a high sucrose diet with scores equal to or greater, than that produced by a reference strain of *S. mutans* (Emilson *et al.*, 1987). This observation underscores the multifactorial nature of the caries process and indicates the problems associated with extrapolating data from animal models to human subjects.

A number of studies have also reported the opposite situation, i.e., cases where caries has developed in the absence of significant numbers of mutans streptococci. These cases generally represent a small number of the carious sites examined, but support the concept that other organisms can be involved in the process. One such study by Loesche and Straffon (1979) reported lesions in which lactobacilli far outnumbered the mutans streptococci in three subjects with a prior history of low caries experience. In another longitudinal study (Lang *et al.*, 1987), 48 school children (7-8 years) from an area with low caries activity were examined over a two-year period focusing on fissure and smooth surface caries. The numbers of *S. mutans* increased 6-9 months prior to detection of 'initial' lesions constituting 11-18% and 10-12% of the streptococci in fissure and smooth sites, respectively. Several sites were observed with high *S. mutans* levels without caries and one of the six fissure sites developed cavitation in the absence of detectable *S. mutans*. The lack of association of mutans streptococci and caries in a small number of caries lesions has also been reported by Hardie and his colleagues in a two-year study of 50 school children in England (Hardie *et al.*, 1977).

Early work in the first half of the twentieth century on the etiology of dental caries focused on the lactobacilli since they were readily isolated from overt caries lesions, undoubtedly because caries lesions are acidic (Dirksen *et al.*, 1962). Such acidic environments would permit these aciduric bacteria to dominate the communities at the expense of the less acid-tolerant flora. More recently, however, a possible role for the lactobacilli in caries initiation was suggested in a number of studies. An early longitudinal study (Ikeda *et al.,* 1973) showed that while the numbers of *S. mutans* increased at all sites where caries occurred, in some cases, this was accompanied by increases in lactobacilli with significant increases in the numbers of these bacteria after overt lesions appeared (i.e., Stage 5, Figure 2). Loesche and Straffon (1979) also demonstrated that in some situations caries could occur when the lactobacilli counts outnumbered those of *S. mutans.* Several lines of evidence also suggest that the lactobacilli are important in the later stages of the caries process since their numbers are low prior to the appearance of the incipient (white spot) lesions and increase to significant levels in cavities. For example, a longitudinal microbial analysis of progressive and non-progressive approximal incipient lesions in children revealed a notable association of *Lactobacillus*, *S. mutans* and *Actinomyces odontolyticus* with the progression to overt cavities (Boyar and Bowden, 1985). Significantly, lactobacilli were associated with 85% of the lesions that progressed to cavitation, but were never found in non-progressive or control sites. Much higher numbers of *S. mutans* and *Lactobacillus* were also observed in white spot lesions and cavities compared to control sites in a 12 month longitudinal study of nursing caries in young children (Milnes and Bowden, 1985). More recently, *S. mutans* and *Lactobacillus*, as well as, *Actinomyces viscosus* (*A. naeslundii* genospecies 2) were also shown to dominate the plaque flora in human root surface lesions (Nyvad and Kilian, 1990) .

Recent clinical studies have focused attention on the 'low pH' non-mutans streptococci (non-MS) in the caries process. In one study (Sansone *et al.,* 1993), counts of lactobacilli, MS and non-MS in pooled plaque samples from coronal incipient 'white spot' lesions and advanced root lesions were compared with sound sites in the same subjects, as well as, with samples from caries-free individuals. In addition, the pH of plaque samples was monitored for 60 minutes following the addition of glucose and the pH minimum achieved by selected isolated colonies of the MS and non-MS were recorded following incubation for three days in glucose-containing medium. The 'resting' (zero time) and 60 minute plaque pH readings from the incipient coronal samples were generally lower, and the MS counts higher, than that of the sound surfaces in the same subjects, although the MS counts were not statistically significant. The lactobacilli counts were generally low; however, those of the non-MS exceeded those of the mutans streptococci and had comparable numbers of isolates capable of lowering the pH below 4.2-4.4. Subjects with advanced root caries had significantly higher numbers of MS than sound surfaces in the same subjects and the numbers of non-MS isolates capable of lowering the pH below 4.4 were higher. This is in contrast to the very low numbers of MS and non-MS observed with samples from caries-free subjects, which exhibited far fewer non-MS capable of lowering the pH below 4.4. Thus, the flora of the caries-positive subjects had a higher proportion of MS and non-MS capable of producing low pH values from glucose, indicating a more acidogenic streptococcal plaque population. This study also showed that in spite of the differences in the MS and non-MS counts, the plaque samples of caries-free and caries-positive subjects had a similar capacity to lower

231

the pH with glucose, an observation also reported by Stephan in 1944. What is less clear is the reason for the comparable numbers of non-MS able to reduce the pH below 4.4 in the samples from both caries lesions and sound surfaces in the same subjects.

A subsequent study (van Houte *et al.,* 1996) used similar techniques to explore the acid-producing potential of a wider selection of oral bacteria from the predominant cultivable flora in sound and carious coronal sites, root surface sites, and caries-free subjects. Almost 26% of the flora from advanced root lesions generated final pH values less than 4.2 in the three-day glucose medium test and this flora included all of the lactobacilli and mutans streptococci, most *Bifidobacterium* and non-MS strains, and approximately 20% of the *Actinomyces.* The coronal caries and sound sites gave similar information. On the other hand, sound root surfaces in subjects free of root caries only had 8.4% of the flora in this category and almost all of these were non-MS. Thus, as the net is thrown wider, more organisms capable of carbohydrate metabolism at low pH can be identified and these organisms may have a legitimate association with the caries process. In addition, several studies have shown an association of *S. mitis* (biovar 1) with root caries suggesting that his species may be a component of the non-MS 'low-pH' group (Bowden *et al.,* 1990; Nyvad and Kilian, 1990).

Animal studies reported in the early 1970s established a role for *Actinomyces* in caries etiology, although a variety of later studies were unable to substantiate a positive association for this genus in human caries etiology (see review: Bowden, 1991). Although *A. odontolyticus* has been associated with the progression from incipient to overt cavities in children (Boyar and Bowden, 1985) and *A. viscosus* (*A. naeslundii* genospecies 2) with nursing caries (Milnes and Bowden, 1985), the more recent data on the role of *Actinomyces* has come from the microbial analysis of human root caries. These studies have provided new information through the application of improved sampling techniques, the use of non-selective media to permit the growth of a wide selection of bacteria, and the application of new taxonomic classifications. This research has demonstrated that root caries has a complex and variable etiology with different species constituting the predominant flora in different lesions with the mutans streptococci, lactobacilli and *Actinomyces* spp. dominant in some lesions, but not in others (Bowden *et al.,* 1990; Nyvad and Kilian, 1990; van Houte *et al.,* 1994). Of the *Actinomyces* spp., *A. naeslundii* genospecies 2 is isolated with a higher frequency from root lesions compared with *A. naeslundii* genospecies 1 (Bowden *et al.,* 1990). Recent data also indicate that other Gram-positive pleomorphic rods (GPPR), probably *Actinomyces,* may be involved in root caries (Brailsford *et al.,* 1998). In this study, the microflora isolated on fastidious anaerobic agar supplemented with blood (FAAB) revealed that *A. naeslundii* was not the predominant *Actinomyces* species in sound and carious plaque, particularly in soft lesions where the GPPR represented 70% of the microflora. As the samples on FAAB were incubated for 7 days, it is conceivable that the higher numbers of putative *Actinomyces* spp. on FAAB isolation may represent slower-growing species associated with the disease (Bowden *et al.,* 1999). A more recent report has indicated that *Actinomyces* species constitute 93% of the GPPR with major contributions from *A. israelii* and *A. gerencseriae* (Brailsford *et al.,* 1999).

A clearer picture of the development and progression of root caries has been advanced in comprehensive cross-sectional studies by Schüpbach and associates (Schüpbach *et al.*, 1995, 1996). In the first study, the surfaces of sound, carious and arrested caries root surfaces on extracted teeth were examined microbiologically using primarily blood agar for microbiological evaluation. The arrested lesions were shown to possess far fewer bacteria than the caries-free and caries-active surfaces. In spite of this factor, *Actinomyces* was the predominant genus on all three surfaces comprising 28.5 to 44.6% of the total colony forming units (CFU), with *A. naeslundii* constituting 21.5% of the total CFU of the surface of the carious root samples. The streptococci were the second most predominant genus (5-11%); however, neither *S. mutans* nor *S. sobrinus* were associated with carious surfaces, nor were *Lactobacillus* species. Interestingly, the paper reported for the first time the observation that species of *Capnocytophaga* and *Prevotella* were associated with root surface caries. In particular, *C. gingivalis*, *C. ochracea* and *P. intermedia* showed elevated levels in active carious surfaces and these species ferment a range of carbohydrates and can degrade protein under acidic conditions (pH $\geq$ 5.0). Thus, in addition to contributing to demineralization of the root surfaces, these organisms may be important in the hydrolysis of dentin collagen matrix.

In a subsequent study (Schüpbach *et al.*, 1996), the same microbiological methods were employed to examine root lesions of different severity in order to assess the flora associated with the progression of the disease. The teeth selected included four grades of lesion from no surface defect (incipient) to grade IV cavitated lesions in which the demineralization had reached the root canal. This study clearly showed that initial lesions below the surface plaque (grades I+II) had a different microflora compared with the advanced (grades III+IV) lesions with the latter having a more diverse community, i.e., 34 vs 17 taxa. Sectioning vertically through the advanced lesions (III+IV) showed that the outer 0.5 mm segment contained >95% of the total viable cells. Comparisons with the results of the previous study on root surfaces (Schüpbach *et al.*, 1995) indicated that the microflora of the initial lesion (grade I) was somewhat similar to that on the intact plaque surface; however, the numbers of streptococci increased from 8.8% of the surface microflora to 44 and 61% of the initial and advanced root lesions, respectively. Of particular interest was the increase of the *S. mutans* populations from 0.001% at the surface to 23% in the initial lesions and 45% in the advanced lesions. Conversely, *A. naeslundii* was higher in the initial lesion (20%) than in the advanced lesions (6.7%). Surprisingly, *Lactobacillus* spp. were higher in the initial lesions (17%) than in the advanced lesions (0.8%), being found in only 3 of 8 of the latter lesions. These results clearly demonstrate that as caries progresses, a succession of bacterial communities are created with the numbers of specific genera and species increasing or decreasing through natural selection by environmental forces. As a consequence, the microbial community at the advancing front of the lesion has a vastly different composition than the original surface plaque.

The microbiological census of caries lesions usually involves a sufficient number of subjects to permit statistical analysis and the results are often reported in isolation frequencies rather than mean counts, since it is well known that the range of counts of specific organisms at the same type of site in different subjects can produce a wide range of numbers. Although this wide range of CFU values for individual organisms at single sites is statistically troublesome when viewing numbers of sites and is often avoided, this grouping can obscure the unique nature of the individual site. For example, Sansone *et al.* (1993) sampled incipient enamel caries sites from 12 subjects and found that the total streptococci varied from 10.3 to 95.5% of the total flora, while the mutans streptococci varied from 0.11 to 39.8% of the total flora and 0.28 to 80.2 % of the total streptococci. Longitudinal studies typically demonstrate a wide variation in cell numbers of specific organisms with sequential sampling at the same site over time. Furthermore, where the microbial census was aimed at a wide range of bacteria, it was apparent that a variety of bacteria can dominate communities at specific diseased sites. This can be illustrated in the work of van Houte and coworkers (1994) with the variation in the organisms that dominate the cultivable flora from individual incipient and advanced root sites in human subjects (Table 2). Observation of this list of 'dominant' organisms indicates a few names, such as *Eubacterium*, *Corynebacterium*, *Gemella* and *Clostridium*, that would not be recognized as members of the 'first team' of cariogenic bacteria seen in Table 1. At the present time, there is little evidence to suggest that these dominant organisms do not have some association with the disease process at these individual sites.

These studies raise a number of questions about the factors that influence the changes in the plaque microflora as the disease progresses. The simplest hypothesis

Table 2. The Dominant Oral Bacteria Associated with individual Incipient and advanced Root Lesions in Human Subjects as a Percentage of the Total Predominant Cultivable Flora (PCF).[a]

Dominant Organism(s)	Number of isolates of the dominant strain(s)	Total number of isolates of all strains (PCF)	Dominant strain(s) as a percent of the PCF
Incipient Root Lesions			
Actinomyces israelii	32	49	65
Actinomyces viscosus	47	48	98
Eubacterium	26	48	54
Corynebacterium	40	48	83
Streptococcus sanguis I + II	34	50	68
Advanced Root Lesions			
Actinomyces naeslundii	31	46	67
Actinomyces naeslundii + *Veillonella*	35	44	80
Bifidobacterium + *Clostridium*	25	44	57
Clostridium	26	48	54
Propionibacterium	17	44	39
Gemella + *A. israelii*	25	46	54

[a] Data derived from van Houte *at al.*, (1994).

suggests that dental caries is promoted in humans by the frequent consumption of refined sugar coupled with a lack of adequate oral hygiene. Carious dental plaque will be highly acidogenic resulting in low plaque pH, particularly at sites with reduced salivary clearance and buffering. This situation, in turn, would result in the selection of acid-tolerant bacteria that become dominant in the community excluding more acid-sensitive bacteria. This proposal would appear to require some modification in view of the above studies showing increased numbers of such genera as *Capnocytophaga*, *Prevotella*, *Eubacterium*, *Corynebacterium*, and *Clostridium* in association with carious root lesions. These bacteria are not known to be acid tolerant, although one cannot eliminate the possibility that a genotypic or phenotypic change occurred in individual cells permitting them to become more dominant in acidic environments. Beneficial microbial interactions among the flora at each site might also have been a factor in promoting the growth of these bacteria by supplying essential nutrients or removing inhibiting substances. Furthermore, the concept that lesion development is always associated with increasing acidity thereby reducing the diversity of the flora, requires some re-evaluation in light of the results of Schüpbach and coworkers (1996), who demonstrated that advanced root lesions (grades III+IV) had a more diverse microflora (34 taxa) than initial lesions (grades I+II)(17 taxa). These results suggest that, in some cases at least, the environment at the advancing front of the lesion may be less acidic than that at earlier stages, permitting the growth of the less aciduric bacteria that have survived within the community.

Microbial Metabolism and Caries

Dental Plaque *in situ*.
To begin a discussion of the microbial metabolism relevant to dental caries, it is probably best to start by reviewing the characteristics of Stephan curves generated by dental plaque when exposed to carbohydrate substrates. The pH profiles generated are influenced by a variety of factors, such as the flow of saliva over the specific site, the nature, consistency and concentration of the carbohydrate in the food ingested, the architecture of the habitat or site, and the nature of the microflora in the plaque at the specific site in question. Three methods have been employed to study such plaque metabolism with each having advantages and disadvantages (Schachtele and Jensen, 1982). These include the use of the 'touch' pH electrodes to analyze *in situ* plaque, plaque sampling, and pH telemetry with indwelling pH electrodes. The use of the 'touch' pH electrode *in situ* was employed in early studies by Stephan (1944), Kleinberg (1961) and others, being particularly useful in field studies with a variety of caries-active and caries-inactive human subjects (Aamdal-Scheie *et al.*, 1996; Fejerskov *et al.*, 1992). Early metabolic studies also employed 'plaque sampling', however, this method suffered from the fact that sampling disrupted the natural plaque structure and eliminated the influence of saliva and the site location as factors, and often required the pooling of plaque from several teeth in order to obtain sufficient plaque material for analysis.

Currently, the most effective method to assess the acidogenicity of dental plaque is by pH telemetry (Graf and Mühlemann, 1966) using small glass electrodes or

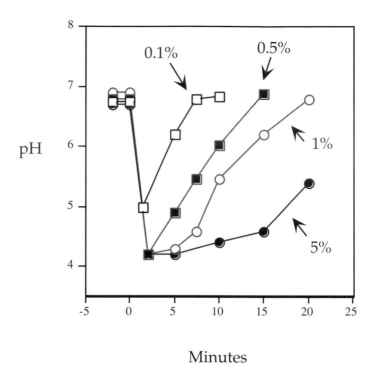

Figure 4. The *in situ* pH change of occlusal three-day old dental plaque of a 32-year old male upon rinsing with varying concentrations of sucrose. The pH was measured by an ion-sensitive, field effect transistor electrode mounted on a partial denture. Sucrose was added at zero time and expectoration was at 2 min. Data derived from Yamada *et al.*, (1980).

ion-sensitive field-effect transistors fixed to the dentition and upon which dental plaque is allowed to accumulate for various periods of time. This latter method has been used extensively to test the cariogenicity of a wide variety of carbohydrates, natural foods, and sugar substitutes, and has the advantage of permitting the continuous monitoring of acid formation during food consumption by undisturbed 'natural' plaque. The indwelling electrode transmits the pH generated by the acid formed by plaque during and after the ingestion of foods and is recorded, usually by computer. An early example of the effect of sucrose concentration on the pH profile of a Stephan curve can be seen in Figure 4 illustrating the pH effect of a two-minute application of 0.1, 0.5, 1 and 5% sucrose (10 ml) to 3-day-old occlusal plaque established on an ion-sensitive transistor in a 32-year-old male subject (Yamada *et al.*, 1980). All concentrations above 0.1% sucrose led to a pH minimum of 4.2 with the time below the critical pH region increasing with the sucrose concentration such that the addition of 5% sucrose prolonged the period of time below the critical pH for about 20 minutes. In addition to readily utilizable sugars, the telemetric technique has demonstrated that a surprisingly long list of foods consumed in the normal fashion are capable of causing the pH of plaque to fall below 5.5 within 30 minutes, with the pH maintained at low periods for several hours (Schachtele and Harlander, 1984; Jensen and Wefel, 1989).

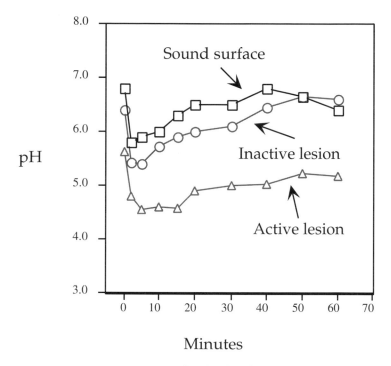

Minutes

Figure 5. The *in situ* plaque pH of sound occlusal surfaces, inactive caries lesions and deep, active occlusal caries lesions after a sucrose rinse as measured by a touch pH electrode. Data derived from Fejerskov *et al.*, (1992).

The maintenance of the pH minimum for such long periods is one point of difference between the original curves by Stephan and Kleinberg and some of the more recent telemetric results. This may reflect the fact that the type of plaque being measured is different from that on the natural surface of the hydroxy- or fluorapatite surface of the tooth (Jensen *et al.*, 1982). Differences in the plaque pH profiles are not unrealistic in keeping with the number of factors affecting Stephan curves by dental plaque at different sites and in different subjects. Such differences also indicate the critical nature of some of these factors in the caries process. For example, in a study of caries-active and -inactive Kenyan children using touch pH electrodes (Fejerskov *et al.*, 1992), maxillary sites showed lower pH levels than mandibular sites with such differences attributed to differences in saliva film velocity that influences both saliva buffering and the accumulation of acids in the plaque matrix at the local sites (Dawes, 1989). The application of a sucrose rinse to sound occlusal surfaces, inactive caries lesions and deep, active occlusal lesions of the Kenyan children produced increasingly lower pH minima (Figure 5) resembling those of the caries-active and -inactive subjects in Stephan's early work (1944).

Such differences in pH profiles are not always exhibited by sound and diseased sites as seen by a recent investigation with elderly Chinese subjects using the same 'touch' electrode technique (Aamdal-Scheie *et al.*, 1996). In this study, no statistically significant differences were observed in the pH profiles of sound and carious root surfaces exposed to sucrose, although maxillary sites again gave more significant pH decreases than mandibular sites. Both sound and carious sites exhibited pronounced pH drops upon the addition of 10 ml of a 10% sucrose solution with prolonged periods of slow pH recovery, such that the pH at 60 minutes was one pH unit below the original 'resting' plaque pH. Interestingly, microbial analysis revealed that the pH response to sucrose was the same in the presence and absence of mutans streptococci at the test sites. The differences between the occlusal sites of the Kenyan children and the root-surface lesions of the Chinese subjects was attributable to the structure of the lesions studied. The occlusal sites were characterized as being deep and inaccessible to clearance by saliva, while the root lesions were often shallow with wide openings, an observation seen in early studies on the pH of overt lesions (Dirksen *et al.,* 1962).

Plaque Acidogenicity.
A consistent feature of the Stephan curves derived from dental plaque with touch electrodes, indwelling electrodes and *in vitro* plaque samples is the rapid reduction in pH immediately following the exposure of the plaque microflora to the carbohydrate substrate in Phase I (Figures 2, 4 and 5). This indicates that supragingival plaque is highly acidogenic, rapidly converting such substrates to acid end-products. This acidogenicity is associated not only with sucrose, glucose and other refined sugars, but also with starch and bread (Mörmann and Mühlemann, 1981), a wide variety of individual food items (Schachtele and Harlander, 1984), and the consumption of complete meals (Jensen and Wefel, 1989). Moreover, this relatively consistent rate of acid formation occurs in spite of the many host, microbial and substrate factors that can affect the pH profiles during food consumption (Schachtele and Jensen, 1982). For the most part, dental plaque is anaerobic resulting in the 'fermentation' of carbohydrates to a mixture of short-chain carboxylic acids reflecting the metabolism of the different acidogenic genera in plaque. A variety of studies have examined the acids formed by the plaque *in situ* by removing and analyzing samples before and after the application of a sugar source, usually a sucrose rinse. The technical difficulties associated with analyzing small amounts of plaque; however, have generally resulted in the need to 'pool' plaque from groups of teeth using similar groups in different jaw quadrants to obtain samples over time after the application of the sugar substrate, or by using weekly samples from the same site for each time point. Consequently, no information is available on the acid profiles generated by single tooth sites when exposed to carbohydrate.

As would be expected from pooled plaque samples, a considerable range of acid concentrations has been observed among subjects at the various time points in each study, and between the various experimental groups. The variation between the individual studies can be ascribed to differences in plaque age, sample weights, sample sites, quadrants, the disease status of the sites, and the fact that not all acids were reported. Nevertheless, in spite of this experimental variation, the results permit a general overview of the acids found in plaque *in situ*: (1) Resting plaque is dominated by acetic acid ($\geq$10 nmol/mg wet weight of plaque)(Distler and Kröncke, 1983; Higham and Elgar, 1989) with high concentrations also reported for formic (Distler and Kröncke, 1983) and propionic acids (Geddes, 1975), and with lesser

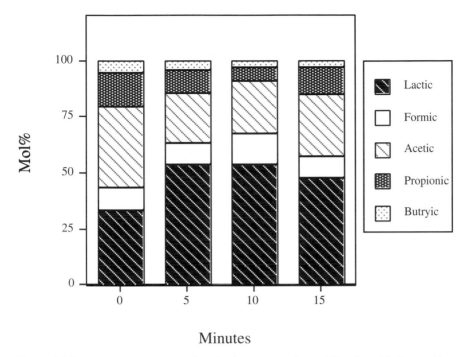

Figure 6. Mean percentage pattern of acids after sucrose rinse of five-day old plaque. Data derived from Distler and Kröncke (1983).

amounts of lactic acid and trace amounts of butyric, succinic and pyruvic acids. (2) The addition of sucrose to plaque *in situ* results in the preferential formation of lactic acid and a reduction in the volatile acids during Phase I of the Stephan curve (Figure 6) (Geddes, 1975; Distler and Kröncke, 1983; Margolis and Moreno, 1992). Furthermore, as demonstrated by Margolis and Moreno (1992), a sucrose rinse in caries-positive subjects resulted in the formation of significantly higher levels of lactic acid over a 30 minute period than in plaque from caries-free subjects (Figure 7). The levels of acetic acid also increased at the 15 and 30 minutes time points. Moreover, the plaque pH of the caries-positive subjects was lower before and after the addition of the sucrose than that of the caries-free individuals. Similar results have been reported by Vratsanos and Mandel (1982) with adult subjects chewing sucrose gum. (3) Phase III of Stephan curves follows the pH minimum and is characterized by the reduction in total acid and increases in the proportion of the volatile acids over lactic acid.

The rapid pH fall in Phase I of the Stephan curve can be ascribed to the formation of lactic acid by the acidogenic plaque microflora under the conditions of high sugar concentration (see below). This is combined with the relatively low buffering capacity of plaque fluid under these conditions, which in the pH 7.0 - 4.0 range, is influenced largely by plaque bacteria and, to a lesser extent, by extracellular plaque fluid. Buffering by plaque fluid is, in part, associated with the ratio of low-pK acids, such as lactic, formic and pyruvic acids, and the high-pK acids, such as acetic,

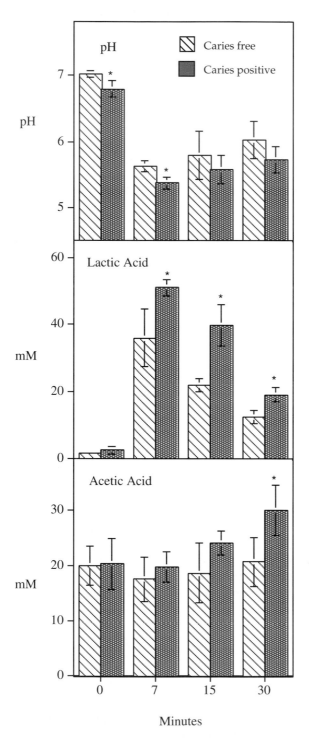

Figure 7. Comparison of the pH and the lactic and acetic acid concentrations of pooled plaque from caries-free and caries-positive subjects receiving a one-minute rinse of 10% sucrose. An asterisk (*) indicates a significant difference (pH, $p<0.001$; lactic, $p<0.025$; acetic, $p<0.05$). Data derived from Margolis and Moreno (1992).

propionic and butyric acids (Margolis and Moreno, 1992; Vratsanos and Mandel, 1982; Margolis *et al.,* 1995). As the pH drops below 6, the weak, high pK acids become undissociated ($A^- + H^+ \rightarrow HA$) and begin to buffer the pH-lowering effects of the low pK acids, with the degree of buffering related to the ratio of low and high pK acids and other buffering components, such as bicabonate and other components of plaque fluid. Thus, high concentrations of high pK acids would resist the lowering of the pH minimum by the strong, low pK acids, although the extent to which this contributes to total plaque buffering capacity is still an open question (Margolis and Moreno, 1992; Shellis and Dibdin, 1988). The activity of undissociated plaque acids and the degree of saturation of calcium and phosphate in plaque fluid with respect to enamel mineral [DS(En)] is important to the process of enamel demineralization. In addition to pH and organic acids, Margolis and Moreno (1992) have analyzed inorganic ions in plaque fluid from caries-free (CF) and caries-positive (CP) subjects and demonstrated that resting CF plaque fluid had a greater remineralizing potential than CP samples. As mentioned previously, the pH of the CP samples was lower and the lactic acid concentrations higher than in CF samples, and this was associated with a lower DS(En) indicating that the caries-positive plaque fluid had a greater potential for demineralization.

Metabolism by the Acidogenic Microflora

Most of the species comprising supragingival plaque, with a few exceptions, are acidogenic. Consequently, the potential role of an organism in the total acid output of a local community is the number of cells at a site, the characteristics of its sugar-metabolizing system(s), and what fraction of the cells present are 'metabolically active' under the prevailing environmental conditions. A factor of considerable importance is the capability of an organism not only to produce acid during carbohydrate metabolism, but to survive and grow at low pH. One can immediately see that the acid-resistant or aciduric properties of an organism will have significance in the caries process as it attempts to survive in the progressively more acidic environments of incipient and overt lesions. While relatively little information is available about the numbers of active cells at a plaque site under these conditions, enumeration at incipient and caries sites often indicates high numbers of *Actinomyces* and *Streptococcus*, with increasing numbers of *Lactobacillus* spp. with overt lesions. In the absence of effective methods for measuring the metabolic activity (i.e., rate of acid formation) of these specific oral bacteria *in vivo*, we must rely on laboratory experiments to assess 'metabolic dominance' under the conditions that are perceived to be typical of the mouth.

Extrapolation from the laboratory to the 'real world' of dental plaque is fraught with difficulties. No laboratory system can duplicate the varying conditions of the tooth surface in the mouth. In addition, the properties of bacteria can vary widely depending on the origin of the strain, the length and nature of the treatment in the laboratory, including the medium, pH and gaseous phase for growth and maintenance. Moreover, organisms, such as *S. mutans*, that are classified as belonging to a specific species by taxonomic methods may exhibit considerable differences in growth and metabolic properties within the species designation, even when tested in the same

Table 3. Comparison of the Glycolytic Rates of Freshly Isolated and Laboratory Strains of Oral Streptococci[a]

Strains	Glycolytic Rate[b]
Fresh Isolates	
S. mutans DC-1	248 ± 13
S. mutans BM-71	212 ± 14
S. mutans PB 1546	84 ± 6
S. gordonii SF-2473	165 ± 7
S. mitis SF-3440	918 ± 109
S. sanguis SF-2731	638 ± 20
Laboratory Strains	
S. mutans Ingbritt	185 ± 12
S. gordonii ATCC 10558	181 ± 8

[a]Data derived from Cvitkovitch and Hamilton (1994).
[b]Nanomoles per mg (dry weight) cells per min.

laboratory under the same conditions (Table 3) (Cvitkovitch and Hamilton, 1994). The transfer of an organism from dental plaque to laboratory medium inevitably results in a selective pressure on subculturing that alters the properties of the original isolate. As mentioned previously, for much of the time oral bacteria in dental plaque must be capable of degrading complex macromolecules for the carbohydrates and amino acids required for growth and survival, since the levels of these components in saliva are low (Carlsson,1986). In the laboratory, on the other hand, the energy source and amino acids are normally supplied in excess concentrations in most media, removing any dependence on macromolecular hydrolysis for nutrients.

With these caveats in mind, there can be little doubt that when it comes to carbohydrate metabolism, the oral streptococci are a dominant force in dental plaque. In fact, most of the virulence properties associated with *S. mutans* in relation to caries involve aspects of carbohydrate metabolism. These characteristics include: (1) high acidogenicity with the ability to rapidly metabolize a wide variety of carbohydrates to lactic, acetic and formic acids; (2) aciduricity with the ability to adapt to growth in moderately acidic (pH ~5.0) environments, and to metabolize carbohydrates at lower pH values; (3) intracellular glycogen synthesis in the presence of high sugar concentrations in order to sustain the cell in periods of low exogenous carbohydrate, and (4) synthesis of extracellular glucans and fructan from sucrose that aid in the colonization of the organism in the plaque biofilm and serve as an endogenous food source. Such intra- and extracellular endogenous energy metabolism may be related to the ability of the organism to survive in biofilms in the presence of inhibitors, such as, sodium fluoride, that can kill the same cells in planktonic phase (Bowden and Hamilton, 1998).

The importance of *S. mutans* in the overall acid production of dental plaque can be envisaged by a comparison of its acidogenicity with species of *Actinomyces* and *Lactobacillus*, two important acid-producing genera in plaque. Cells of three representative strains, *S. mutans* Ingbritt, *A. naeslundii* GN 431/175 (genospecies 2) and *L. casei* RB1014, were grown to steady state in continuous culture under essentially identical conditions, and the glycolytic rates and acids formed from glucose were measured. As seen in Table 4, *S. mutans* Ingbritt exhibited a glycolytic rate 20-fold higher than that of *A. naeslundii* GN 431/175, suggesting that *S. mutans* would be more competitive with respect to the sugar utilization and acid formation, even in the presence of high numbers of *Actinomyces*. On the other hand, the glycolytic rate of *L. casei* RB1014 was only slightly lower than that of *S. mutans* Ingbritt, suggesting that it would compete with the latter organism. In normal non-carious dental plaque this would not be the case since the numbers of lactobacilli are significantly lower than the numbers of *S. mutans* and the total streptococci. Lactobacilli would be competitive only in acidic environments, such as overt lesions, since the terminal pH for growth and the pH at which killing occurs is lower than that for *S. mutans* even though the latter organism can adapt to acid tolerance (Svensäter *et al.*, 1997).

The comparable glycolytic rates of *S. mutans* Ingbritt and *L. casei* RB1014 are undoubtedly due to the fact that both transport sugars via the high affinity ($K_s = 10$ µM) phosphoenolpyruvate (PEP) phosphotransferase sugar transport system (PTS)(Figure 8), whereas *Actinomyces* species have low PTS activity (Hamilton and Ellwood, 1983; Hamilton, 1987). The PTS is the principal system in oral streptococci and lactobacilli for the transport of glucose and a variety of sugars, including sucrose, mannose, fructose, lactose and maltose. In the PTS, phosphate is transferred from PEP in the glycolytic pathway via the general PTS proteins, HPr and Enzyme I (EI), to the sugar-specific, membrane-bound protein Enzyme II (EII) and then to the incoming sugar. The domains that make up the EIIs can vary depending on the organism and the sugar to be transported, appearing either as a single protein or as separate polypeptides with IIA and IIB possessing the first and second

Table 4. Comparison of the Glycolytic Rate, Acids Formed and the pH Tolerance of *Streptococcus mutans* Ingbritt, *Actinomyces naeslundii* GN 431/175[a], and *Lactobacillus casei* RB1014 Grown in Continuous Culture[b]

Organism	Glycolytic Rate[c]	Acids Formed From Glucose	Terminal pH During Growth With Excess Glucose[d]	Killing pH[d]
Streptococcus mutans Ingbritt	244	Lactate, Acetate, Formate	4.28	3.0
Actinomyces naeslundii GN 431/175	12	Lactate, Acetate, Formate, Succinate	4.68	3.2
Lactobacillus casei RB1014	195	Lactate, Acetate, Formate, Succinate, Butyrate, Fumarate	3.81	2.3

[a]Formerly *A. viscosus* GN 431/175.
[b]Conditions: dilution rate = 0.05 per hour at pH 6.5 with a glucose limitation. Data derived from Hamilton, (1987).
[c]Nanomoles per mg (dry weight) cells per min.
[d]Derived from Svensäter *et al.*, (1997).

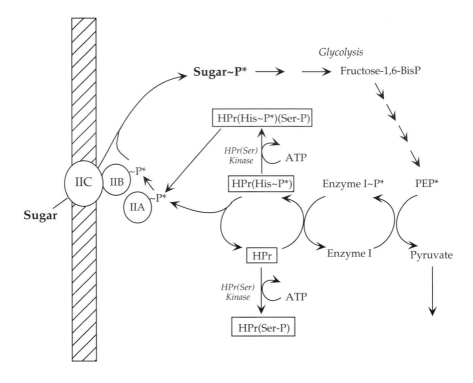

Figure 8. General structure of the phosphoenolpyruvate (PEP) phosphotransferase sugar transport system in oral streptococci showing the phosphorylation of the general protein, HPr, by PEP generating the phosphoryl donor HPr(His~P), and by ATP forming the donors, HPr(Ser-P) and HPr(His~P)(Ser-P).

phosphorylation sites, respectively, while IIC provides the non-phosphorylated, membrane channel and sugar-binding site. The PTS also has a variety of regulatory functions in cell physiology in addition to sugar transport with IIA^{Glc} acting as the key element in Gram-negative bacteria. The protein, IIAB^{Man}, appears to have a similar function in oral streptococci since lactose metabolism is regulated by the mannose-PTS when the cells are grown on glucose by a process that is thought to involve the regulation of gene expression. However, for the most part, regulation in Gram-positive bacteria is associated with the general PTS protein, HPr (see review: Vadeboncoeur and Pelletier, 1997).

During PTS activity, HPr is phosphorylated at histidine residue 15, creating HPr(His~P), which in turn transfers the phosphoryl group to IIA and ultimately to the incoming sugar. In Gram-positive bacteria, but not Gram-negative organisms, HPr can also be phosphorylated at serine 64 by ATP to form HPr(Ser-P), a reaction catalyzed by HPr(Ser) kinase (Figure 8). Originally, the formation of HPr(Ser-P) was postulated to result in the regulation of transport via the PTS; however, more recent data have indicated that HPr(Ser-P) may be involved in regulating both non-PTS lactose transport in *Lactobacillus brevis* and PTS transport systems in *Lactococcus lactis* by inducer exclusion. The control in the latter organism would ensure that glucose is utilized preferentially over other PTS substrates which are metabolized more slowly. Furthermore, studies with *Bacillus subtilis* have indicated that HPr(Ser-P) is involved in PTS-mediated catabolite repression, whereby PTS sugars, such as glucose, prevent the transcription of certain sugar catabolic genes. This postulated mechanism suggests that with high levels of glucose, increased concentrations of the glycolytic intermediate, fructose-1,6-bisphosphate (FBP), would activate HPr(Ser) kinase to increase the cellular levels of HPr(Ser-P). This promotes the association of HPr(Ser-P) with the transcriptional regulatory protein, CcpA, which interacts with a short DNA sequence CRE (catabolite-responsive element) to prevent transcription of the target operon. As the concentration of glucose decreases, and thereby FBP, the level of HPr(Ser-P) also declines removing the inhibition of the target permeases permitting transport of other sugars.

Recent research with *S. mutans* and *S. salivarius* suggest that while HPr is involved in catabolite repression, HPr(Ser-P) would appear not to regulate lactose permease as it does in other Gram-positive bacteria, nor does it appear to inhibit PTS transport activity (Vadeboncoeur and Pelletier, 1997). One important difference is that the activity of HPr(Ser) kinase is not regulated by FBP in oral streptococci as in other bacteria indicating that activation at high glucose concentrations (i.e., high FBP) would not influence activity. Moreover, several key observations were derived from the measurement of the four forms of HPr [i.e., HPr, HPr(His~P), HPr(Ser-P) and HPr(His~P)(Ser-P)] in chemostat-grown cells of *S. mutans* Ingbritt grown under glucose limiting (10 and 50 mM) and glucose excess (100 and 200 mM) conditions (Thevenot *et al.,* 1995). While the intracellular concentration of HPr and HPr(His~P) varied little between the various glucose levels, HPr(Ser-P) and the doubly phosphorylated compound [HPr(His~P)(Ser-P)] increased in relation to the glucose concentration, although the HPr(Ser-P) level declined at 200 mM. These results indicated that the concentrations of HPr and HPr(His~P) were sufficiently high to permit maximum PTS activity at low glucose concentrations. As the glucose was increased, the level of HPr(His~P)(Ser-P) in the steady state cells was high enough to supplement the HPr(His~P)-directed phosphorylation of the IIA component, thereby, increasing total PTS transport activity (Figure 8).

In addition to the PTS, *S. mutans* possesses at least two other sugar transport systems: the multiple sugar metabolism (Msm) transport system and a glucose permease. Discovered by genetic means, the Msm system is a novel binding-protein-dependent system with homology to the osmotic-shock transport systems in the enteric bacteria (Russell *et al.,* 1992). The *S. mutans* Msm operon comprises eight genes, including *msmK*, which codes for an ATP-binding protein, the first such protein identified in Gram-positive bacteria. This system is primarily responsible for the transport of raffinose, melibiose and isomaltosaccharides, but is also capable of transporting sucrose, glucose and fructose (Tao *et al.,* 1993). A second non-PTS glucose transport system in *S. mutans* was indicated from early studies with chemostat-grown cells (Hamilton, 1987) and membrane vesicles prepared from the organism (Buckley and Hamilton, 1994). This work demonstrated that this non-PTS transport process functioned at high sugar concentrations, low pH and high growth rates, conditions repressing the glucose-PTS. However, this was not confirmed until the genes for the general PTS proteins, Enzyme 1 or HPr, in *S.*

mutans were cloned (Boyd *et al.,* 1994) making it possible to inactivate the gene (*ptsI*) for Enzyme 1 and, as a consequence, blocking all PTS activity (Cvitkovitch *et al.,* 1995a). Using the resultant PTS-defective mutant (DC-10), it was possible to show that the Msm transport system in *S. mutans* is under the control of the PTS (Cvitkovitch *et al.,* 1995b).

As mentioned previously, the ability of the non-mutans streptococci to generate acid at low pH is now considered a factor in the formation of caries (Sansone *et al.,* 1993). The number of the strains in this group is often high in carious plaque and significant numbers have the capability of lowering the pH below 4.4 and would be classified as being aciduric. It has long been known that oral bacteria can metabolize sugar at a lower pH than that for growth, a factor of considerable significance in the mouth. Early studies showed that mutans streptococci can metabolize glucose and sucrose at pH 4.0, but can initiate growth only at pH 5.0 or higher (Harper and Loesche, 1984). Consequently, the aciduric non-mutans streptococci possess acid-tolerant characteristics typical of the acid-tolerant mutans streptococci. Moreover, as seen in Table 3, the glycolytic rates of freshly-isolated non-MS strains can exceed those of fresh strains of *S. mutans*, as well as laboratory strains cultivated in the laboratory for long periods of time (Cvitkovitch and Hamilton, 1994). Thus, these bacteria must be considered an integral part of the acidogenic driving force of plaque in relation to caries formation and, as yet, very little is known about the individual species involved and their metabolic characteristics.

One interesting feature of the acidogenic plaque flora is the ability to deal with the 'feast' and 'famine' conditions with respect to readily utilizable carbohydrate. Resting plaque fluid is very low in free sugars and the input of dietary sugar can increase the carbohydrate concentration approximately 10,000-fold, a condition that could lead to 'substrate accelerated death' (Carlsson, 1986). To overcome the problem of high sugar concentrations and assuming anaerobic conditions, organisms such as *S. mutans* regulate the glycolytic pathway to increase the flow of carbon via lactic dehydrogenase resulting in the formation of lactic acid (Figure 9), a feature mentioned previously as being a characteristic of Phase 1 of the Stephan Curve. As the sugar concentration falls with microbial degradation and clearance, an increasing amount of sugar carbon is diverted to the pyruvate:formate lyase system with the generation of acetic and formic acids and ethanol. This latter pathway is preferred under low sugar conditions since it is more energy efficient. Thus, *S. mutans* can be described as being both homofermentative at high sugar concentrations and heterofermentative at low sugar concentrations. Strains of *Lactobacillus* can be homofermentative and/ or heterofermentative, while species of *Actinomyces* are heterofermentative.

As one would expect of complex plaque ecological systems, the rate and end-product profile formed during carbohydrate metabolism is affected by a variety of factors, such as oxygen, pH, microbial interactions, the presence of inhibitors and saliva, as well as a plethora of microbial interations (Marquis, 1995; Bowden and Hamilton, 1998). Also important to plaque composition and metabolic activity is the frequency, consistency and concentration of sugar available. Considerable information is available on the relationship of sugar and caries (Chapters 6 and 7, Nikiforuk 1985) and, with the exception of malnutrition during infancy (Alvarez *et al.,* 1993), dietary sugar is one of the few ingredients in human diets known to

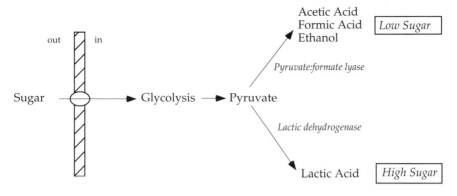

Figure 9. Acid end-products generated by *S. mutans* in the presence of high and low glucose concentrations.

influence the composition and metabolic properties of dental plaque. The increased frequency of sugar consumption naturally results in low plaque pH values and such conditions are known to result in population shifts *in vivo* to more aciduric bacteria, although these changes are usually site-specific. For example, in one human study (Minah *et al.*, 1985), sucrose consumption was positively correlated with the numbers of *S. mutans*, total lactobacilli and *Veillonella* at a maxillary approximal site, with yeast and *S. mutans* at a molar site, and with *A. viscosus* (*A. naeslundii* genospecies 2) in a mandibular approximal site.

Acid Tolerance

No chapter on plaque ecology in relation to dental caries would be complete without some discussion of the capacity of oral bacteria to tolerate acidic environments. Even in the absence of significant caries activity, bacteria in dental plaque are subjected to continual cycles of 'acid shock' created by the formation of metabolic acid end-products following the consumption of dietary carbohydrate. Inspection of Stephan's 1944 data (Figure 2) reveals that caries-active individuals differ from caries-free subjects in three respects: the resting and minimum plaque pH values are lower and the time at low pH values is longer. Clearly, the bacteria in plaque of subjects with high caries activity had the ability not only to survive the long-term acidic environment, but were able to function metabolically to produce acid once supplied with sugar. The low resting plaque pH in caries-active individuals is not surprising since, as mentioned previously, the pH of caries lesions can be near or below 4.0, and exposure of human plaque to acid is known to alter the composition of the normal oral flora to increase the proportion of aciduric bacteria (Svanberg, 1980).

This change in composition has been confirmed in a variety of continuous culture studies using mixtures of oral bacteria. In a three-component mixture, *S. mutans* was seen to dominate *S. sanguis* and *S. mitior* (*S. mitis* group) as the pH of the medium was reduced from pH 7.0 to 4.5 (Bowden and Hamilton, 1987). In a second study, a mixture of nine oral bacteria was pulsed daily with glucose over a 10 day period in the presence and absence of pH control (Bradshaw *et al.,* 1989). During glucose pulses without pH control, *S. mutans, Veillonella dispar* and *L. casei* increased dramatically, whereas *S. sanguis* and the Gram-negative components declined in comparison with the same culture pulsed with glucose at a constant pH of 7.0. Lactic acid was the major acid end-product of the culture without pH control and the successive glucose pulses were shown to increase the glycolytic rate of the consortium. For example, the time to reach pH 5.0 on day 1 was 3.3 hours; however, this was reduced to only 45 minutes on day 10. In addition, the hydrogen ion concentration increased 10-fold over this period, such that the pH after 6 hours of glucose metabolism on day 1 was pH 4.8 and this was reduced to pH 3.8 on day 10. These observations reflect not only the change in composition of the culture to a predominance of the more acid-tolerant strains, but also indicate a degree of "carbohydrate training" whereby the acidogenic bacteria, such as *S. mutans,* adapt to increased glycolytic activity at low pH (Belli and Marguis, 1991: Hamilton and Buckley, 1991).

Microbial survival in acidic environments depends on the ability of an organism to maintain intracellular pH homeostasis. *S. mutans* and other oral bacteria maintain such homeostasis by at least two mechanisms: proton extrusion from the cell via a membrane-associated, proton-translocating ATPase (H^+/ATPase) and by acid end-product efflux (Marquis, 1995; Bowden and Hamilton, 1998). These function to maintain the intracellular pH above that of the external environment as the external pH falls with the formation of metabolic acid end-products. Unlike enteric bacteria, streptococci do not attempt to maintain a constant intracellular pH, but simply strive to ensure a relatively stable pH difference (ΔpH) between the intracellular compartment and the external environment, with the maintenance of the ΔpH dependent on the presence of an energy source and potassium ions. Exposure to more sustained acid environments can induce changes in cell physiology that supplement the normal homeostatic mechanism to permit growth at low pH (Belli and Marquis,, 1991; Hamilton and Buckely, 1991). Studies with oral streptococci have shown that this 'acid-tolerance response' (ATR) results in (1) increased glycolytic activity to provide ATP for proton efflux, (2) shifts to lower pH optimum for glucose transport, the glycolytic pathway and proton impermeability, (3) decreased activity by the acid-sensitive, membrane-associated, sugar-specific Enzyme IIs of the PTS sugar transport system and increased activity by the more acid-tolerant, non-PTS transport activity, (4) increased specific activity of the H^+/ATPase responsible for proton efflux from cells in low pH environments, (5) increased capacity to maintain ΔpH at lower pH values, and (6) shifts to predominately homofermentative metabolism as indicated by increases in lactate formation (Bowden and Hamilton, 1998).

Bacterial survival during shifts to low pH depends very much on the time over which the pH change occurs. This was demonstrated in experiments with glucose-limited continuous cultures of *S. mutans* 2452 in which the culture pH was dropped from 7.0 to 4.8 either by the rapid addition of HCl over a 10 min period or by the accumulation of metabolic acid-end products over a 24 hour period (Hamilton, 1986). The culture acidified with HCl 'washed out' of the chemostat indicating that the

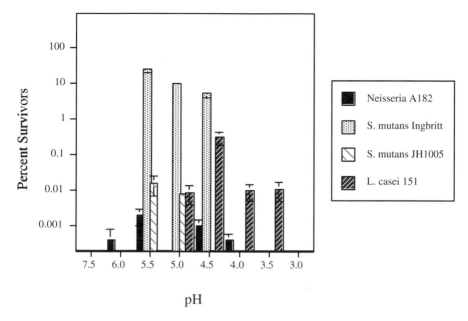

Figure 10. Induction of an acid-tolerance response by various oral bacteria that enhances survival during a three hour period at a pH killing control cells maintained at pH 7.5. Log-phase cells were grown at pH 7.5 and incubated for two hours at pH values between 7.5 and 3.0 prior to acidification to the killing pH.

cells were unable to grow during such rapid acidification. On the other hand, gradual acidification over a 24 hour period with the acids generated by the culture itself resulted in survival of a significant proportion of the cell population. Clearly, the latter cells had adapted to acid tolerance during the 24 hour period that permitted growth and metabolism at pH 4.8 that was not possible during the 10 minute acidification period. In applying this information to dental plaque, one must consider that organisms in the biofilm are not 'washed out' when they cease to grow, but remain components of the biofilm. Clearly, the rapid addition of acid will kill biofilm cells; however, some cells will survive such an acid shock and persist, particularly if the environmental pH increases.

This survival will depend on the characteristics of the organism in the community since not all acidogenic oral bacteria are capable of inducing an acid tolerance response (ATR), and those that exhibit an ATR do so to various degrees. This can be seen in a recent study (Svensäter *et al.,* 1997) with 21 species of six acidogenic genera. Adaptation was tested by incubating log-phase cells, grown at pH 7.5, for two hours in fresh medium buffered at low, but non-lethal pH values (6.0 – 3.5), and then measuring whether such treatment resulted in enhanced survival at a pH (4.0 - 3.0) that killed control cells maintained at pH 7.5. This experiment revealed three groups of organisms: (1) the strongly-responding organisms, such as strains of *S. mutans, S. gordonii, S. salivarius* and *L. casei,* that generated a high number of survivors, (2) weakly-responding strains that generated low numbers of survivors, and (3) strains exhibiting no ATR. Figure 10 demonstrates the relationship of the weakly adaptive strain, *Neisseria* A182, and the strongly adaptive organisms, *L. casei* 151 and the *S. mutans* strains Ingbritt and JH1005, with considerable differences observed between the latter two strains, confirming intraspecies variation. This study also showed that the terminal pH an organism achieves during growth in medium with excess carbohydrate is related to its ability to induce an ATR, with organisms generating low terminal pH values more adaptable than those that produce higher pH values.

The inducible ATR requires protein synthesis since no survival is observed at the killing pH when chloramphenicol is present during the adaptation phase (Svensäter *et al.*, 1997). In order to assess the number of proteins involved in the ATR in various oral bacteria, protein synthesis was analyzed during adaptation by pulsing the cells with [^{14}C]-amino acids at intervals over the two hour period following the pH change from 7.5 to 5.5, extracting the cells and subjecting them to one-dimensional polyacrylamide gel electrophoresis (1DE) and autoradiography. Analysis of the 1DE gels revealed the upregulation of a variety of proteins and the diminished synthesis of others, with the changes of all but a few proteins transitory over the two hour period (Hamilton and Svensäter, 1998). The most complex pattern was observed with the strongly adaptive strain, *S. mutans* LT11, which showed the upregulation of 36 proteins, with 25 appearing in the first 30 minutes of the pH change. The other adaptive oral streptococci and lactobacilli showed increases in fewer proteins and even those strains that were classified as 'non-adaptive' showed the enhanced formation of 3-9 proteins. Analysis of these latter strains indicated that protein synthesis was not sustained during the entire two hour period suggesting that the cells were unable to maintain pH homeostasis.

More comprehensive protein analysis by two-dimensional PAGE (2DE) gives a much clearer impression of the 'global effects' of a pH change to a low acid environment. When the same experiment with [^{14}C]-amino acids was carried out with a fresh strain of *S. mutans* (H7), isolated at low pH from a caries lesion, 64 proteins were shown to be upregulated by at least two-fold within the first 30 minutes of the pH change from 7.5 to 5.5, while a further 49 proteins were downregulated (Svensäter *et al.*, 2000). This number of proteins is not surprising in view of the number of physiological changes that can be observed upon sustained exposure to acid. Overall, the ATR exhibited by the *S. mutans* strains involved a greater number of 'acid-responsive proteins' than that exhibited by *Lactobacillus* species, which are acid tolerant to lower pH values. This suggests that lactobacilli possess a greater complement of 'constitutive' acid-tolerant proteins present during all stages of growth, whereas *S. mutans* strains require a more complex adaptive process to withstand acidic environments.

Currently, there is virtually no information on the identity of the 'acid-responsive' proteins observed in the 2DE gels, although identification is possible by 'proteome analysis', a concept embracing the analysis of the entire protein complement expressed by the genome of a cell or tissue. Some progress in the identification of proteins of importance in acid tolerance has come from the use of molecular biological techniques, specifically the random inactivation of genes by transposon mutagenesis. The first study (Yamashita *et al.*, 1993) utilized transposon Tn*916* to isolate a mutant of *S. mutans* GS5 that was defective in acidurance, as well as being sensitive to high osmolarity and high temperatures. The transposon had inactivated a gene (*dgk*) whose product had significant homology to diacylglycerol kinase, which catalyzes the ATP-dependent phosphorylation of *sn*-1,2-diglyceride resulting in the formation of phosphatidic acid, as well as the phosphorylation of other diacylglycerol-like molecules. This suggests that membrane integrity had been compromised resulting in alterations in the cellular response to environmental stress. Recently, sequence analysis of the region downstream of the *dgk* resulted in the identification of the *sgp* gene coding for the *Streptococcus* GTP-binding protein (SGP), an essential GTPase implicated in the stress response of *S. mutans* (Baev *et al.,* 1999). The total amount of SGP was found to increase with the age of the culture, at elevated temperatures and at acidic pH, with a substantial amount of the protein associated with the membrane under these conditions. The construction of a strain capable of chromosomal *sgp* antisense RNA expression permitted physiological studies with cells depleted of SGP. This work demonstrated a role for SGP in the regulation of the GTP/GDP ratio under different growth conditions, and also implicated the glucan-binding protein A in stress response, since levels of the protein increased significantly in SGP-depleted cells.

In other recent work, *S. mutans* JH1005 has been transformed with transposon Tn917 leading to the identification of three acid-sensitive genes: *fhs*, encoding formyltetrahydrofolate synthetase; *dfp*, a gene in pantothenate synthesis, and *ffh*, a homolog of the eucaryotic signal recognition particle, which is a chaperonin involved in protein translocation and membrane biogenesis (Gutierrez *et al.,* 1996). Later work revealed that the mutation in the *ffh* gene resulted in decreased levels of *ffh* mRNA, lower levels of the proton-extruding ATPase (H$^+$/ATPase), and an inability to induce a log-phase acid tolerance response (Gutierrez *et al.,* 1999). Recently, transcriptional analysis of the heat-shock (HS) system in *S. mutans* GS-5 revealed that the gene, *dnaK* encoding the protein DnaK, was in an operon-like region preceded by two HS genes, *hrcA* and *grpE*, with transcription initiated from a σ^A-type HS-responsive promoter, P1, and to a lesser extent from a σ^B-type promoter, P2 (Jayaraman *et al.,* 1997). Of significance to acid tolerance was the observed increased levels of *dnaK* mRNA and DnaK protein in steady-state cells growing in continuous culture when subjected to acid shock and following adaptation to acid. This indicated enhanced transcription from P1 implicating this promoter in general stress regulation.

Recombination events in bacterial cells are very much affected by acidic environments because of the need for DNA repair and, as a consequence, cells devoid of RecA activity have enhanced sensitivity to organic acids. Quivey and coworkers (1995) studied the characteristics of a RecA-deficient mutant (UR100) of *S. mutans* UA159 and demonstrated that such a mutant was killed more rapidly at pH 2.5 than the wild-type cell. In spite of this, UR100 was able to adapt to growth at pH 5.0 in continuous culture and such pH 5.0-adapted cells regained the same survival characteristics at pH 2.5 as UA159 demonstrating that adaptation had reduced the effect of the RecA deficiency. Combined with the observation that adapted cells of the mutant were more resistant to UV irradiation, the overall results suggested that the organism possessed a RecA-independent, acid-inducible repair system. Subsequent work identified an AP endonuclease in *S. mutans* that was induced at low pH, was independent of the RecA protein and showed other functional similarities to DNA repair enzymes in other bacteria (Hahn *et al.,* 1999). Clearly, these studies provide a hint of the complexity of the mechanisms employed by bacteria to enhance their survival under adverse environmental conditions.

Other Environmental 'Stress' Factors

In addition to acid, oral bacteria in the plaque biofilm can also be stressed by the presence or absence of nutrients, oxygen, competing bacteria and inhibiting substances, with each individual organism within a community having specific requirements for growth and survival. Compared with many natural ecosystems, the nutrient supply in the mouth is more abundant and localized, although starved micro-environments exist at individual sites, such as occlusal fissures, caries lesions and the condensed layer of mature plaque next to the tooth surface. Since saliva is the main nutrient source for supragingival plaque, survival of an individual organism depends on its ability to collaborate with neighbours in the micro-community to degrade the complex salivary glycoproteins, proteins and peptides. As shown in early studies of tube-fed subjects (Littleton *et al.,* 1967), plaque bacteria can exist for long periods of time ($\geq$ 30 months) without the oral consumption of food. As one might expect, the plaque of such individuals was much less acidogenic when supplied with sucrose, glucose and fructose than that of subjects fed orally in the normal manner. Of interest, however, was the fact that while counts of lactobacilli and filamentous bacteria were low and observed in only one of the seven subjects, the streptococci were present in all tube-fed individuals and averaged 2.5×10^6 per mg of plaque, which was only 12-fold lower than that of persons fed orally. Thus, the absence of diet reduced plaque acid formation, but the major acidogenic population, the streptococci, remained in relatively high numbers in a 'glycolytically-restrained' condition. This 'physiological flexibility' of the oral streptococci is the basis for their ability to persist in the oral cavity under conditions of carbohydrate 'starvation', as well as being able to adjust rapidly to high sugar concentrations in dietary foods.

The remarkable ability of oral streptococci to persist in the mouth under extreme conditions of stress was also seen in a recent study of the oral flora of children undergoing bone marrow transplantation (Lucas *et al.,* 1997). The children were treated with chemotherapy and total body irradiation, in addition to intensive oral care, including antifungal drugs and rinsing with 0.2% chlorhexidine solution four times daily. The majority (16/20) were pyrexial 5-6 days following transplantation and were administered antibiotic regimens effective against Gram-negative and Gram-positive pathogens. While there was a significant decrease in the total streptococcal flora 7 days post-transplantation, the predominant species were members of the '*S. oralis* group' (*S. mitis* and *S. oralis*) which increased from 12.1% at baseline to 48.4% of the total anaerobic count at the 7th day. Thus, with the 16 pyrexial children, the '*S. oralis* group' was able to persist in spite of 36-48 hours of intravenous antibiotic administration. Since the children were fed by total parenteral nutrition prior to the 7-day sample, this persistence was attributed to the ability of this group to specially bind to, and degrade, the carbohydrate sidechains of glycoproteins in saliva and in human serum for growth.

The presence or absence of oxygen is a major environmental factor influencing the growth, metabolism and survival of oral bacteria. The early observations of Ritz (1967) demonstrated that if plaque was allowed to develop undisturbed over time, the aerobic populations in plaque decreased, the obligate anaerobes increased while the facultative bacteria, such as the streptococci, remained relatively constant. The fact that the oxidation/reduction potential (E_h) of supragingival plaque has been shown to be as low as -141 mV (Kenny and Ash, 1969) has resulted in the long-held concept that the key to survival for strict anaerobes in plaque is their association with microaerophilic or facultative bacteria that can maintain a low redox potential. This concept has been confirmed in a mixed culture oral biofilm study which showed that obligate anaerobes not only persisted, but grew in biofilms in the presence of oxygen even when the dissolved oxygen tension was 40-50% and the E_h was as high as +100 mV (Bradshaw *et al.*, 1996).

The proximity of oxygen to many supragingival plaque sites suggests that even the obligate anaerobes in the plaque microflora are exposed to considerable levels of oxygen. Marquis (1995) has suggested that all plaque bacteria metabolize oxygen to varying degrees with the differentiation into the various oxygen-sensitive classes based upon the ability of an organism to defend against the toxic effects of various reactive oxygen byproducts. Newer information has shown that organisms previously thought to be strict anaerobes can tolerate oxygen and grow in both aerobic and anaerobic environments. Survival against oxidative stress is dependent on the ability of an organism to metabolize oxygen to harmless end-products and/or to remove toxic oxygen by-products through the action of protective enzymes, such as superoxide dismutase, peroxidase and catalase. Aerobes utilize oxygen-linked respiratory chains and are superbly equipped with defensive enzymes, whereas obligate or strict anaerobes lack respiratory chains and are normally devoid of such enzymes. Facultative anaerobes, such as the oral streptococci, do not require oxygen for growth, although they take up oxygen at rates similar to aerobic organisms, producing H_2O_2 or water. Although this uptake is not linked to ATP formation, it contributes to lowering the O_2 concentration and E_h in plaque (Marquis, 1995).

Saliva is a factor in oxygen survival by oral bacteria. In addition to the antimicrobial components, lysozyme, lactoferrin, antibodies, and histidine-rich peptides, saliva contains salivary peroxidase, which catalyzes the oxidation of thiocyanate (SCN^-) by hydrogen peroxide (H_2O_2), generated during oxidative metabolism of oral bacteria, to the weak oxidizing agent, hypothiocyanite ($OSCN^-$), that has been shown to inhibit the glycolytic enzyme, glyceraldehyde 3-phosphate dehydrogenase, in oral streptococci (Carlsson *et al.*, 1983; Thomas *et al.*, 1994). Thus, this process has the potential of inhibiting the establishment of alien organisms in the mouth, control members of the indigenous plaque flora, and to protect host epithelial cells from the harmful effects of H_2O_2. Experiments with *S. mutans* GS-5 and *S. sobrinus* OMZ-176 have indicated that H_2O_2 + salivary peroxidase and SCN^- was much more inhibitory than H_2O_2 alone, suggesting that hypothiocyanite is an effective inhibitor of microbial metabolism.

In addition to acid stress, cell products such as bacteriocins are antagonistic to the growth and activity of certain oral bacteria. For example, the mutacins formed by mutans streptococci have a relatively wide spectrum of antibacterial activity. However, this may be restricted to micro-environments in plaque since both sensitive and mutacinogenic strains can be isolated from the same plaque samples. The mutacin produced by *S. mutans* T8 is a lanthionine-containing polypeptide that is sensitive to trypsin (Novak *et al.,* 1994); consequently, such compounds in plaque would be subject to degradation by proteases, which would reduce activity at local sites. Also, mutacin-producing strains have been shown to require a minimal amount of carbohydrate (0.09%) in order to produce mutacins, suggesting that these inhibitors are not a factor in microbial interactions in the absence of dietary carbohydrate (Bowden and Hamilton, 1998).

Like the study of acid tolerance, the physiological response of *S. mutans* to other stress factors, such as salt, heat, oxidation and starvation, has also been examined by incubating log-phase cells with [^{14}C]-amino acids during the first 30 minutes of the application of stress, followed by extraction and 2D electrophoretic separation of the cellular proteins (Svensäter *et al.,* 2000). The addition of NaCl (0.2 M), heat (42°C), hydrogen peroxide (2 mM), or dilution of medium components, resulted in global changes to protein synthesis in a fashion analogous to that seen with acid shock. Using consistent 2-fold changes as a threshold, oxidative stress (H_2O_2) upregulated 69 proteins, 15 of which were oxidation-specific, and downregulated 24 proteins when compared with the control cells. Reduction in the concentration of salts, vitamins and amino acids of the basal medium resulted in the increased formation of 58 proteins, with 11 starvation-specific proteins and 20 showing decreased synthesis. Some 52 and 40 proteins were enhanced by salt and heat stress with 10 and 6 of these proteins specific to the stress, respectively. Interestingly, when the acid protein profiles are included, it is apparent that a significant number of proteins were increased by more than one stress condition, in fact, six proteins were enhanced by all five stress conditions and could be classified as general stress proteins. Of potential relevance to the caries process was the observation that prior adaptation to the other stresses, except heat (42°C), protected the cells during a subsequent acid challenge at pH 3.5. For example, 5-fold dilution of the basal medium and glucose increased the numbers of acid-resistant survivors 12-fold, while elimination of the glucose entirely caused a 7-fold enhancement compared with the control cells, indicating a relationship between the acid and starvation-stress responses. Clearly, the response of *S. mutans* to adverse environmental conditons results in complex and diverse alterations in protein synthesis that serve to promote cell survival.

Base Formation

In most cases, the production of acid by dental plaque during the consumption of dietary carbohydrate overwhelms buffering and clearance by saliva in Phase I of the Stephan curve (Figure 2-5). The factors involved in the pH minimum (Phase II) and the 'pH rise' portion (Phase III) of the curve are complex, depending on the concentration of the carbohydrate, the amount and properties of the microflora, and the composition and flow rate of saliva. The pH minimum can be seen as the point of exhaustion of the carbohydrate source, the cessation of microbial metabolism due to the low pH, or an equilibrium between the low rate of acid formation and clearance/buffering by saliva. In Phase III the salivary factors are augmented by two microbial processes that increase plaque pH: acid utilization whereby strong acids are metabolized to weaker acids and gaseous products, and base formation from the metabolism of urea, proteins and peptides with the formation of ammonia and amines.

Acid utilization by oral bacteria in dental plaque has not been studied extensively and the degree to which this process modifies plaque pH is not known, although there has been some interest in acid conversion to other products as a feature of plaque 'food chains'. The best example of acid conversion is the metabolism of the strong acid, lactic acid by species of *Veillonella* to the weaker acids, propionic and acetic acids, and carbon dioxide and hydrogen gas (Carlsson, 1986). Species of *Veillonella* fit well into plaque ecology since they are anaerobic and form a food chain with the dominant acidogenic bacteria that generate lactic acid, such as the streptococci. Such a mutualistic arrangement benefits both populations since the veillonellae are presented with an energy source and the conversion of lactic acid to weaker acids decreases the acid of the micro-environment resulting in the stimulation of streptococcal glycolysis (Hamilton and Ng, 1983). Therfore, it is not surprising that *Veillonella* can be positively associated with human sucrose consumption and increases in *S. mutans* and lactobacilli (Minah *et al.,* 1985), as well as, *in vitro* continuous culture studies with mixtures of oral bacteria subjected to glucose pulses resulting in low pH environments (Bradshaw *et al.,* 1989). Lactate can also be used by members of the *Propionibacterium*, *Clostridium* and *Eubacterium*, with *E. alactolyticum* also able to convert acetic acid to butyric and caproic acids causing decreases in environmental acidity (Carlsson, 1986).

Urea is the most readily available and simplest base-generating substrate in the mouth, being present at a concentration of 2 - 5 mM in saliva with concentrations several times higher in gingival crevicular fluid (Golub *et al.,* 1971). Urea in saliva and crevicular fluid is rapidly degraded to ammonia and carbon dioxide by the action of microbial urease (Equation 2) and, while plaque has significant ureolytic activity, the identification of the ureolytic bacteria has not been completed. Ureolytic species of *Actinomyces* and *Streptococcus*, along with *Staphylococcus epidermidis* and *Bifidobacterium adolescentis*, have been found in low numbers in some subjects (Gallagher *et al.,* 1984), while ureolytic *Haemophilus parainfluenzae* has been found in high numbers at sites with high pH and ammonia (Salako and Kleinberg, 1989).

$$NH_2 - CO - NH_2 + H_2O \xrightarrow{\text{urease}} 2NH_3 + CO_2 \qquad \text{Equation 2}$$

One oral organism known to produce urea is *S. salivarius*, which is present in high numbers in saliva. Even though urease is an intracellular enzyme requiring urea to be transported into the cell to be degraded, the resulting ammonia is removed from the cell rapidly, since the acidification of cell suspensions in the presence of urea results in rapid pH increases (Kleinberg, 1967). Moreover, the intracellular formation of ammonia maintains internal pH during periods of falling extracellular pH and this pH difference can enhance survival even at pH values that will progressively kill cells. The urease level in *S. salivarius* is regulated by pH and increases in specific activity with changes in pH from 7.0 to 5.5. The magnitude of these changes differs depending on whether the cells are grown in batch or continuous culture; however, the activity of the enzyme is severely compromised at pH values below 4.5 (Sissons and Hancock, 1993). Recent continuous culture studies have demonstrated that urease activity in *S. salivarius* is also increased by increasing growth rate and the presence of excess glucose under conditions of nitrogen limitation (Chen and Burne, 1996).

Recent studies by Clancy and Burne (1997) have involved the transformation of *S. mutans* (strain GS5) with the nickel-dependent urease genes from *S. salivarius*. By regulating the nickel concentration in the growth medium, it was possible to test the effect of increasing urease activity on the pH achieved during glycolysis with glucose as the carbon source. With the transformed strain, the simultaneous catabolism of glucose with as little as 2 mM urea resulted in final pH values 0.5 units higher than with glucose alone, and increasing the nickel concentration to 25 μM nickel chloride with 10 mM urea was sufficient to eliminate acidification by glucose. These experiments demonstrated that it is theoretically possible to genetically engineer strains of *S. mutans* to reduce their cariogenicity by incorporating enzymes that will enhance base production from natural substrates, thereby, neutralizing the acid formed during glycolysis.

Another plaque organism that is known to possess urease activity is *A. naeslundii* genospecies 1. Very recent work with strain WVU45 (Morou-Bermudez and Burne, 1999) has identified the nucleotide sequence of the urease gene cluster, which is comprised of seven open-reading frames. This led to the construction of a urease-defective mutant by insertional inactivation of the *ureC* gene. Acid-sensitivity studies demonstrated that in the presence of 25 mM urea, the wild-type strain had 100-fold higher numbers of survivors at pH 4.0 than the mutant strain, and the terminal pH was 3 units higher than that of the same cells incubated without urea. However, urea had no protective effect at pH 3.0, confirming that urease activity decreases as the external pH values falls. These experiments also demonstrated that in addition to pH homeostasis, urea would satisfy the nitrogen requirements of the organism, suggesting that such organisms would have a selective advantage in dental plaque over organisms devoid of ureolytic activity.

The other source of alkaline products in dental plaque is salivary or crevicular protein. Historically, nitrogen metabolism by the bacteria in supragingival plaque has received relatively little attention, although this has begun to change in recent years with increasing interest in the ability of oral bacteria to utilize host-derived glycoproteins and proteins for growth. The removal of the oligosaccharide side-chains from complex glycoproteins increases the susceptibility of the protein core to proteolysis leading to the formation of polypeptide chains, smaller peptides and ultimately the unit amino acids (Figure 11). Further metabolism of small peptides and amino acids by plaque bacteria contributes to the acid-base balance in dental plaque through the formation of ammonia or amines depending on the pH of the environment of the local habitat. This process requires the synergistic action of the plaque consortium and, as a consequence, base formation from protein degradation is normally a slow process that cannot compete with the rate of acid formation from dietary carbohydrate. Nevertheless, this type of metabolism can sustain the oral

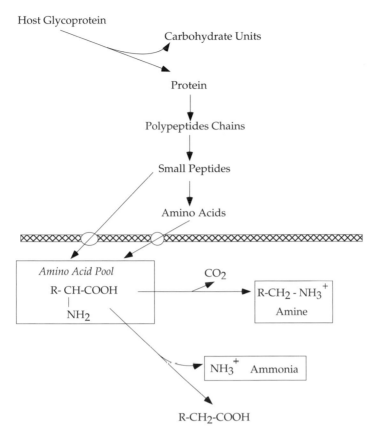

Figure 11. General pathway of host glycoprotein degradation by the concerted action of the oral microflora in dental plaque.

microflora for prolonged periods in the absence of food consumption by their host.

Coupled to protein degradation and base formation is the role of arginine degradation by the arginine deiminase system (ADS) of oral bacteria. The ADS degrades arginine to ornithine, CO_2 and NH_3, and this activity has been reported for a variety of oral strains including *A. naeslundii*, *Treponema denticola*, *L. fermentum*, and various streptococci, such as. *S. sanguis*, *S. gordonii*, and *S. rattus*, but not *S. mutans*. Of interest to plaque acid-base physiology is the observation that ADS activity has been reported at pH values below the minimum for growth and glycolysis (Marquis *et al.*, 1987). For example, *S. sanguis* NCTC 10904 exhibited a pH minimum of 5.2 for growth and glycolysis, but was able to carry out arginolysis at pH 4.0. An interesting cyclic survival strategy has been postulated for such bacteria in dental plaque exposed to high concentrations of sugar. Even in the presence of arginine, the ADS is normally repressed by glucose with the regulation postulated to occur via the glucose-PTS transport system. However, as glycolysis proceeds to generate acid, the cells enter the stationary phase at low pH, a condition that represses the PTS. Under these conditions and in the presence of arginine, the ADS is derepressed even though sugar may be available. Increased ADS activity results in the increase in the pH, allowing glycolysis to be 'reactivated' resulting in repression of the ADS until the pH has again reached a low level, whereupon the cycle can be repeated by the derepression of the ADS provided an adequate carbon source is available.

Caries Lesions -To Be or Not to Be

Hopefully, the above general background of the plaque biofilm, the oral microflora, and the forces promoting and inhibiting the caries process will now make it possible to examine in a little more detail the conundrum associated with the approximal caries lesions presented in Figure 1 [A] and [B], i.e., how can a caries lesion appear at an approximal site of one tooth and not at a site only a millimeter away on the adjacent tooth? In keeping with the complexity of the caries process alluded to above and the individual age and eruption patterns of the various teeth in these photographs, it will not be difficult to see that no one single answer will suffice in both these situations. Initially, it would be useful to list a set of variables that might contribute to the caries process at the carious site which do not exist at the non-caries site. These variables can be grouped under each of the three main components of the ecosystem: namely, the characterisitcs of the habitat itself, and those associated with the environment (abiotic) and the biological (biotic) components of the system (Table 5).

With respect to the lesion in Figure 1[A], the status of the habitat was critical to the caries process since the subject was 11 years old and the carious tooth (#35) erupted 4-5 years after the caries-free tooth (#36). Thus, the latter tooth had undergone

Table 5. Factors that May Contribute to the Incidence of Caries at One Approximal Site and Not at a Second Approximal Site on the Adjacent Tooth

Ecological Component	Factors
Habitat	Age of the individual. Differences in time of tooth eruption leading to differences in mineralization. Differences in the morphology of the tooth surface that promotes plaque accumulation. Differences in the amount of fluorapatite in enamel. Differences in exposure to saliva. Food trapping.
Environment	Differences in the retention of sugar and acid end-products at the caries site. Differences in saliva flow.
Microflora	More bacteria (thicker plaque) are associated with the caries site. Flora at the caries site is dominated by aciduric bacteria. The amount and composition of the flora of the two sites may be the same except that localized anti-microbial factors from saliva or other bacteria may limit the flora at the non-caries site. The amount of plaque at both sites is the same, but local microbial interactions limit the aciduric bacteria at the non-caries site.

a much longer period of secondary maturation during which fluoride was undoubtedly taken up into the apatite lattice of the tooth resulting in the formation of fluorapatite making the tooth more resistant to demineralization by acid than tooth #35. The appearance of the lesion in the younger tooth also suggests that the total sugar challenge in the 5-11 year period had increased compared with the early five year period, coupled perhaps with a reduction in effective oral hygiene procedures. This would result in the accumulation of more and older plaque at the approximal sites leading to the increased retention of sugar and acid end-products. The lower pH profile so generated would promote the proliferation of the acidogenic/aciduric flora, such as lactobacilli, *S. mutans* and the low pH non-mutans streptococci, leading to increasingly lower pH profiles. These changes in the composition of the plaque microflora would overpower the capacity of the community to generate basic end-products from urea and protein driving the equilibrium of Equation 1 in the direction of demineralization.

The approximal mesial and distal lesions on tooth #45 in Figure 1 [B] present a different and more perplexing scenario since the subject was 28 years old and the question of differential eruption times would appear not to be a factor in this case. The evidence suggests that tooth #45 is caries prone; however, it is conceivable that with time the sites on the adjacent teeth (#44 and #46) would also develop carious lesions. Certainly, the appearance of restorations in all three teeth indicate that the total dentition has a history of being challenged by the continual consumption of a high carbohydrate diet accompanied by the retention of acidogenic/aciduric plaque microflora due to inadequate toothbrushing and flossing. It is difficult to see differential effects of saliva flow at the approximal sites being a factor, although the radiographs do not provide an adequate picture of the tooth architecture that would influence the action of saliva. The morphology of the carious sites could promote the accumulation of more plaque than the caries-free sites and differences in the mineral composition of tooth #45 cannot be excluded.

A major element in the caries process is the number, type and characteristics of the bacteria of the local habitat. The evolution of a caries-producing ecosystem results in the ultimate dominance by a group of acidogenic 'pathogens' that can function physiologically at low pH. Within the time frame of caries development, each small site undergoes a constant series of pH reductions during the consumption of carbohydrate, most of which reach into the 'critical pH' zone of enamel demineralization. The key element is the amount of time the tooth is exposed to the critical pH. As a consequence, carious plaque might be defined as a group of bacteria generating sufficient acid from carbohydrate substrates over time to sustain the equilibrium in Equation 1 predominantely in the direction of demineralization. Non-carious plaque would, on the other hand, support demineralization only for brief periods during carbohydrate consumption, but would sustain the equilibrium in the direction of remineralization.

An important question is what are the key factors driving plaque development towards carious or non-carious plaque at a tooth site during the early stages of development? Clearly, plaque development in subjects with high caries activity will be strongly influenced by the acidogenic and aciduric bacteria in saliva and the reservoirs on the teeth. In such subjects, one can imagine a relatively 'straight line' process to lesion development in the presence of a relatively constant supply of dietary carbohydrate, and in the absence of adequate oral hygiene procedures. Similarly, subjects with greatly diminished saliva flow (xerostomia) would follow the same 'straight line' process in the absence of an aggressive oral hyiene regime. In these cases, the environment favours demineralization and one might even suggest that caries development was preordained. However, in cases where the environment does not immediately favour lesion formation, one might imagine that plaque develops more or less randomly with an increasingly diverse collection of bacteria that is subject to constant modification in response to environmental forces. Potentially cariogenic bacteria in plaque would occupy their normal niche being constrained by competition with other bacteria. Only during an alteration in the environment, such as the increased consumption of carbohydrate, or a physiological change to increase its competitiveness, would cariogenic bacteria begin to dominate the plaque microflora. The previously mentioned results of Schüpbach *et al.,* (1996) demonstrated that the evolution of root surface lesions involved a series of different microbial communities. Succession in the lesion was probably influenced by the environment and by interaction among the members of the lesion community.

Control of Caries

No chapter on the ecological basis for dental caries would be complete without a short discussion on caries control, or at least the philosophy of caries control. Much has been written, of course, on the various strategies that can and are being applied to reduce the incidence of caries from the use of chemicals to modify both the composition of tooth enamel and the activity of microflora in dental plaque, to those designed to modify the behavior of the patient. Since normal plaque at non-diseased sites is protective, 'caries control' implies a modulation of the amount of dental plaque and the activity of the indigenous flora. Some authors have proposed the development of the "specific pathogen-free human" by genetic modification of the

---

Table 6. Procedures that Might be Applied to Control Caries that are Consistent with Good Ecological Principles.

Ecological Component	Procedure
Habitat	Alter the chemical composition of enamel to make it more resistant to demineralization (e.g. fluorapatite). The use of sealants to protect surfaces from caries. The use of plaque-inhibitory varnishes.
Environment	Reduce the consumption of sugar-containing foods. Eat carbohydrate-containing foods under conditions of high saliva flow. Use sugar substitutes and anti-caries substances such as xylitol.
Microflora	Use effective oral hygiene procedures in a consistent fashion. Apply anti-microbial agents that modify the activity of the acidogenic oral flora, but do not eliminate it, thereby, creating minimum ecological disruption. The use of replacement therapy targeting the principal cariogenic bacteria. The use of vaccines against epitopes on surface molecules of principal cariogenic bacteria.

host, the oral microflora, and by the use of anti-idiotype vaccines (Taubman *et al.*, 1989). In contrast, this chapter concludes with a discussion of procedures that are currently relevant and have a good basis in ecology. Table 6 lists some of the measures that might be applied and which are based on sound ecological principles, the latter referring to procedures that result in minimal ecological disruption of the indigenous microflora.

A good example is the use of fluoride, both through drinking water and topical application in hygiene products. The addition of 1 ppm fluoride to community water sources is probably the single most effective public health measure that has resulted in a general reduction in dental caries over the past 20 years. The primary anti-caries effect of fluoride is the conversion of hydroxyapatite in tooth enamel to the more acid-resistant fluorapatite with a secondary effect being the partial inhibition of carbohydrate metabolism by the plaque microflora (Hamilton and Bowden, 1988). The success of fluoride as an anti-caries agent serves as a 'gold standard' by which other methods of caries control can be measured and serves to remind us that effective procedures must have an ecological basis. The medical and dental literature is replete with references to situations where the use of anti-microbial agents, usually antibiotics, have resulted in elimination of the indigenous flora only to be replaced by a more dangerous or obnoxious microflora. A good example of the protective nature of the indigenous oral flora are cases of candidiasis, or the overgrowth of the normal flora by *Candida albicans*, following the administration of antibiotics that reduce the Gram-positive microflora in the mouth. In the case of fluoride, the conversion of enamel hydroxyapatite to fluorapatite is a beneficial modification of the tooth structure and the inhibition of glycolysis by fluoride does not eliminate the acidogenic flora, but serves only to reduce the amount of acid that is generated in the presence of exogenous sugar.

The other measures listed in Table 6 are accepted procedures for caries control, with the exception of the use of replacement therapy. Theoretically, this latter procedure is valid on the basis of ecological principles since it would replace the pathogen native to dental plaque with an altered strain that would be less virulent. To be effective, the replacement strain would have to adhere and grow in dental plaque and assume the same niche as the original native strain. The concept of replacement therapy in caries prevention was tested in early animal studies with lactic dehydrogenase (LDH)-defective mutants of mutans streptococci since this enzyme is responsible for formation of lactic acid at high sugar concentrations (Figure 9). Johnson and Hillman (1982) demonstrated that the prior establishment of such a strain in rats resulted in a 10-10,000-fold increase in the minimum infectious dose for persistent colonization in the rat by the wild-type strain. Subsequent work by Fitzgerald and coworkers (1989) demonstrated the reduced cariogenicity of a LDH-defective serotype c strain of *S. mutans* that also exhibited bacteriocin-like activity inhibiting other cariogenic mutans streptococci. These early studies involved the isolation of spontaneous or chemically-derived mutants and, as such, were hampered by questions of reversion and the potential for mutiple genetic lesions with the use of some chemical mutagenic agents.

More recent research has taken advantage of the newer molecular biological techniques to generate mutants with stable characteristics. Using such methods, it was soon apparent that the insertional inactivation of LDH in *S. mutans* was a lethal event (Hillman *et al.*, 1994) unless the cells were grown anaerobically on limiting glucose. While pyruvate can be metabolized by pyruvate dehydrogenase and the pyruvate:formate lyase system (Carlsson, 1986), these reactions in *S. mutans* are insufficient to re-oxidize all of the cellular NADH to NAD at high glucose concentrations and the accumulation of NADH results in growth inhibition. To overcome this problem, a recombinant mutant (CH4ts) of *S. mutans* NG8 was generated possessing a temperature-sensitive LDH that permitted growth at 30°C, but not 42°, unless it was grown on limiting glucose. To increase the NADH-oxidizing capacity of mutant CH4ts, the alcohol dehydrogenase II gene from *Zymomonas mobilis* was fused to and placed under the control of the *spaP* gene of the mutant. The mutant grew as well as, if not better, than the wild-type strain under a variety of conditions, producing from sugar substrates less lactic acid, but more ethanol (Hillman *et al.*, 1996). More recently, a mutant strain of *S. mutans*, named BCS3-L1, was derived entirely deficient in LDH activity and engineered to produce high levels of the peptide antibiotic, mutacin 1140, which provides the organism with selective advantage over mutacin-sensitive strains of *S. mutans* (Hillman, 1999). Strain BCS3-L1 was genetically stable, possessed strong colonization properties and was significantly less cariogenic than the wild-type strain in gnotobiotic rats. As a consequence, provided such a mutant could be successfully implanted in human dental plaque and occupy the niche of native strains of *S. mutans*, the potential exists for the development of a less acidogenic microbial community through replacement therapy.

A second approach involves the construction of recombinant strains of *S. mutans* that are capable of base production to counteract the acidification of the environment during carbohydrate metabolism. An example of this strategy is work by Clancy and Burne (1997), mentioned previously, in which *S. mutans* GS-5 was transformed with a plasmid containing the urease genes, *ureA-G*, from *S. salivarius* 57.1, producing a stable recombinant strain capable of degrading urea to ammonia, a property that was not exhibited by the wild-type strain. This research demonstrated that by increasing the urease activity of the cells, it was possible in the presence of physiological concentrations of urea to reduce the depth and duration of the pH fall during glycolysis with glucose as the substrate. Thus, replacement therapy utilizing such a strain would serve to increase the ureolytic capacity of dental plaque and increase the minimum pH observed with carbohydrate substrates provided that the strains would implant in dental plaque comprised of the normal indigenous flora.

A considerable long-term research effort has been directed at the development of a vaccine against caries by utilizing specific surface components of mutans streptococci. One target for immunological intervention are the glucosyltransferases (GTF), enzymes responsible for the synthesis of extracellular glucan from sucrose. Glucans are associated with plaque accumulation by interacting with glucan-binding proteins (GBPs) on mutans streptococci and other streptococci. GTF is one such protein, possessing both catalytic and glucan binding properties, although not all binding proteins have enzymatic activity. The N-terminal region of GTF contains the catalytic domain, while the glucan-binding site is in the C-terminal third of the protein (Abo *et al.*, 1991). Synthetic peptide constructs of these two regions have been used to immunize rats to determine whether the induced antibodies would protect against dental caries in the rat model system. In one study (Taubman *et al.,* 1995), serum immunoglobulin (IgG) antibodies to both peptides were shown to inhibit insoluble glucan synthesis by *S. sobrinus* GTF, while the catalytic peptide antibody inhibited soluble glucan synthesis by the *S. mutans* GTF. Importantly, both peptides provided protective immunity against dental caries in the rat system.

Further detailed analysis of the glucan-binding proteins showed that a highly immunogenic 59 kDa protein (GBP$_{59}$) was secreted by fresh and laboratory strains of *S. mutans* and was observed in the saliva of children during the initial colonization by mutans streptococci (Smith and Taubman, 1996). Immunization of rats with the GBP$_{59}$ from *S. mutans* was shown to interfere both with the accumulation of the organism and the incidence of dental caries caused by the organism in rats. More recently, serum IgG and salivary IgA antibodies, induced by two synthetic peptides whose sequences contained the b5 and b7 strands of the putative catalytic region of the *S. mutans* GTF, were shown to interfere with GTF-mediated glucan synthesis (Smith *et al.,* 1999). The epitopes comprising these two peptides had significant immungenicity and would be good candidates to test for anti-caries activity in the rat model system.

Studies on the interference in adherence of *S. mutans* to enamel have also involved testing muscosal antibodies raised to the surface fibrillar protein, streptococcal antigen Ag I/II (SA I/II). This protein interacts with salivary glycoprotein in the acquired pellicle or plaque matrix to promote the adhesion of *S. mutans* to the tooth surface (see review: Hajishengallis and Michalek, 1999). Coupled with the considerable advances in the understanding of the mucosal immune system and the delivery of antigens to enhance antibody formation, the use of such vaccines is feasible, particularly with young children at the time of tooth eruption prior to the "window of infectivity" of the mutans streptococci (19 to 31 months)(Caulfield *et al.,*1993). Interference with the re-colonization of mutans streptococci on human teeth has been demonstrated in an early study using topically-applied mouse monoclonal IgG antibodies to AgI/II after the treatment of the subjects with chlorhexidine (Ma and Lehner, 1990). This experiment showed not only the inhibition of colonization, but a substantive effect of treatment since the teeth were free of *S. mutans* for a two-year period. Since the antibody was present for only a brief period, the authors suggested that the niche normally occupied by *S. mutans* was filled by other members of the plaque microflora during the early stages of treatment, thereby, preventing re-colonization by *S. mutans*.

More recent studies have examined the binding activity of SA I/II (Ag I/II) in greater detail. Munro *et al.,* (1993) initially demonstrated that binding was associated with the region of the protein comprising residues 816 to 1213 (fragment 3). More recently, Todryk *et al.,* (1996) showed that the topical application of monoclonal antibodies that recognized this fragment prevented the colonization of *S. mutans* in primates. Subsequently, inhibition of *S. mutans* binding to salivary receptors was demonstrated *in vitro* using a synthetic peptide (p1025) corresponding to residues 1025-144 of this fragment (Kelly *et al.,* 1999). Two residues, Q1025 and E1037, within the peptide, identified by site-directed mutagenesis, were shown to be involved in surface binding. The direct application of p1025 to human teeth prevented the colonization of *S. mutans*, but not *Actinomyces*, suggesting that a variety of peptide inhibitors might be generated against various bacteria to prevent their colonization of oral surfaces.

Future Directions

Even with the passage of over 100 years, our understanding of the etiology of dental caries continues to evolve. Organisms such as *S. mutans* and *Lactobacillus* spp. have long been associated with the disease; however, the application of more comprehensive and sophisticated microbiological analysis in cross-sectional and longitudinal human studies over the past 20 years has expanded the list of the putative pathogens. Much more information is needed on the contribution of the individual species grouped within the "low pH-non-mutans streptococci" and the *Actinomyces*. And what of the role, if any, of those organisms not normally associated with caries etiology that assume dominant positions in incipient and advanced root lesions, such as those listed in Table 2. Cariogenic bacteria are generally assumed to be acidogenic with the capacity to metabolize carbohdyrate substrates at low pH values and to possess a degree of acid tolerance that permits them to grow and survive in such environments. If this definition applies to those organisms in Table 2 then one must assume that some genotypic or phenotypic change had occurred to alter cell physiology in a substantial manner to permit growth and survival under acidic conditions.

A variety of environmental signals are received by bacteria in complex communities, including those from other bacteria in the habitat. The signal triggering a change increasing fitness to acid may have been received and acted upon by the entire community, a species population within the community, or even a single cell within a population. Since bacteria in dental plaque are sequestered in the biofilm, we can assume that within a species population, diversity will increase with distinct clones descending from a single ancestor, much like the wedge-shaped, radically oriented sectors seen within colonies of a single organism on agar plates (Shapiro, 1997). Such self-generated changes within bacteria to increase fitness would likely result in alterations in dedicated signalling and regulatory molecules and ultimately gene expression. We already have evidence of differences between members of the same species as exemplified in Table 3 by the variation in glycolytic rates by fresh isolates of *S. mutans*. Since the domination of lesions by organisms other than *S. mutans*, *Lactobacillus* spp. and the non-MS may be a relatively rare event, it is conceivable that there will be a species variation in the potential to receive and act on 'fitness' signals.

These future studies, of course, require that we have a much better understanding of the characteristics and properties of oral biofilms, and the degree to which such microbial communities permit 'biochemical flexibility' and enhance survival from external and internal forces. An array of sophisticated techniques are now being applied to the study of biofilms, including fluorescence microscopy, scanning laser confocal microscopy, attenuated total reflectance infrared spectroscopy, two-photon scanning microscopy and a variety of molecular biological, fluorometric and other methods. Current results clearly show that biofilm formation involves the activation of specific genes (Burne *et al.,* 1997; Davies *et al.,* 1998). Thus, future research will undoubtedly involve the increasing application of molecular biology to oral biofilms, such as the use of reporter genes, differential display polymerase chain reaction, *in vivo* expression technology, and proteomics, to name a few. Such methods will not only identify biofilm-expressed genes, but permit the identification of the important genes involved in global signalling that enhances survival to environmental stress, such as acid, oxidation and starvation, as well as, the response of biofilm cells to the application of biocides. These methods could also be used to test the effects of interspecies interactions in mixed biofilm model systems. The combined information would provide a clearer picture of dental plaque as an integrated multicellular unit operating with a complex network of intercellular communication that can be involved in gene expression, differentiation and other cellular processes. From this knowledge should come the identification of key molecular events in the plaque microflora that trigger the pathogenic process leading to dental caries.

References

Aamdal-Scheie, A., W. -M. Luan, G. Dahlén, and O. Fejerskov. 1996. Plaque pH and microflora of dental plaque on sound and carious root surfaces. J. Dent. Res. 75: 1901-1908.

Abo, H., T. Masumura, T. Kodama, H. Ohta, K. Fukui, K. Kato, and H. Kagawa. 1991. Peptides sequences for sucrose splitting and glucan binding with *Streptococcus sobrinus* glucosyltransferase (water-insoluble glucan synthetase). J. Bacteriol. 173: 989-996.

Alvarez, J. O., Caceda, J., T. W. Woolley, K. W. Carley, N. Baiocchi, L. Caravedo and J. M. Navia. 1993. A longitudinal study of dental caries in the primary teeth of children who suffered from infant malnutrition. J. Dent. Res. 72: 1573-1576.

Baev, D., R. England, and H. K. Kuramitsu. 1999. Stress-induced membrane association of the *Streptococcus mutans* GTP-binding protein, an essential G protein, and investigation of its physiological role by utilizing an antisense RNA strategy. Infect. Immun. 67: 4510-4516.

Beckers, H. J. A., and J. S. van der Hoeven. 1982. Growth rates of *Actinomyces viscosus* and *Streptococcus mutans* during early colonization of tooth surfaces in gnotobiotic rats. Infect. Immun. 35: 583-587.

Beckers, H. J. A., and J. S. van der Hoeven. 1984. The effects of mutual interaction and host diet on the growth rates of the bacteria *Actinomyces viscosus* and *Streptococcus mutans* during colonization of tooth surfaces in di-associated gnotobiotic rats. Arch. Oral Biol. 29: 231-236.

Beighton, D., K. Smith, and H. Hayday. 1986. The growth of bacteria and the production of exoglycosidic enzymes in the dental plaque of macaque monkeys. Arch. Oral Biol. 31: 829-835.

Belli, W. A, and R. E. Marquis. 1991. Adaptation of *Streptococcus mutans* and *Enterococcus hirae* to acid stress in continuous culture. Appl. Environ. Microbiol. 57: 1134-1138.

Berry, C. W., and C. A. Henry. 1977. The effect of adsorption on the acid production of caries and noncaries-producing streptococci. J. Dent. Res. 56: 1193-1200.

Bowden, G. H. W. 1991. Which bacteria are cariogenic in humans? P. 266-286. *In:* N. W. Johnson (ed.), Risk markers for oral diseases. 1. Dental caries. Cambridge University Press, Cambridge.

Bowden, G. H. 1995. The role of microbiology in models of dental caries: Reaction paper. Adv. Dent. Res. 9: 255-269.

Bowden, G. H., and I. R. Hamilton. 1987. Environmental pH as a factor in the competition between strains of the oral streptococci, *Streptococcus mutans, S. sanguis*, and "*S. mitior*" growing in continuous culture. Can. J. Microbiol. 33: 824-827.

Bowden, G. H. W., and I. R. Hamilton. 1998. Survival of oral bacteria. Crit. Rev. Oral Biol Med. 9: 54-85.

Bowden, G. H. W., J. Ekstrand, B. McNaughton, and S. J. Challacombe. 1990. The association of selected bacteria with the lesions of root surface caries. Oral Microbiol. Immunol. 5: 346-351.

Bowden, G. H. W., N. Nolette, H. Ryding, and B. M. Cleghorn. 1999. The diversity and distribution of the predominant ribotypes of *Actinomyces naeslundii* genospecies 1 and 2 in samples from enamel and healthy, carious root surfaces of teeth. J. Dent. Res. 78: 1800-1809.

Boyar, R. M., and G. H. Bowden. 1985 The microflora associated with the progression of incipient carious lesions in teeth of children living in a water-fluoridated area. Caries Res. 19: 298-306.

Boyar, R. M., A. Thylstrup, L. Holmen, and G. H. Bowden. 1989 The microflora associated with the development of initial enamel decalcification below orthodontic bands in vivo in children living in a water-fluoridated area. J. Dent. Res. 68: 1734-1738.

Boyd, D. A., D. G. Cvitkovitch and I. R. Hamilton. 1994. Sequence and expression of the genes for HPr (*ptsH*) and Enzyme I (*ptsI*) of the phosphoenolpyruvate-dependent phosphotransferase system from *Streptococcus mutans*. Infect. Immun. 62: 1156-1165.

Bradshaw, D. J., A. S. McKee, and P. D. Marsh. 1989. Effects of carbohydrate pulses and pH on population shifts within oral microbial communities *in vitro*. J. Dent. Res. 68: 1298-1302.

Bradshaw, D. J., K. A. Homer, P. D. Marsh and D. Beighton. 1994. Metabolic cooperation in oral microbial communities during growth on mucin. Microbiology 140: 3407-3412.

Bradshaw, D. J., P. D. Marsh, C. Allison, and K. M. Schilling. 1996. Effect of oxygen, inoculum composition and flow rate on development of mixed culture oral biolfilms. Microbiology 142: 623-629.

Brailsford, S. R., E. Lynch and D. Beighton. 1998. The isolation of *Actinomyces naeslundii* from sound and root surfaces and root caries lesions. Caries Res. 32: 100-106.

Brailsford, S. R., R. B. Tregaskis, H. S. Leftwich and D. Beighton. 1999. The predominant *Actinomyces* spp. isolated from infected dentin of active root caries lesions. J. Dent. Res. 78: 1525-1534.

Buckley, N. D., and I. R. Hamilton. 1994. Vesicles prepared from *Streptococcus mutans* demonstrate the presence of a second glucose transport system.

Microbiology 140: 2639-2648.

Burne, R. A., Y.-Y, M. Chen, and J. E. C. Penders. 1997. Analysis of gene expression in *Streptococcus mutans* in biofilms. Adv. Dent. Res. 11: 100-109.

Carlsson, J. 1986. Metabolic activities of oral bacteria. P.74-106. *In* A. Thylstrup and O. Fejerskov (ed.) Textbook of Cariology. Munksgaard, Copenhagen.

Carlsson, J., Y. Iwami , and T. Yamada. 1983. Hydrogen peroxide excretion by oral streptococci and effect of lactoperoxidase-thiocyanate-hydrogen peroxide. Infect. Immun. 40: 70-80.

Carlsson, P., I. A. Gandour, B. Olsson B. Rickardsson and K. Abbas. 1987. High prevalence of mutans streptococci in a population with extremely low prevalence of dental caries. Oral Microbiol. Immunol. 2: 121-124.

Caufield, P. W., G. R. Cutter and A. P. Dasanayake. 1993. Initial acquisition of mutans streptococci by infants: Evidence for a discrete window of infectivity. J. Dent. Res. 72: 37-45.

Chen, Y-Y. M., and R. A. Burne. 1997. Analysis of *Streptococcus salivarius* urease expression using continuous chemostat culture. FEMS Microbiol. Lett. 135: 223-229.

Clancy A., and R. A. Burne. 1997. Construction and characterization of a recombinant ureolytic *Streptococcus mutans* and its use to demonstrate the relationship of urease activity to pH modulating capacity. FEMS Microbiol. Lett. 151: 205-211.

Cvitokovitch, D. G., and I. R. Hamilton. 1994. Biochemical change exhibited by oral streptococci resulting from laboratory subculturing. Oral Microbiol. Immunol. 9: 209-217.

Cvitokovitch, D. G., D. A. Boyd, T. Thevenot, and I. R. Hamilton. 1995a. Glucose transport by a mutant of *Streptococcus mutans* unable to accumulate sugars via the phosphoenolpyruvate-dependent phosphotransferase system J. Bacteriol. 177: 2251-2258.

Cvitokovitch, D. G., D. A. Boyd, and I. R. Hamilton. 1995b. Regulation of sugar sugar uptake via the multiple sugar metabolism operon by the phosphoenolpyruvate-dependent sugar phosphotransferase transport system of *Streptococcus mutans*, p. 351-356. *In* J. J. Ferretti, M. S. Gilmore, T. R. Klaenhammer and F. Brown (eds.), Genetics of streptococci, enterococci and lactococci. Dev. Biol. Stand., Basel

Davies, D. G., M. R. Parsek, J. P. Pearson, B. H. Iglewski, J. W. Costerton, and E. P. Greenberg. 1998. The involvement of cell-to-cell signals in the development of a bacterial biofilm. Science 280: 295-298.

Dawes, C. 1989. An analysis of factors influencing diffusion from dental plaque into a moving film of saliva and the implications for caries. J. Dent. Res. 68: 1483-1488.

Dawes, C. and L. M. D. Macpherson. 1993. The distribution of saliva and sucrose around the mouth during the use of chewing gum and the implications for the site-specificity of caries and calculus deposition. J. Dent. Res. 72: 852-857.

De Jong, M. H. and J. S. van der Hoeven. 1987. The growth of oral bacteria on saliva. J. Dent. Res. 66: 498-505.

Dirksen, T. R., M. F. Little, B. G. Bibby and S. L. Crump. 1962. The pH of carious cavities. I. Effect of glucose and phosphate buffer on cavity pH. Arch. Oral Biol. 7: 49-58.

Distler, W., and A. Kröncke. 1986. Formic acid in human single-site resting plaque-quantitative and qualitative aspects. Caries Res. 20: 1-6.

Distler, W., and A. Kröncke. 1983. The acid pattern in human dental plaque. J. Dent. Res. 62: 87-91.

Dreizen, S. and L. R. Brown. 1976. Xerostomia and dental caries, p. 263-273. *In* H. M. Stiles, W. J. Loesche and T. O'Brien. Microbial aspects of dental caries. Information Retrieval, Washington, D. C.

Emilson, C. G., P. Carlsson, and D. Bratthall. 1987. Strains of mutans streptococci isolated in a population with extremely low caries prevalence are cariogenic in the hamster model. Oral Microbiol. Immunol. 2: 183-186.

Fejerskov, O., A. A. Scheie, and F. Manji. 1992. The effect of sucrose on plaque pH in the primary and permanent dentition of caries-inactive and –active Kenyan children. J. Dent. Res. 71: 25-31.

Fitzgerald, R. J., B. O. Adams, H. J. Sandham and S. Abhyankar. 1989. Cariogenicicity of a lactate dehydrogenase-deficient mutant of *Streptococcus mutans* serotype c in gnotobiotic rats. Infect. Immun. 57: 823-826.

Gallagher, I. H. C., E. I. F. Pearce and E. M. Hancock. 1984. The ureolytic microflora of immature dental plaque before and after rinsing with a urea-based mineralizing solution. J. Dent. Res. 63: 1037-1039.

Geddes, D. A. M. 1975. Acids produced by human dental plaque metabolism *in situ*. Caries Res. 9: 98-109.

Gibbons, R. J. 1964. Bacteriology of dental caries. J. Dent Res. 46: 1021-1028.

Gibbons, R. J. 1989. Bacterial adhesion to oral surfaces: A model for infectious diseases. J. Dent Res. 68: 750-760.

Golub, L. M., S. M. Borden, and I. Kleinberg. 1971. Urea content of gingival crevicular fluid and its relation to periodontal disease in humans. J. Periodontol. 6: 243-251.

Graf, H., and H. R. Mühlemann. 1966. Telemetry of plaque pH from interdental area. Helv. Odont. Acta 10: 94-101.

Gutierrez, J. A., P. J. Crowley, D. P. Brown, J. D. Hillman, P. Youngman, and A. S. Bleiweis. 1996. Insertional mutagenesis and recovery of interrupted genes of *Streptococcus mutan* by using Tn*917*: preliminary characterization of mutants displaying acid sensitivity and nutritional requirements. J. Bacteriol. 178: 4166-4175.

Gutierrez, J. A., P. J. Crowley, D. G. Cvitkovitch, L. J. Brady, I. R. Hamilton, J. D. Hillman, and A. S. Bleiweis. 1999. *Streptococcus mutans ffh*, a gene encoding a homoloque of the 54 kDa subunit of the signal recognition particle, is involved in resistance to acid stress. Microbiology 145: 357-365.

Hahn, K., R. C. Faustoferri, and R. G. Quivey. 1999. Induction of an AP endonuclease activity in *Streptococcus mutans* during growth at low pH. Mol. Microbiol. 31: 1489-1498.

Hamada, S. and H. D. Slade. 1980. Biology, immunology and cariogenicity of *Streptococcus mutans*. Microbiol. Rev. 44: 331-384.

Hamilton, I. R. 1986. Growth, metabolism and acid production by *Streptococcus mutans*, p. 255-262. *In*: S. Hamada S. M. Michalek, H. Kiyono. L. Menaker and J. R. McGhee (ed.), Molecular microbiology and immunobiology of *Streptococcus mutans*. Elsevier, Amsterdam.

Hamilton, I. R. 1987. Effect of changing environment on sugar transport and metabolism by oral bacteria, p. 94-133. In J. Reizer and A. Peterkofsky (ed.), Sugar transport and metabolism in gram-positive bacteria. Ellis Horwood Ltd., Chichester.

Hamilton, I. R. and G. H. Bowden. 1988. Effect of fluoride on oral microorganisms., p. 77-103. *In* J. Ekstrand, O. Fejerskov and L. M. Silverstone (ed.), Fluoride in dentistry. Munksgaard, Copenhagen.

Hamilton, I. R. and G. H. Bowden. 2000. Oral microbiology. *In*: J. Lederberg (ed.), Encyclopedia of Microbiology, Academic Press, San Diego. (In Press).

Hamilton, I. R., and N. D. Buckley. 1991. Adaptation by *Streptococcus mutans* to acid tolerance. Oral Microbiol. Immunol. 6: 65-71.

Hamilton, I. R. and D. C. Ellwood. 1983. Carbohydrate metabolism by *Actinomyces viscosus* growing in continuous culture. Infect. Immun. 42: 19-26.

Hamilton, I. R. and S. K. C. Ng. 1983. Stimulation of glycolysis through lactate consumption in a resting cell mixture of *Streptococcus salivarius* and *Veillonella parvula*. FEMS Microbiol, Letters 20: 61 65.

Hamilton, I. R., and G. Svensäter. 1998. Acid-regulated proteins induced by *Streptococcus mutans* and other bacteria during acid shock. Oral Microbiol. Immunol. 13: 292-300.

Hajishengallis, G., and S. M. Michalek. 1999. Current status of a mucosal vaccine against dental caries. Oral Microbiol. Immunol. 14: 1-20

Hardie, J. M., P. L. Thompson, R. J. South, P. D. Marsh, G. H. Bowden, A. S. McKee and G. L. Slack. 1977. A longitudinal epidemiological study on dental plaque and the development of caries. Interim results after two years. J. Dent. Res. 56: C90-C98.

Harper, D. S. , and W. J. Loesche. 1984. Growth and acid tolerance of human dental plaque bacteria. Arch Oral Biol. 29: 843-848.

Higham, S. M., and W. M. Edgar. 1989. Human dental plaque pH, and the organic acid and free amino acid profiles in plaque fluid, after sucrose rinsing. Arch. Oral Biol. 34: 329-334.

Hillman, J. D. 1999. Replacement therapy of dental caries, p. 587-599. In H. N. Newman and M. Wilson (eds.), Dental plaque revisited: Oral biofilms in health and disease. BioLine, Cardiff

Hillman, J. D., A. Chen and J. L. Snoep. 1996. Genetic and physiological analysis of the lethal effect of L-(+)-lactic dehydrogenase deficiency in *Streptococcus mutans*: Complementation by alcohol dehydrogenase from *Zymomonase mobilis*. Infect. Immun. 64: 4319-4323.

Hillman, J. D., A. Chen, M. Duncan and S-W Lee. 1994. Evidence that L-(+)-lactic dehydrogenase deficiency is lethal in *Streptococcus mutans*. Infect. Immun. 62: 60-64.

Hudson. M. C., and R. Curtiss. 1990. Regulation of expression of the *Streptococcus mutans* genes important to virulence. Infect. Immun. 58: 464-470.

Ikeda, T., H. J. Sandham and E. L. Bradley, Jr. 1973. Changes in *Streptococcus mutans* and lactobacilli in plaque in relation to the initiation of dental caries in negro children. Arch. Oral Biol. 18: 555-566.

Jayaraman, G. C., J. E. Penders and R. A. Burne. 1997. Transcriptional analysis of the *Streptococcus mutan hrcA, grpE* and *dnaK* genes and regulation of expression in response to heat shock and environmental acidification. Mol. Microbiol. 25: 329-341.

Jensen, M. E., and J. S. Wefel. 1989. Human plaque pH responses to meals and the effects of chewing gum. Br. Dent. J. 167: 204-208.

Jensen, M. E., P. J. Polansky, and C. F. Schachtele. 1982. Plaque sampling and telemetry for monitoring acid production on human buccal tooth surfaces. Arch. Oral Biol. 27: 21-31.

Johnson, K. P., and J. D. Hillman. 1982. Competitive properties of lactate dehydrogenase mutants of the oral bacterium *Streptococcus mutans* in the rats. Arch. Oral Biol. 27: 513-516.

Kelly, C. G., J. S. Younson, B. Y. Hikmat, S. M. Todryk, M. Czisch, P. I. Haris, I. R. Flindall, C. Newby, A. I. Mallet, J. K. Ma, and T. Lehner. 1999. A synthetic peptide adhesion epitope as a novel antimicrobial agent. Nat. Biochnol. 17: 42-47.

Kenny, E. B., and M. M. Ash Jr. 1969. Oxidation reduction potential of developing plaque, periodontal pockets and gingival sulci. J. Periodontol. 40: 630-633.

Kleinberg. I. 1961. Studies on dental plaque. I. The effect of different concentrations of glucose on the pH of dental plaque *in vivo*. J. Dent. Res. 40: 1087-1111.

Kleinberg, I. 1967. Effect of urea concentration on human plaque pH levels *in situ*. Arch. Oral Biol. 12: 1475-1484.

Kolenbrander, P. E., and J. London. 1992. Ecological significance of coaggregation among oral bacteria, p. 183-217. *In* K. C. Marshall (ed.) Advances in microbial ecology, Vol 12. Plenum Press, New York.

Köhler, B., and D. Bratthall. 1978. Intrafamilial levels of *Streptococcus mutans* and some aspects of the bacterial transmission. Scand. J. Dent. Res. 86: 35-42.

Köhler, B., D. Bratthall, and B. Krasse. 1983. Preventive measures in mothers influence the establishment of the bacterium *Streptococcus mutans* in their infants. Arch. Oral Biol. 28: 225-231.

Lang, N. P., P. R. Hotz, F. A. Gusberti and A. Joss. 1987. Longitudinal clinical and microbiological study on the relationship between infection with *Streptococcus mutans* and the development of caries in humans. Oral Microbiol. Immunol. 2: 39-47.

Larsen, M. J., and C. Bruun. 1994. Caries chemistry and fluoride-Mechanism of action, p. 231-257. *In* A. Thylstrup and O. Fejerskov (ed.) Textbook of Clinical Cariology. Munksgaard, Copenhagen.

Littleton, N. W., R. M. McCabe, and C. H. Carter. 1967. Studies of oral health in persons nourished by stomach tube. II Acidogenic properties and selected bacterial components of plaque material. Arch. Oral Biol. 12: 601-609.

Loesche, W. J. 1982. Dental caries. A treatable infection. Charles C. Thomas, Springfield, Ill.

Loesche, W. J. 1986. Role of *Streptococcus mutans* in human dental decay. Microbiol. Rev. 50: 353-380.

Loesche, W. J., and L. H. Straffon. 1979. Longitudinal investigation of the role of *Streptococcus mutans* in human fissure decay. Infect. Immun. 26: 498-507.

Lucas, V. S., D. Beighton, G. J. Roberts and S. J. Challacombe. 1997. Changes in the oral streptococcal flora of children undergoing allogenic bone marrow transplantation. J. Infect. 35: 135-141.

Ma, J. K-C., and T. Lehner. 1990. Prevention of colonization of *Streptococcus mutans* by topical application of monoclonal antibodies in human subjects. Arch. Oral Biol. 35 (suppl.): 115s-122s.

Margolis, H. C., and E. C. Moreno. 1992. Composition of pooled plaque fluid from caries-free and caries-positive individuals following sucrose exposure. J. Dent. Res. 71: 1776-1784.

Margolis, H. C., E. C. Moreno, and B. J. Murphy. 1995. Importance of high pKA acids in cariogenic potential of plaque. J. Dent. Res. 64: 786-792.

Marquis, R. E. 1995. Oxygen metabolism, oxidative stress and acid-base physiology of dental plaque biofilms. J. Ind. Microbiol. 15: 198-207.

Marquis, R. E., G. R. Bender, D. R. Murray, and A. Wong. 1987. Arginine deiminase sysem and bacterial adaptation to acid environments. App. Environ. Microbiol. 53: 198-200.

Marsh, P. D. 1995. The role of microbiological models of dental caries. Adv. Dent Res. 9: 244-254.

Miller, W. D. 1890. The microorganisms of the human mouth. S. S. White Dental Mfg Co., Philadelphia, Penn.

Milnes, A. R. 1987. A longitudinal investigation of the microflora associated with developing lesions of nursing caries. Ph.D. thesis. University of Manitoba, Winnipeg.

Milnes, A. R. and G. H. W. Bowden. 1985. The microflora associated with developing lesions of nursing caries. Caries Res. 19: 289-297.

Minah, G. E., S. E. Solomon, and K. Chu. 1985. The association between dietary sucrose consumption and microbial population shifts at six oral sites in man. Arch. Oral Biol. 30: 397-401.

Mörmann, J. E., and H. R. Mühlemann. 1981. Oral starch degradation and its influence on acid production in human dental plaque. Caries Res. 15: 166-175.

Morou-Bermudez, E., and R. A Burne. 1999. Genetic and physiologic characterization of urease from *Actinomyces naeslundii*. Infect. Immun. 67: 504-512.

Munro, G. H., P. Evans, S. Todryk, P. Buckett, C. G. Kelly and T. Lehner. 1993. A protein fragment of streptococcal cell surface antigen I/II which prevents adhesion of *Streptococcus mutans*. Infect. Immun. 61: 4590-4598.

Nikiforuk, G. 1985. Understanding dental caries. I. Etiology and Mechanisms. Karger, Basel.

Novak, J., P. W. Caulfield and E. J. Miller. 1994. Isolation and biochemical characterization of a novel lantibiotic mutacin from *Streptococcus mutans*. J. Bacteriol. 176: 4316-4320.

Nyvad, B., and M. Kilian. 1987. Microbiology of the early colonization of human enamel and root surfaces *in vivo*. Scand. J. Dent. Res. 95: 369-380.

Nyvad, B., and M. Kilian. 1990. Microflora associated with experimental root surface caries in human. Infect. Immun. 58: 1628-1633.

Quivey, R. G., R. C. Faustoferri, K. A. Clancy and R. E. Marquis. 1995. Acid adaptation in *Streptococcus mutans* UA159 alleviates sensitization to environmental stress due to RecA deficiency. FEMS Microbiol. Letters 126: 257-262.

271

Ritz, H. L. 1967. Microbial population shifts in developing human dental plaque. Arch. Oral Biol. 12:1561-1568.

Ripa, L. W. 1978. Nursing habits and dental decay in infants: nursing bottle caries. J. Dent. Child. 45: 274-275.

Russell, R. R. B., J. Aduse-Opoku, I. C. Sutcliffe, L. Tao, and J. J. Ferretti. 1992. A binding protein-dependent transport system in *Streptococcus mutans* responsible for multiple sugar metabolism. J. Biol. Chem. 267: 4631-4637.

Salako, N. O. and I. Kleinberg. 1989. Incidence of selected ureolytic bacteria in human dental plaque from sites with different salivary access. Arch. Oral Biol. 34: 787-791.

Sansone, C., J. van Houte, K. Joshipura, R. Kent and H. C. Margolis. 1993. The association of mutans streptococci and non-mutans streptococci capable of acidogenesis at a low pH with dental caries on enamel and root surfaces. J. Dent. Res. 72: 508-516.

Scannapieco, F. A. 1994. Saliva-bacterium interactions in oral microbial ecology. Crit. Rev. Oral Biol. Med. 5: 203-248.

Schachtele, C. F., and S. K. Harlander. 1984. Will the diets of the future be less cariogenic? J. Can. Dent. Assoc. 50: 213-220.

Schachtele, C. F., and M. E. Jensen. 1982. Comparisons of methods for monitoring changes in the pH of human dental plaque. J. Dent. Res. 61: 1117-1125.

Schüpbach, P., V. Osterwalder and B. Guggenheim. 1995. Human root caries: Microbiota in plaque covering sound, carious and arrested carious root surfaces. Caries Res. 29: 382-395.

Schüpbach, P., V. Osterwalder and B. Guggenheim. 1996. Human root caries: Microbiota of a limited number of root caries lesions. Caries Res. 30: 52-64.

Shapiro, J. A. 1997. Multicellularity: the role, not the exception: Lessons from *Escherichia coli* colonies, p. 14-49. *In* J. A. Shapiro and M. Dworking (ed.), Bacteria as multicellular organisms. Oxford University Press, Oxford.

Shellis, R. P., and G. H. Dibdin. 1988. Analysis of the buffering systems in dental plaque. J. Dent. Res. 67: 438-446.

Silverstone, L. M., N. W. Johnson, J. M. Hardie, and R. A. D. Williams. 1981. Dental caries: Aetiology, pathology and prevention. MacMillan Press, London.

Sissons, C. H., and E. M. Hancock. 1993. Urease activity in *Streptococcus salivarius* at low pH. Arch. Oral Biol. 38: 507-516.

Smith. D. J., and M. A. Taubman. 1996. Experimental immunization of rats with *Streptococcus mutans* 59-kilodalton glucan-binding protein protects against dental caries. Infect. Immun. 64: 3069-3073.

Smith. D. J., R. L. Heschel, W. F. King, and M. A. Taubman. 1999. Antibody to glycosyltransferase induced by synthetic peptides associated with catalytic regions of α-amylases. Infect. Immun. 67: 2638-2642.

Stephan, R. M. 1944. Intra-oral hydrogen-ion concentration associated with dental caries activity. J. Dent. Res. 23: 257-266.

Svanberg, M. 1980. *Streptococcus mutans* in plaque after mouthrinsing with buffers of varying pH value. Scand. J. Dent. Res. 88: 76-78.

Svensäter, G., B. Sjögreen, and I. R. Hamilton. 2000. Multiple stress responses in *Streptococcus mutans* and the induction of the general and stress-specific proteins.

Microbiology 146: 107-177.

Svensäter, G., U-B Larsson, E. C. G. Greif, D. G. Cvitkovitch, and I. R. Hamilton. 1997. Acid tolerance and survival by oral bacteria. Oral Microbiol. Immunol. 12: 266-273.

Tao, L., I. C. Sutcliffe, R. R. B. Russell and J. J. Ferretti. 1993. Transport of sugars, including sucrose, by the *msm* transport system of *Streptococcus mutans*. J. Dent. Res. 72: 1386-1390.

Taubman, M. A., R. J. Genco and J. D. Hillman. 1989. The specific pathogen-free human: A new frontier in oral infectious disease research. Adv. Dent. Res. 3: 58-68.

Taubman, M. A., C, J. Holmberg and D. J. Smith. 1995. Immunization of rats with synthetic peptide constructs from the glucan-binding or catalytic region of mutans streptococci glucosyltransferase protects against dental caries. Infect. Immun. 63: 3088-3093.

Thevenot, T., D. Brochu, C. Vadeoncoeur and I. R. Hamilton. 1995. Regulation of ATP-dependent P-(Ser)-HPr formation in *Streptococcus mutans* and *Streptococcus salivarius*. J. Bacteriol. 177: 2751-2759.

Thomas, E. L, T. W. Milligan, R. E. Joyner, and M. M. Jefferson. 1994. Antibacterial activity of hydrogen peroxide and the lactoperoxidase-hydrogen peroxide-thiocynate system against oral streptococci. Infect. Immun. 62: 529-535.

Thylstrup, A., and O. Fejerskov. 1994. Clinical and pathological features of dental caries, p. 111-157. *In* A. Thylstrup and O. Fejerskov (ed.) Textbook of Clinical Cariology. Munksgaard, Copenhagen.

Todryk, S. M., C. G. Kelly, G. H. Munro, and T. Lehner. 1996. Induction of immune responses to functional determinants of a cell surface antigen. Immunology 87: 55-63..

Vadeboncoeur, C., and M. Pelletier. 1997. The phosphoenolpyruvate:sugar phosphotransferase system of oral streptococci and its role in the control of sugar metabolism. FEMS Microbiol. Rev. 19: 187-207.

van der Hoeven, J. S., M. H. de Jong, A. Van Nieuw Amerogen. 1989. Growth of oral microflora on saliva from different glands. Microbial Ecol. Health Dis. 31: 129-133.

van Houte, J., J. Lopman and R. Kent. 1994. The predominant cultivable flora of sound and carious human root surfaces. J. Dent. Res. 73: 1727-1734.

van Houte, J., J. Lopman and R. Kent. 1996. The final pH of bacteria comprising the predominant flora on sound and carious human root and enamel surfaces. J. Dent. Res. 75: 1008-1014.

van Loveren, C., M. M. E. Straetemans, and J. P. Buijs. 1999. Similarity of mutans streptococci in mothers and children who acquired mutans streptococci after the age of 5, p. 281. *In* Abstracts 46th ORCA congress. Caries Res. 33: 281.

Vratsanos, S. M., and I. D. Mandel. 1982. Comparable plaque acidogenesis of caries-resistant and caries-susceptible adults. J. Dent. Res. 61: 465-468.

Wexler, D. L., M. C. Hudson, and R. A. Burne. 1993. *Streptococcus mutans* fructosyltransferase (*ftf*) and glucosyltransferase (*gft*BC) operon fusion strains in continuous culture. Infect. Immun. 61:1259-1267.

Wimpenny, J. 1995. Biofilms: structure and organization. Microbial Ecol. Health

Dis. 8: 305-308.

Yamada, T., K. Igarashi and M. Mitsutomi. 1980. Evaluation of cariogenicity of glycosylsucrose by a new method to measure pH under dental plaque *in situ*. J. Dent. Res. 59: 2157-2162.

Yamashita, Y., T. Takehara , and H. Kuramitsu. 1993. Molecular characterization of a *Streptococcus mutans* mutant altered in environmental stress responses. J. Bacteriol. 175: 6220-6228.

From: *Oral Bacterial Ecology: The Molecular Basis*
ISBN 1-898486-22-0 ©2000 Horizon Scientific Press, Wymondham, U.K.

6

PERIODONTITIS AS AN ECOLOGICAL IMBALANCE

Daniel Grenier and Denis Mayrand

Contents

Introduction

There is now much evidence to suggest that the physiology, biological properties, and pathogenicity of a bacterium are largely influenced by the environment. As a matter of fact, signal transduction systems that recognize and respond to specific environmental cues are often utilized by pathogens to ensure that genes required for survival, multiplication and virulence are appropriately expressed during the infection (DiRita and Mekalanos, 1989; Guiney, 1997; Straley and Perry, 1995). The situation is much more complex when an infectious disease results from a mixed infection, i.e. from the interaction of several bacterial species each contributing to cause the disease. Mixed anaerobic infections are generally found to have several characteristics: i) the infectious agents are members of the normal indigenous bacterial microbiota and hence may be regarded as opportunistic pathogens, ii) as individuals, the pathogenic members of such infections are only minimally virulent under most conditions, and iii) predisposing factors may play an important if not essential role in the initiation of such infections, since a change in the environmental conditions enables some members of the ecosystem to become prominent and/or to manifest some of their pathogenic properties.

Virulence promotion by mixed bacterial populations has three underlying features: bacterial interactions, expression of virulence factors and host-bacteria interactions. The oral cavity and more particularly the subgingival environment represent a perfect example of microbial ecology where different bacterial populations cohabitate. Periodontal diseases, which are considered as mixed anaerobic infections, develop when the equilibrium of the community is broken in favor of the pathogenic bacteria that most individuals harbor in low numbers in their subgingival sites. The mechanisms that regulate and influence the proportions of pathogenic and commensal or beneficial bacteria in subgingival plaque are not fully understood. However, it is likely that host, bacterial and environmental factors are of utmost importance for determining the subgingival microbiota composition and modulating the expression of bacterial virulence factors.

Periodontal Diseases and Periodontopathogens

Periodontal diseases are probably the most common chronic inflammatory disorder in adults and may lead to tooth loss in the absence of appropriate treatments. Periodontal diseases are customarily separated into infections affecting only the gingiva (gingivitis) and those affecting the underlying, tooth-supporting tissues of the periodontium (periodontitis) including the periodontal ligament and the alveolar bone. Gingivitis is defined as the result of a non-specific inflammatory reaction in response to an increase in the mass of bacteria (either Gram-negative or Gram-positive) at or under the gingival margin. On the other hand, periodontitis is initiated by an overgrowth of specific Gram-negative bacterial species found in the same site. This results in the formation of a periodontal pocket which greatly favors further accumulation and a shift in the proportional composition of the microbiota. The progression of periodontitis is episodic, with active and inactive phases of tissue destruction. This reflects the opposing actions of the bacterial challenge and the host immune response.

It has been demonstrated that over 250 different cultivable bacterial species may be present in subgingival plaque samples (Moore *et al.,* 1982). Among these bacteria only some, either alone or in combination, have the biochemical flexibility to dominate the oral microbiota and to express a periodontopathogenic potential to initiate disease when a critical concentration is reached. Generally, the development of periodontitis is accompanied by an increase of the proportion of specific Gram-negative anaerobic bacterial species commonly referred to as periodontopathogens. Criteria have been proposed for considering bacteria as periodontopathogens i.e. contributing to the etiology of periodontal diseases (Haffajee and Socransky, 1994; Socransky, 1979): i) the bacteria must be present in high numbers or proportions in progressing lesions but absent or present in low numbers in healthy sites, ii) a clinical improvement should be observed following elimination of the bacteria or modification of their virulence factors, iii) the bacteria must be present prior to disease progression, iv) the bacteria must stimulate the host defense system, v) virulence factors must be produced by the bacteria, and vi) the bacteria must induce lesions in an appropriate animal model.

Based on frequency of isolation in lesion sites, *Actinobacillus actinomycetemcomitans* is now considered as a key etiologic agent of early-onset periodontitis (Slots *et al.,* 1980; Wilson and Henderson, 1995; Wolff *et al.,* 1985). The fact that this bacterium is rarely predominant among the total microbiota suggests that it may have a high virulence potential and thus a low threshold level for inducing disease. A group of about ten bacterial species, including *Porphyromonas gingivalis, Prevotella intermedia/nigrescens, Bacteroides forsythus, Treponema denticola, Fusobacterium nucleatum, Peptostreptococcus micros, Campylobacter rectus, Eikenella corrodens* and *Selenomonas sputigena* have been associated with adult chronic periodontitis, the most common form of periodontitis (Dzink *et al.,* 1988; Haffajee and Socransky, 1994; Socransky *et al.,* 1988; Tanner, 1991). These bacterial species behave in a cooperative or synergistic fashion to fulfil the requirements for pathogenicity and produce periodontitis, a mixed anaerobic infection. It is likely that some other bacteria not yet suspected as playing a role in the initiation and/or progression of periodontitis may also be actively involved. These bacteria may be either uncultivable or simply absent or present in low numbers in affected sites, being only required for the initiation of the disease. As a matter of fact, Choi *et al.* (1994) using rRNA sequence analysis reported the presence of approximately 20 new species of *Treponema* that have not been cultured and classified so far. The use of this molecular approach may be of great value to elucidate the role of uncultivable oral bacteria in the pathogenesis of periodontal infections.

Periodontal infections are initiated when the above periodontopathogens increase in proportions and replace the commensal microbiota. This often results from a disequilibrium in the host which may be induced by several ways: (i) a modification of the environmental conditions of the site caused by either bacterial interactions or accumulation of dental plaque, (ii) a decrease in the proportion of beneficial bacteria caused by bacterial interactions or the use of systemic antibiotics, and (iii) a decreased efficacy of the host immune system.

Although bacteria are the primary factor in the etiology of periodontitis, tissue destruction is also a consequence of the host response (Birkedal-Hansen, 1993; Kornman *et al.,* 1997). In fact, bacteria and their products may (i) trigger resident and immigrant host cells to express tissue degrading enzymes, and (ii) provoke an immune response that results in the release of cytokines by lymphocytes, macrophages, and other cell lineages. These cytokines can subsequently activate one or more degradative pathways: matrix metalloproteinase, plasminogen-dependent, phagocytic and polymorphonuclear-serine proteinase pathways and ultimately osteoclastic bone resorption. In a recent study, Loomer *et al.* (1998) used an osteogenic/osteoclastic cell *in vitro* co-culture model and postulated that periodontal bone loss caused by a *P. gingivalis* cell extract is the sum of net resorption and net formation. Therefore, periodontopathogens not only stimulate osteoclasts to resorb but also suppress osteogenesis and uncouple normal communication

277

between bone forming and bone resorbing cells (uncoupled homeostasis).

To better understand the entire pathogenic process of a mixed anaerobic infection such as periodontitis, it is critical to study the basic features of the relationships bacteria/bacteria and host/bacteria, as well as the effects of environmental parameters on virulence factors and stress responses of periodontopathogens.

Bacterial Interactions

As mentioned above, dental plaque which accumulates in the gingival sulcus contains a great variety of bacteria that may act positively or negatively. These interactions may exist between individual cells within one population or between members of different bacterial populations forming a diversified community. The sum of positive and negative interactions among the different microbial populations is responsible for maintaining the ecological balance within the community. However, when the equilibrium of the community is broken in favor of the pathogens, periodontal disease may develop. Homeostasis of the dental plaque can be upset endogenously by immunological deficiencies (genetic defects) or exogenously by external agents such as drugs, infections or malnutrition (Marsh, 1989). Certain non-immune factors including the use of oral contraceptives or antibiotics, can also cause homeostatic mechanisms to break down, allowing pathogenic species to emerge (Marsh, 1989).

Positive interactions occurring between oral bacteria include mutualism, commensalism and synergism. Mutualism is defined as a relationship between two species where both of them benefit from the association. Commensalism occurs when only one species benefits whereas the other one obtains nothing from the association. Lastly, synergism is found when the interaction between the two microbial species has a greater effect than the sum of both species taken individually. The presence of surface components which promote coaggregation of two bacterial species may greatly favor positive interactions (Kolenbrander and London, 1992; London and Kolenbrander, 1996). However, a positive interaction does not necessarily imply that the two species need to be in close contact. In fact, a bacterial species may stimulate the growth of another species simply by secreting in the environment a nutritional factor needed by the other species or by modifying the physico-chemical environment of the subgingival site, thereby, permitting the establishment or overgrowth of another species.

Recent studies have revealed that numerous bacterial species of non-oral origin can communicate via excreted signaling peptides or pheromones, the cell-to-cell signaling phenomenon which results in positive interactions (Wirth *et al.,* 1996). The production of these signaling peptides is a means by which a single bacterial cell can obtain information from others of the same species. It is thought that pheromonal signals produced by bacteria might be received and transduced in a way similar to other environmental signals. Communication via such signals is important for the survival of bacteria, for increasing bacterial virulence or for facilitating the transfer of conjugal plasmids (Clewell, 1993; Wirth *et al.,* 1996). For instance, a pheromone belonging to the class of N-acyl-L-homoserine lactones has been found to regulate the expression of elastase and other virulence factors in *Pseudomonas aeruginosa* (Latifi *et al.,* 1995). Although pheromone-like peptides have been found in some oral streptococci (Lunsford, 1998), production of pheromones and similar signaling peptides by periodontopathogens and their role in the pathogenesis of periodontal disease have not been investigated yet and represent an area of interest for future research.

Negative interactions occurring between bacteria found in subgingival sites include competition and antagonism. Competition is represented by two populations competing for common nutrients and optimal conditions in order to multiply and survive. Normally, this results in a lower cell density or growth rate than if the two populations were separated. This type of relationship occurs when two species try to occupy a particular site or to obtain any similar essential nutrients. Usually, bacteria harboring adhesins, possessing efficient mechanisms of nutrient acquisition or exhibiting a high growth rate will be favored in this type of interaction. When a bacterial population secretes products (hydrogen peroxide, bacteriocins, organic acids) which inhibit other populations or negatively alter environmental conditions (pH, oxidation-reduction potential), interactions are said to be antagonistic. Bacteria producing such compounds or conditions will have an ecological advantage over the others.

Specific examples of positive and negative interactions involving the principal periodontopathogens and which may modulate their concentration in the periodontal pocket and consequently influence the progression of periodontitis are presented

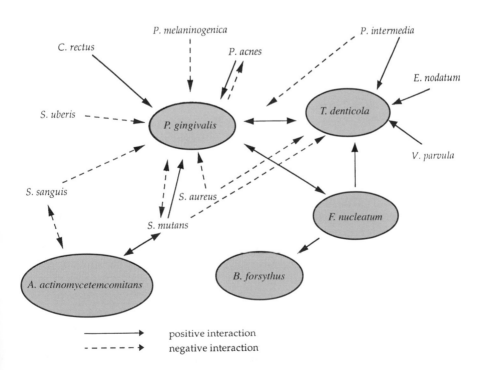

Figure 1. Positive (commensalism, mutualism, synergism) and negative (antagonism, competition) interactions involving the principal periodontopathogens.

(Figure 1). Interestingly, several of the *in vitro* observations are supported by microbiological analysis of subgingival plaque samples or experimental infections in animal models. A mutualistic relationship between *P. gingivalis* and *T. denticola* was studied by Grenier (1992). In a complex culture medium lacking hemin, both species could not grow separately but exhibited significant growth when co-cultivated. Growth factors produced by the two species were shown to be organic acids i.e. isobutyric acid produced by *P. gingivalis* and succinic acid by *T. denticola*. It was also observed that the two species coaggregated in the mixed culture. Since *T. denticola* does not produce a large quantity of succinic acid, coaggregation may be a way to efficiently supply the growth factor to *P. gingivalis*. The capacity of *P. gingivalis* to support growth of *T. denticola* was also reported by another research group (Nilius *et al.*, 1993). The mutualistic relationship between *P. gingivalis* and *T. denticola* is strongly supported by *in vivo* observations. Indeed, Simonson *et al.* (1992) showed that the presence of *P. gingivalis* is a prerequisite for the subsequent detection of *T. denticola* in periodontal pockets. Lastly, a synergistic virulence of these two periodontopathogens in a murine lesion model was demonstrated (Kesavalu *et al.*, 1998). It was also shown using protease deficient mutants of *P. gingivalis* that the trypsin-like activity (possibly the arg-gingipain activity) is of utmost importance for virulence expression of the bacterial mixture in the animal model.

A commensalism relationship between *P. gingivalis* and *C. rectus* has been described (Grenier and Mayrand, 1986). The growth of *P. gingivalis* appeared to be stimulated by a protoheme-related compound, most probably a cytochrome b derivative that was released in the external environment following cell lysis of *C. rectus*. This *in vitro* observation concerning the relationship between *P. gingivalis* and *C. rectus* was confirmed by *in vivo* data. Indeed, Riviere *et al.* (1996) recently assessed by estimated odds ratios which bacteria were associated with *P. gingivalis* and found that associations of *P. gingivalis* with *C. rectus* were the strongest, irrespective of periodontal status.

Analysis of subgingival plaque samples from actively-progressing lesions and inactive periodontal sites by Socransky *et al.* (1988) revealed that *P. gingivalis* was positively associated with a number of bacteria including *F. nucleatum*. Using an experimental model of mixed infections in mice, a synergistic relationship was demonstrated between *P. gingivalis* and *F. nucleatum* (Baumgartner *et al.*, 1992; Ebersole *et al.*, 1997; Feuille *et al.*, 1996). The pathogenic potential of the bacterial mixture was significantly greater than that of each individual species. Feuille *et al.* (1996) studied the various aspects of virulence of *P. gingivalis* and *F. nucleatum* in a murine lesion model by altering bacterial ratios, time of infection and virulence factors such as coaggregation and proteolytic activity. The authors showed that the two bacteria can synergistically initiate soft tissue destruction, that coaggregation may be important in the synergy, that viable *P. gingivalis* is needed to produce a lesion and that inhibiting the trypsin-like protease activity by inhibitors or mutagenesis eliminated the synergistic interactions. The same group also showed that priming the mice with low levels of *P. gingivalis* had little or no effect on lesion development and that active immunization with whole *P. gingivalis* cells elicited a significant increase in serum Ig antibody (Ebersole *et al.*, 1997). A synergistic relationship between *P. gingivalis* and *F. nucleatum* was also found *in vitro* by Gharbia *et al.* (1989) who showed that the proteolytic capacity of the combination was 30% higher than the sum of the individual activities. It was suggested that the synergistic effect could be related to the high peptidase activities of *F. nucleatum,* which may expose susceptible amino acids in protein substrates permitting a more efficient action of *P. gingivalis* proteases. Lastly, Grenier (1994) reported that the *in vitro* growth of *F. nucleatum* was stimulated by adding into the medium proteolytic enzymes (trypsin and pancreatic chymotrypsin) or a proteolytic fraction prepared from cells of *P. gingivalis.*

Apart from the positive interaction described above with *P. gingivalis,* other interactions involving *T. denticola* have been reported. A study by Socransky (1964) showed that a mixture of bacteria, not identified at the species level, was able to provide essential factors for the growth of *T. denticola.* The growth stimulation was associated with the production of isobutyrate and polyamines as well as with a decrease of the oxidation-reduction potential. More recently, Ter Steeg *et al.* (1988) have shown that inoculation of subgingival plaque samples into human serum led to the accumulation of *T. denticola.* Subsequent studies led this group to identify four oral species (*P. intermedia, Eubacterium nodatum, Veillonella parvula, F. nucleatum*) capable of stimulating the growth of the spirochete (Ter Steeg and van der Hoeven, 1990). The authors have also been able to suggest different putative stimulatory mechanisms. *P. intermedia* and *E. nodatum* may have been able to cleave carbohydrate chains from glycoproteins rendering the molecules susceptible to the proteolysis by *T. denticola. E. nodatum* was able to supply a low molecular weight, heat stable growth factor to *T. denticola* whereas *V. parvula* might have provided peptidase activities complementary to those of *T. denticola.* The role of *F. nucleatum* in the growth stimulation could not be identified.

A mutual relationship between *B. forsythus* and *F. nucleatum* has been shown *in vitro* (Dzink *et al.,* 1986). This observation was supported by Takemoto *et al.* (1997) who found that *B. forsythus,* by itself, did not induce abscess formation in rabbits but caused abscess formation when coinoculated with either *F. nucleatum* or *P. gingivalis.* Socransky *et al.* (1988) identified positive associations between the isolation of *B. forsythus* and some species found in affected periodontal sites. *B. forsythus* was often found in samples containing *Streptococcus mitis, Streptococcus morbillorum, Prevotella melaninogenica, F. nucleatum* and *C. rectus.*

Growth inhibitory factors produced by oral bacteria also play a critical role in the ecology of subgingival areas and likely in the evolution of periodontitis. Van Winkelhoff *et al.* (1987) found that *P. gingivalis* possessed the most extensive inhibitory capacity among all black-pigmented anaerobic bacteria tested. It showed inhibitory activity against *P. intermedia, Porphyromonas endodontalis, Prevotella loescheii* and *P. melaninogenica.* The exact nature of the inhibitory compounds were not investigated. A study by Nakamura *et al.* (1978) revealed that the pigment (hematin) produced by *P. gingivalis* had the capacity to inhibit Gram-positive bacteria including *Streptococcus mutans, Streptococcus mitis, Actinomyces viscosus, Actinomyces naeslundii, Actinomyces israelii, Corynebacterium matruchotii, Corynebacterium parvum* and *Propionibacterium acnes.* The pigment had no effect on growth of the Gram-negative bacteria tested, and was found to be produced and active in a guinea pig model (Nakamura *et al.,* 1980).

A number of bacteria, including *Staphylococcus aureus, S. mutans, P. melaninogenica* and *P. intermedia* have been found to interfere with the growth of *P. gingivalis* (Grenier, 1996; Takazoe *et al.,* 1984). Recently, Johansson *et al.* (1994) showed that growth of *P. gingivalis* was reduced to various extent by UV-killed bacterial species (*A. actinomycetemcomitans, F. nucleatum, S. aureus, S. mitis, Staphylococcus epidermidis*) isolated from subgingival plaque. The bacterial interference was significant but the authors could not define the mechanism responsible for the effect. Studies of antagonism carried out by Hillman and Socransky (1982) showed that the growth of *P. gingivalis* was affected by *Streptococcus sanguis* and *Streptococcus uberis.* Two bacteriocins active against *P. gingivalis* have been isolated respectively from *P. melaninogenica* (melaninocin) (Nakamura *et al.,* 1981) and *S. sanguis* (sanguicin) (Fujimura, 1979).

A recent study by Grenier (1996) has shown that *S. aureus* and *S. mutans*, both isolated from healthy subgingival sites, had the capacity to inhibit growth of *T. denticola.* The inhibitory activity of *S. aureus* was associated with the production of a bacteriocin-like compound whereas the growth inhibition caused by *S. mutans* was due to the production of lactic acid.

Several reports have indicated a significant negative association between *S. sanguis* and *A. actinomycetemcomitans* (Hillman and Socransky, 1982; Hillman and Socransky, 1989; Hillman *et al.,* 1985; Socransky *et al.,* 1988). It was found that the presence of *S. sanguis* correlated with the absence of *A. actinomycetemcomitans* in the sites. It seems that the capacity of *S. sanguis* to inhibit *A. actinomycetemcomitans* is due to the production of hydrogen peroxide. Interestingly, *A. actinomycetemcomitans* is catalase activity positive but it still may be overwhelmed by the amount of hydrogen peroxide produced by streptococci. It was also observed that the levels of *A. actinomycetemcomitans* present in the mouth of gnotobiotic rats could be diminished by a factor of 25 within a few weeks following a massive infection with *S. sanguis* (Shivers *et al.,* 1987). On the other hand, the levels of the *A. actinomycetemcomitans* remained unchanged when a mutant strain of *S. sanguis* deleted of its capacity to produce hydrogen peroxide was inoculated. Another antagonistic relationship of interest is the one between *S. mutans* and *A. actinomycetemcomitans.* It is well known that *A. actinomycetemcomitans* is frequently found in patients with juvenile periodontitis but not in normal subjects with a healthy peridontium (Slots *et al.,* 1980; Wilson and Henderson, 1995; Wolff *et al.,* 1985). Its presence in affected sites was correlated with lower numbers of *S. mutans* in those sites (Hillman *et al.,* 1985). The decrease in the population of *S. mutans* was thought to be related to the production of a bacteriocin called actinobacillin (Stevens *et al.,* 1987). Hammond *et al.* (1987) have purified and characterized the biochemical properties of the *A. actinomycetemcomitans* bacteriocin. The bacteriocin was also found to be active against *S. sanguis.* Overall, the above interactions found for *A. actinomycetemcomitans* and oral streptococci agree with *in vivo* observations made by Socransky *et al.* (1997).

For anaerobic bacteria such as the periodontopathogens, the oxidation-reduction potential represents a key parameter for survival. In fact, growth of obligate anaerobes is only possible below a certain level of oxidation-reduction potential. Generally, the more oxygen-sensitive anaerobes require a lower oxidation-reduction potential to initiate growth. For *P. gingivalis*, the oxidation-reduction potential is likely to represent a critical environmental determinant affecting its growth characteristics and mediating its dynamic character as a potential pathogen. In a recent study, it was found that under conditions of positive values of oxidation-reduction potential obtained by adding 3 and 30 mM hydrogen peroxide, the growth of *P. gingivalis* was only possible in a complex culture medium containing small peptides and amino acids (Leke *et al.,* 1999). No growth occurred in a defined basal medium containing bovine serum albumin as the only source of nitrogen and energy. It was also found that some oral bacterial species could either decrease (*S. mutans, Peptostreptococcus micros, F. nucleatum*) or increase (*S. sanguis*) the oxidation-reduction potential. This suggests that these bacterial species may modulate the composition of subgingival plaque by modifying the oxidation-reduction potential of the environment.

It is evident from this brief discussion and as summarized in Figure 1 that there are numerous bacterial interactions which may lead to virulence promotion among species in subgingival sites. Three basic types of mechanisms can be proposed to explain virulence promotion by mixed bacterial populations: the "addition" model, the "cascade" model and the "sequential" model (Figure 2). Of course, a number of combinations of these basic mechanisms may also occur. These rather simplistic models may in part explain different types of mixed infections; first, from the point of view of time of appearance of clinical symptoms and secondly, from the mechanistic events which occur. The "addition" model may reflect an acute situation where the infection is the net result of additive effects of bacteria that all possess the required abilities to penetrate host tissues, to survive the host defense system and to cause tissue damage. The sheer number of these bacteria or the increased proportion of specific bacteria enables them to resist the host defense system. The "cascade" model might reflect a two or multiple stage reaction which can either be acute or manifest itself within days or a few weeks from initiation. On the other hand, the

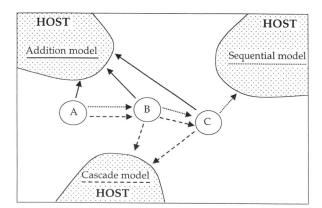

Figure 2. Proposed mechanisms of virulence promotion in mixed anaerobic infections such as periodontitis. A, B and C represent different bacterial species. In the cascade and sequential model, bacterial interactions may result in an increase of growth and/or virulence of bacteria.

"sequential" model can reflect a more chronic situation, and the clinical symptoms may take months or even years to manifest themselves. In the cascade and sequential models bacterial interactions may result in an increase of growth and/or virulence of bacteria. Some periodontal disease entities fall within the sequential model. Results from ligature-induced periodontitis in monkeys (Kiel *et al.,* 1983; Kornman *et al.,* 1981) along with studies of combinations of oral bacteria in periapical tissues of monkeys (Fabricius *et al.,* 1982) indicate that this latter model may be common for mixed bacterial infections in the oral cavity.

In summary, there is ample evidence to indicate that virulence can be enhanced within mixed bacterial populations. What remains to be determined is the exact pathway(s) effecting the promotion of virulence. Mixed infections in which each and every single bacterial component does not show overt pathogenicity present a dilemma. On the one hand, dominant species within established periodontal lesions, though endowed with a large set of pathogenic attributes, nevertheless require factors from other bacteria or the host to fully express their pathogenicity. On the other hand, within the mixed bacterial population, species devoid of overt pathogenicity may actively participate in the infectious process to establish the clinical lesion. In other words, the ability of oral bacteria to invade or otherwise provoke disease not only depends on the inherent characteristics of the bacterial species itself but may result as a consequence of bacterial interactions.

Virulence Factors of Periodontopathogens

To be recognized as a periodontopathogen, a bacterial species must possess virulence factors. These pathogenic determinants enable bacteria to elicit the key stages of the infective process of periodontitis. Specifically, to colonize the host and multiply, to avoid destruction or neutralization by the host's defenses, and to cause tissue damage. During the course of an infection, virulence factors may act alone or in combination. Important virulence factors produced by periodontopathogens include adhesins, lipopolysaccharides (LPS), hemolysins, proteinases and outer membrane vesicles. The following summarizes the key roles that may be accomplished by these bacterial components (Table 1).

Adhesins
The ability of periodontopathogens to colonize subgingival sites represents the first stage of the disease process. Periodontopathogens exhibit specificity for their respective colonization sites. Such specificity is directed by adhesin-receptor cognate pairs which largely influence the distribution of bacteria in periodontal pockets. Adhesins and their ecological roles are covered in detail in a previous chapter (Lamont and Jenkinson, chapter 3). Adhesins associated with either the outer membrane or the fimbriae can promote attachment of most periodontopathogens (*P. gingivalis*, *T. denticola*, *A. actinomycetemcomitans*) to tooth surfaces, gingival cells, basement membrane components, erythrocytes or oral bacteria (Whittaker *et al.,* 1996). The extent of their functional diversity and the approaches used to identify and characterize them have recently been described (London, 1999). Additional roles in pathogenicity have been attributed to adhesins of periodontopathogens. Hanazawa

Table 1. Virulence Factors of Periodontopathogens and their Functional Roles in Pathogenicity*

Functional roles	Virulence factors
Adherence	Outer membrane vesicle, adhesin, LPS, protease
Bacterial growth	Protease, hemolysin
Host defense perturbation	Outer membrane vesicle, LPS, protease, leukotoxin, Fc-binding protein
Tissue destruction	Outer membrane vesicle, LPS, protease, cytotoxin, adhesin

*See the following reviews for more detail on virulence factors: Daly *et al.,* 1990; Grenier and Mayrand, 1993; Henderson *et al.,* 1996; Ishihara, K., and K. Okuda. 1999; London, 1999; Mäkinen and Mäkinen, 1996; Mayrand and Grenier, 1989; Meyer and Fives-Taylor, 1997; Smalley *et al.,* 1993; Whittaker *et al.,* 1996; Wilson and Henderson, 1995.

et al. (1995) showed that *P. gingivalis* fimbriae stimulate bone resorption *in vitro*. Adhesion of bacteria to host cells may also trigger cytokine synthesis, and consequently bacterial adhesins are a new class of modulins which include other bacterial factors such as LPSs (Henderson *et al.*, 1996). As a matter of fact, purified fimbriae from *P. gingivalis* were reported to induce proinflammatory cytokines in human peripheral blood monocytes and mouse peritoneal macrophages (Hanazawa *et al.*, 1991; Ogawa *et al.*, 1994).

Lipopolysaccharides

Endotoxins or LPS are major constituents of the outer membrane of Gram-negative bacteria and consist of three components: the O-specific chain or O-polysaccharide, the outer and inner polysaccharide core, and the hydrophobic lipid A, which represents the toxic part of the molecule. Bacterial LPS are called modulins since they are potent inducers of proinflammatory mediators such as cytokines (Henderson *et al.*, 1996; Rietschel *et al.*, 1994), and may thus initiate host-mediated damage by stimulating gingival inflammation, increasing the secretion of collagenases from host cells, reducing collagen formation, and inducing localized bone resorption (Bartold, 1991; Daly *et al.*, 1990). The mechanism by which LPS stimulate macrophages and fibroblasts to produce cytokines has not been clearly identified but may involve a LPS surface receptor (see chapter 4). Shapira *et al.* (1998) reported that the activation of monocytes and inflammatory macrophages by lipopolysaccharides of *P. gingivalis* was strain dependent. This may explain, at least in part, the variability observed in the virulence potential of various strains of *P. gingivalis* (Grenier and Mayrand, 1987; Neiders *et al.*, 1989). LPS from several periodontopathogens were recently reported to prime neutrophils for enhanced respiratory burst (Aida *et al.*, 1995). This phenomenon may increase periodontal tissue damage by enhancing the generation of oxygen metabolites in infected sites. Lastly, the lipid A component of LPS was also reported to be involved in hemin binding by *P. gingivalis* and hemoglobin binding by *A. actinomycetemcomitans*, and may thus participate in the mechanism of iron acquisition (Grenier 1991; Grenier *et al.*, 1997). LPS from several periodontopathogens were also found to attach to saliva- and serum-coated hydroxyapatite beads, suggesting that they may participate in the bacterial colonization of subgingival sites (Okuda, *et al.*, 1991).

Proteinases

Proteinases produced by periodontopathogens may play multiple roles in the pathogenesis of periodontitis: adherence, bacterial growth, perturbation of the host defense system and tissue destruction (Grenier and Mayrand, 1993; Kuramitsu, 1998; Ishihara and Okuda, 1999). Although proteinases may adversely influence several mechanisms responsible for attachment *in vivo* (Childs and Gibbons, 1990; Hesketh *et al.,* 1987) they may also favor bacterial adherence either directly or following exposure of hidden binding sites. The mechanism by which cryptitopes, the hidden segments of molecules, may become exposed and participate in adhesion of oral bacteria has been initially described by Gibbons (1989). Childs and Gibbons (1988, 1990) as well as Naito and Gibbons (1988) reported that pretreatment of epithelial cells and fibronectin-collagen complexes with pancreatic trypsin could significantly enhance the adherence of *P. gingivalis*. Similar observations were noted for the attachment of *T. denticola* to fibroblasts (Weinberg and Holt, 1990). Since *P. gingivalis* and *T. denticola* produce enzymes with trypsin-like specificity (Grenier and Mayrand, 1993; Mäkinen and Mäkinen, 1996), it is logical to speculate that the above phenomenon could take place in subgingival sites. Several groups of invest-igators have also presented evidence for the direct involvement of cell-associated *P. gingivalis* cysteine proteinases in adherence to host cells or other oral bacteria (Grenier, 1992; Hoover, *et al.,* 1992; Li *et al.,* 1991). Recent molecular and genetic data support the proposal that proteinases and related gene products produced by *P. gingivalis* may act as adhesins distinct from other surface components with more specialized adherence functions (Curtis *et al.,* 1996; Nakayama *et al.,* 1995; Tokuda *et al.,* 1996). Nakayama *et al.* (1995) reported that hemagglutination and arginine-specific cysteine proteinase (RGP) activities are coded for by the same genes.

A number of important periodontopathogens, including *P. gingivalis* and *F. nucleatum*, are asaccharolytic bacteria whose metabolism is dependent on the uptake of small peptides and amino acids (Gharbia *et al.,* 1989; Shah and Gharbia, 1993). Proteolytic degradation of periodontal tissue components and host crevicular fluid proteins with production of lower molecular weight peptides may thus represent a critical requirement for the establishment and proliferation of these bacteria. Moreover, it has been suggested that the ability of bacterial proteinases to cleave host iron-binding proteins such as hemoglobin and transferrin, may be important for the acquisition of iron by bacteria (Carlsson *et al.,* 1984). Interestingly, a spontaneous proteinase deficient mutant of *P. gingivalis* could neither degrade transferrin nor use it as the source of iron for growth (Brochu *et al.,* 1999).

Proteinases from periodontopathogens may also allow bacteria to counteract the highly efficient host immune system by degradation of immunoglobulins, complement factors and plasma proteinase inhibitors (Fletcher *et al.,* 1994; Grenier, 1996; Kilian, 1981; Sundqvist *et al.,* 1985). Cleavage of immunoglobulins may protect periodontopathogens by reducing the beneficial effects of the immunoglobulin-mediated host defense system whereas inactivation of complement proteins (C3, C4, C5 and C5a factors) may enhance pathogenicity since complement is directly involved in the phagocytosis and killing of bacteria. The destruction of plasma proteinase inhibitors such as α_1-antitrypsin, α_2-macroglobulin and α_2-antiplasmin) may enhance inflammatory responses, vascular permeability and fibrinolysis. A new mechanism by which proteinases may reduce the efficacy of the host defense system has been proposed by Fletcher *et al.* (1997). They showed that *P. gingivalis* hydrolyzes cytokines including interleukin-1β, which may cause a reduction in the local concentration of pro-inflammatory cytokines resulting in a decrease in the host's ability to mount an effective response to periodontopathogens present in the periodontal pocket.

A significant reduction in gingival collagen fiber density is associated with the progression of adult chronic periodontitis. Proteinases from periodontopathogens may act directly on collagen as well as augment the activity of host tissue collagenases and thus may account for some of the destruction observed. Moreover, degradation of basement membrane components (collagen and fibronectin) may allow these periodontopathogens to (i) damage the tissue barrier and favor the diffusion of toxic bacterial products, and/or (ii) invade (along with other periodontopathogens) the subepithelial basement membrane and gain access to connective tissues. Proteinases from periodontopathogens may also have an important impact on the cytoskeleton and the crucial functions of the cytoskeleton which allow epithelial cells and fibroblasts to maintain homeostasis in the periodontium. This aspect has been recently reviewed by Ellen (1999).

The most proteolytic bacteria that have been associated with periodontitis include *P. gingivalis, T. denticola, P. intermedia/nigrescens* and *B. forsythus*. Proteolytic enzymes produced by *P. gingivalis* are either extracellular or cell-bound. A close examination of the numerous reports on purified *P. gingivalis* proteinases suggests that there are three types of enzymatic activities: arginine and lysine-specific cysteine proteinases (trivially called "trypsin-like proteinases"), X-prolyl dipeptidyl peptidase (class of serine proteases), and collagenases (Grenier and Mayrand, 1993). However, it appears that most of the proteolytic activity exhibited by *P. gingivalis* is due to its arg- and lys-gingipain cysteine proteinases. An analysis of available data suggests that three different genes code for arginine-X- (Arg-gingipain 1 and 2 [RGP-1 and RGP-2]) and lysine-X- (Lys-gingipain 1 [KGP]) specific cysteine proteinases (Potempa *et al.,* 1995). Several of these cysteine proteinases are processed forms of much larger protein precursors. Whether the processing is autocatalytic or results from the activity of other proteinases has not yet been clearly determined. Since cysteine proteinases of *P. gingivalis* are likely a critical virulence factor as well as an important antigen, antibodies to these enzymes may function in a protective capacity. Interestingly, immunization with purified Arg-gingipain 1 or 2, followed by a challenge with a virulent *P. gingivalis* strain, was found to protect against *P. gingivalis* colonization and invasion in a mouse chamber model (Genco *et al.,* 1998).

T. denticola is also known as a highly proteolytic bacterium and produces two major activities: chymotrypsin- and trypsin-like (Ishihara and Okuda, 1999; Mäkinen and Mäkinen, 1996). Most of the activities appear to be cell surface-associated. The chymotrypsin-like enzyme, also termed dentilisin, is a serine proteinase with a wide specificity being active on fibronectin, fibrinogen, type IV collagen and IgG (Ishihara *et al.,* 1996; Uitto *et al.,* 1988). Sequencing of the gene revealed that it has some similarity with subtilisin, which is the *Bacillus subtilis* serine proteinase (Ishihara *et al.,* 1996). Lastly, although much less data are available, proteinases are also produced by *P. intermedia/nigrescens* (Grenier *et al.,* 1994; Jansen *et al.,* 1995) and *B. forsythus* (Grenier, 1995).

Outer Membrane Vesicles
Since several determinants of virulence are surface components, the bacterial cell envelope of many bacterial species plays a crucial role in pathogenicity. The shedding of bacterial cell surface components, such as the vesicles, could then contribute to the pathogenic process and deserve to be considered as important virulence factors. Compared with the whole bacterial cell, biologically active substances such as proteinases, LPS and adhesins are present in larger amounts in vesicles. This is due to the fact that the ratio surface:volume is significantly higher for vesicles than for cells. Because of their small dimensions (approximately 50 nm) vesicles would more easily reach inaccessible areas or diffuse through anatomical barriers and thus initiate tissue invasion by whole bacterial cells. Outer membrane vesicles may also be involved in protective phenomena. First, they could compete for antibodies directed toward bacteria and thus impede the specific antibacterial immune defense. Secondly, it has been reported that *P. gingivalis* vesicles can completely abolish the bactericidal activity of human serum directed against other oral bacteria (Grenier and Bélanger, 1991). The LPS and the proteolytic activity associated with the vesicles were suggested to be responsible for this phenomenon. Lastly, *P. gingivalis* vesicles were found to bind chlorhexidine and decrease susceptibility of chlorhexidine-sensitive bacteria (Grenier *et al.*, 1995). Zhou *et al.* (1998) proposed that the genesis of extracellular vesicles by Gram-negative bacteria is a result of cell wall turnover. According to this hypothesis, peptidoglycan turnover results in the swelling of the outer membrane, causing the outer membrane to bulge and finally to bleb. Thereafter, the blebs are released in the environment by mechanical motion. Among the principal periodontopathogens, vesicles have been shown to be produced in great quantities by *P. gingivalis*, *T. denticola* and *A. actinomycetemcomitans* (Mayrand and Grenier, 1989; Nowotny *et al.*, 1982; Rosen *et al.* 1995; Smalley *et al.*, 1993).

Other Virulence Factors of Ecological Significance
Bacterial hemolysins may function to lyse erythrocytes present in the periodontal pocket, resulting in the liberation of hemoglobin that represents a potential source of iron for bacterial growth. Hemolysins have been reported to be produced by *P. gingivalis* (Chu *et al.*, 1991; Deshpande and Khan, 1999), *A. actinomycetemcomitans* (Kimizuka *et al.*, 1996), and *T. denticola* (Grenier, 1991). The capacity of bacteria to interact non-immunologically with immunoglobulins via the Fc portion may prevent attachment of opsonising antibodies to polymorphonuclear leukocytes Fc receptors, thereby inhibiting phagocytosis. The binding of immunoglobulins may also enable the bacteria to mimic the host and hence to hide from its immune system. Among periodontopathogens, *P. intermedia/nigrescens*, *F. nucleatum* and *A. actinomycetemcomitans* have the ability to bind Fc components of IgG (Grenier and Michaud, 1994; Labbé and Grenier, 1995; Tolo and Helgeland, 1991). *A. actinomycetemcomitans* is the only oral bacterium that has been reported to elaborate a specific leukotoxin capable of killing polymorphonuclear cells and monocytes (Zambon *et al.*, 1983). The leukotoxin appears to kill cells by inducing apoptosis (Korostoff *et al.*, 1998). Kolodrubetz *et al.* (1989) first cloned and expressed the leukotoxin gene. Bacterial extracts and metabolic products (ammonia, fatty acids) of *A. actinomycetemcomitans*, *T. denticola* and *P. gingivalis* can also exert toxic

effects toward several types of host cells (Stevens and Hammond, 1988; Taichman *et al.,* 1984; van Steenbergen *et al.,* 1986).

Modulation of Periodontal Microbiota and Expression of Virulence Factors by Environmental Parameters

In many pathogens, virulence factors are produced under coordinate regulation (Miller *et al.,* 1989). Generally, a common regulon which globally regulates the simultaneous expression of a series of virulence factors is present in pathogenic microorganisms. The regulon seems to sense and be regulated by the environment. Based on such a mechanism of regulation, individuals may harbor latent pathogens in their subgingival plaque for years. However, a change in the local environment of the subgingival sites may turn on or activate the virulence factor regulon leading to disease initiation.

Many triggers are capable of modifying biological responses of a bacterium. These triggers take the form of environmental and chemical stimuli such as pH, temperature, metabolic inhibitors, nutrient limitation (iron or nitrogen limitation) and oxidative stresses, and may enable some members of the ecosystem to become either more prominent and/or to manifest some of their pathogenic properties. Such conditions which operate *in vivo* but can be tested *in vitro* may affect the survival of oral bacteria (Bowden and Hamilton, 1998) and also induce the accumulation of damaged or denatured proteins, resulting in the induction of stress proteins (discussed below).

Iron is a constituent of important metabolic enzymes and is essential for the growth of almost all microorganisms. Lactoferrin, transferrin and hemoglobin are known constituents of gingival crevicular fluid (Cimasoni, 1983) and most likely represent the major sources of iron for periodontopathogens. Total iron in gingival crevicular fluid of adult periodontitis patients was reported to be in the range of 26 – 170 µM (Mukherjee, 1985). In a variety of pathogens, virulence factors are regulated by iron (Litwin and Calderwood, 1993). Most of the studies concerning this aspect in periodontopathogens have been conducted in *P. gingivalis*, although the precise role of iron in the regulation of specific virulence genes has not been clearly defined. Contradictory results were reported for the effect of hemin-restricted growth conditions on virulence of *P. gingivalis* in an animal model. Bramanti *et al.* (1993) reported that hemin limitation results in increased virulence of *P. gingivalis* whereas McKee *et al.* (1986) and Marsh *et al.* (1994) observed a diminished virulence. This may simply be related to the animal models used, which are not ideal for studying periodontal disease.

Hemin conditions have also been reported to affect virulence determinants of *P. gingivalis.* It was found that *P. gingivalis* cells grown under hemin-limited conditions elaborated 2 to 3 fold more hemolytic activity compared with cells grown under hemin excess (Chu *et al.,* 1991). On the other hand, Marsh *et al.* (1994) reported that trypsin-like activity (likely the arg-gingipain activity) of *P. gingivalis* was 3.5 times higher in cells grown in a condition of hemin excess. Moreover, extracellular outer membrane vesicles from hemin-limited *P. gingivalis* cultures were produced in greater numbers (McKee *et al.,* 1986) and possessed higher trypsin-like and

289

hemagglutinating activities as compared with vesicles from cultures grown in hemin excess (Smalley *et al.,* 1991). Contradictory results were obtained by Barua *et al.* (1990) who reported that iron availability had no effect on proteinase production in *P. gingivalis.* C. A. Genco's group (Genco, 1995; Genco *et al.,* 1995; Simpson *et al.,* 1994) showed that the decreased transport of hemin by a mutant of *P. gingivalis* created by transpositional mutagenesis resulted in the increased expression of several virulence factors (vesicles, hemolytic and proteinase activities) which may be coordinately regulated by hemin. It is also possible that the increased expression of the virulence factors may simply be a consequence of poor growth of the mutant.

Kesavalu *et al.* (1999) evaluated the effect of environmental iron on the virulence of *T. denticola* in mice. Growth of the spirochetes under iron-limiting conditions had no effect on abscess induction in comparison with bacteria grown under normal growth conditions. However, increasing systemic iron availability by injection of iron dextran led to significantly larger lesions in mice as well as systemic manifestations of the infectious challenge.

The metabolic activity of asaccharolytic bacteria such as *P. gingivalis* and *F. nucleatum* found in the periodontal pocket may be associated with a rise in the pH of gingival crevicular fluid to values of 7.5 to 8.0. Such alkaline conditions may favor the growth of several periodontopathogens and enhance the activity of virulence factors. For instance, the optimal pH for growth of *A. actinomycetemcomitans* (Sreenivasan *et al.,* 1993) and *P. gingivalis* (McDermid *et al.,* 1988; Takahashi and Schachtele, 1990) is in the range of 7.0 to 8.0. In addition, an alkaline environmental condition was found to increase the proteolytic activity of *P. gingivalis* (McDermid *et al.,* 1988; Takahashi and Schachtele, 1990).

Other environmental parameters have been found to influence the proteinase activities of *P. gingivalis.* A number of studies have revealed that the expression of some proteinases is increased under conditions of peptide limitation (Lu and McBride, 1988; Park *et al.,* 1997; Tokuda *et al.,* 1998). On the other hand, an increase in growth temperature (37°C to 41°C) of *P. gingivalis* has been reported to significantly reduce the level of arg- and lys-gingipain activities as well as their hemagglutination, while not affecting bacterial growth (Percival *et al.,* 1999). The presence of regulatory or processing genes in *P. gingivalis* could explain the simultaneous effects caused by a modification of a single environmental parameter.

A. actinomycetemcomitans is known to produce a leukotoxin that may play a critical role in the pathogenic process of periodontitis (Zambon *et al.,* 1983). Anaerobic culture conditions were found to induce leukotoxin production by highly toxic strains (Hritz *et al.,* 1996). Spitznagel *et al.* (1991) first showed that the regulation of leukotoxin mRNA can be affected by both genotype as well as environment. They found that *A. actinomycetemcomitans* leukotoxin is regulated by environmental signals (oxygen tension) that can be found in the gingival crevice (Spitznagel *et al.,* 1995). It was also reported that the bacteria exhibited more fimbriae when grown anaerobically than when cultured in CO_2 (Scannapieco *et al.,* 1987). On the other hand, growth in the presence of CO_2 resulted in a much greater amount of amorphous extracellular material surrounding the cells (Scannapieco *et al.,* 1987).

Lowering of the oxidation-reduction potential clearly favors anaerobic bacteria and may be brought about in various ways. Obstruction and stasis in areas containing an indigenous microbiota typically lead to a low oxidation-reduction potential and overgrowth of anaerobic bacteria. Tissue destruction, trauma or surgery is also an important condition underlying decreased oxidation-reduction potential. Lastly, the presence and growth of aerobic or facultative bacteria also leads to a significant reduction in oxidation-reduction potential. It is surprising that despite the fact that oxidation-reduction potential is likely a critical factor in anaerobic infections, very few studies have been devoted to the effect of this environmental parameter on the biology and physiology of periodontopathogens. During the course of periodontitis, the formation of periodontal pockets is associated with a significant decrease of oxidation-reduction potential thus resulting in a reduced environment favorable for the multiplication of strictly anaerobic periodontopathogens. Kenney and Ash (1969) reported values in the range of +14 to –157 mV in these sites whereas the oxidation reduction potential of healthy gingival sulcus is approximately +74 mV. The growth of obligate anaerobic periodontopathogens in a mixed population containing facultative anaerobes may allow them to survive in an environment containing oxygen. Bradshaw *et al.* (1996) reported that anaerobes not only persist but also grow in biofilms in the presence of oxygen, even when the dissolved oxygen tension was 40-50% and the oxidation-reduction potential was as high as +100 mV.

Cysteine, which is known to decrease the oxidation-reduction potential was found to significantly enhance *in vitro* growth of *T. denticola* as well as virulence of these spirochetes in a model of localized inflammatory abscesses in mice (Kesavalu *et al.*, 1999). Recently, the influence of oxidized (positive Eh) and reduced (negative Eh) conditions on growth and selected properties of *P. gingivalis* was evaluated (Lekc *et al.*, 1999). Results indicated that under positive Eh values, growth, hemagglutination and arg-gingipain activity were negatively affected. This suggests that an increase in the Eh condition in the periodontal pocket may be a way to create an environment incompatible with growth and virulence of anaerobic periodontopathogens. A number of clinical studies have already proposed a therapeutic approach for periodontitis patients based on the application in periodontal pockets of an oxidizing agent that can increase the oxidation-reduction potential (Gibson *et al.*, 1994; Wennstrom *et al.*, 1987; Wolff *et al.*, 1989). Conclusions from these studies have been variable, most likely because the substances were eliminated from the diseased sites too rapidly to be effective. Incorporation of the oxidizing agents into a slow release carrier showed more convincing clinical improvements (Ower *et al.*, 1995).

Stress Response and Pathogenicity

When submitted to a wide range of environmental stresses, bacteria (like most living organisms) respond by inducing or accelerating the synthesis of a specific set of proteins which by convention are termed stress proteins. Environmental parameters known to induce the synthesis of these proteins include temperature, oxidative, nutritional and chemical stresses (Lathigra *et al.*, 1991; Lindquist, 1986; Watson, 1990). For instance, in *Escherichia coli*, a temperature shift from 30°C to 37°C results in a 2 to 20-fold increase in synthesis of at least 17 proteins (Dowds, 1994). Similar responses resulting in expression of specific «heat shock proteins» (HSPs) can also be induced by stresses other than heat. HSPs are divided into five families which are composed of similarly sized proteins that are named according to the size of the protein in kiloDaltons: HSP90, HSP70 (DnaK), HSP60 (GroEL), low molecular weight and GroES (10 kDa) (Lathigra *et al.*, 1991; Lindquist, 1986; Watson, 1990). Stress proteins appear to be biologically important since they have been implicated in thermotolerance, immunodominance and autoimmunity phenomena (Dowds, 1994; Kaufmann and Schoel, 1994; Zügel and Kaufmann, 1999). Bacteria under stress produce abnormal, misfolded or damaged proteins and synthesis of the stress proteins is required to protect proteins by reversible association (Visick and Clarke, 1995). Being highly immunoreactive, they are likely to play a role in the immune response to many bacterial pathogens (Lathigra *et al.*, 1991; Watson, 1990). Specific stress proteins have also been found to possess a variety of biological activities. For instance, a superoxide dismutase activity has been ascribed to a HSP in *E. coli* (Privalle and Fridovich, 1987) whereas a urease activity was ascribed to a HSP in *Helicobacter pylori* (Evans *et al.*, 1992). Recent interest in HSPs has led to the suggestion that HSPs of pathogenic organisms may play a key role in infection and thus represent virulence factors (Fernandez *et al.*, 1996; Lathigra *et al.*, 1991). Interestingly, it has been reported that exposure of *Leishmania braziliensis* promastigotes to heat shock increases their infectivity and pathogenicity (Smejkal *et al.*, 1988). Furthermore, mutations in a HSP gene (*htrA*) in *Salmonella typhimurium* were associated with attenuation in the virulence of the bacteria (Lathigra *et al.*, 1991). All the above findings represent strong experimental evidence that HSPs may play a role in microbial virulence.

Environmental stress factors affecting oral bacteria have been recently reviewed by Bowden and Hamilton (1998). The stress response to heat-shock has also been studied in several periodontopathogens including *P. gingivalis, A. actinomycetemcomitans, P. intermedia, B. forsythus* and *C. rectus* (Hinode *et al.*, 1998; Kadri *et al.*, 1998; Koga *et al.*, 1993; Lokensgard *et al.*, 1994; Lu and McBride, 1994; Vayssier *et al.*, 1994). In general, all these bacteria were found to express a classical heat-shock response resulting in an increased synthesis of both HSP60 and HSP70. Moreover, a temperature stress (37 to 39°C) in *P. gingivalis* resulted in a reduction of biosynthesis of fimbriae and an increase in the amount of superoxide dismutase (Amano *et al.*, 1994). Future studies should reveal the effects of environmental factors, such as those reported for inflamed periodontal pockets, on virulence properties and host immune responses of other periodontopathogens.

Host-Bacteria Interactions

Periodontopathogens present in subgingival sites induce host cellular and humoral responses. In most cases, this response results in the elimination or local control of the pathogens and significant periodontal disease does not establish or progress. However, the protective responses to periodontal pathogens may be overcome via the action of bacterial virulence factors, and consequently pathogens in subgingival plaque may be able to reach a critical concentration required for initiation or progression of periodontitis. The continuous challenge of the host defense system by periodontopathogens and their products, most particularly the LPS, initiates a number of host-mediated destructive processes (Birkedal-Hansen, 1993; Kornman *et al.*, 1997). Firstly, a large number of polymorphonuclear leukocytes accumulate in the periodontal pocket and release latent matrix metalloproteinases (MMPs) or active hydrolytic enzymes. Proteinases from periodontopathogens may activate latent MMPs as well as degrade host tissue inhibitor of metalloproteinases (TIMP) resulting in a significant increased amount of active MMPs (Sorsa *et al.*, 1992). These enzymes are capable of degrading the major host tissue components including collagen, fibronectin and laminin. Secondly, T lymphocytes are continually stimulated by periodontopathogens, a phenomenon that results in a production of increased amount of cytokines, including the multifunctional cytokine interleukin-1 (IL-1). Among others, cytokines have a cytotoxic effect on fibroblasts, and attract macrophages in the periodontal pocket prior to activating them to release latent or active hydrolytic enzymes. Macrophages can also be stimulated directly by LPS and fimbriae of periodontopathogens (Hanazawa *et al.*, 1991; Shapira *et al.*, 1998). This complex inflammatory phenomenon results in the destruction of the tooth-supporting tissues, mostly mediated by host collagenases (MMPs).

Inflammation is a localized protective response elicited by injury and infection. Whereas the absence of inflammatory responses leads to a compromised host (see chapter 4), excessive or sustained inflammation results in chronic inflammatory diseases such as periodontitis. As mentioned above, pro-inflammatory cytokines, and more particularly IL-1, are key factors in the initiation and development of the inflammatory cascade. In addition, IL-1, which has pleiotropic effects on almost every cell type, can participate in attacking local tissues in periodontal disease by inducing collagenase production by fibroblasts, and promoting cartilage and bone resorption (Dinarello, 1994; Ohshima *et al.*, 1994; Tatakis, 1993). More specifically, IL-1β is a pro-inflammatory cytokine produced as an inactive precursor (pro-IL-1β) by various cell types including monocytes, macrophages and keratinocytes (Dinarello, 1994). This precursor, with a molecular weight of 31 kDa, must be cleaved enzymatically into a 17.5 kDa fragment to exert its biological activities (Black *et al.*, 1988; Mosley *et al.*, 1987).

Bacterial factors (LPS, peptidoglycan, polysaccharide, fimbriae, etc) that are known to induce the production of IL-1β have been identified (Hamada *et al.*, 1988; Saito *et al.*, 1996; Takada *et al.*, 1991; Wilson *et al.*, 1996; Yoshimura *et al.*, 1997). More specifically, several research groups have reported that LPS from periodontopathogens can induce the release of IL-1β by polymorphonuclear leukocytes, macrophages and fibroblasts (Henderson and Wilson, 1998; Takada *et al.*, 1991; Wilson *et al.*, 1996). However, once released, the cytokine must be cleaved into the active form to exert its key role in the inflammatory process. In a previous study, the ability of *T. denticola* to cleave the inactive pro-IL-1β precursor with the production of biologically active fragments was demonstrated (Beauséjour *et al.*, 1997). As *T. denticola* is present in high proportions in affected periodontal sites, it is possible that it can gain access to the epithelial and phagocyte IL-1β precursor, and generate biologically active fragments. Indeed, *T. denticola* can attach to epithelial cells and increase permeability of the epithelium thus favoring invasion of periodontal tissues (Uitto *et al.*, 1995). Interestingly, the presence of spirochetes has been previously demonstrated in histological tissue sections of gingiva (Mikx *et al.,* 1990; Saglie *et al.,* 1985). The cleavage of the IL-1β precursor could occur via secreted enzymes or via a reaction at the bacterial cell surface, as *T. denticola* has been shown to possess both surface and secreted forms of proteinases (Mäkinen and Mäkinen, 1996). If so, *T. denticola* could exert a pro-inflammatory role in periodontal diseases, and thus display a doubly deleterious effect on periodontal tissues by virtue of its own enzymes, and by activating a tissue destructive inflammation cascade. This hypothesis is supported by the fact that IL-1β was detected in the gingival crevicular fluid of patients with periodontitis at higher concentrations than in patients of a healthy control group (Hou *et al.,* 1995; Preiss and Meyle, 1994). It is also in agreement with the concept proposed by Henderson and Wilson (1998) that the normal microbiota has evolved cytokine-modulating molecules to live in harmony with host mucosal surfaces. They also introduced the concept of cytokine polymorphisms as a contributory factor to pathogenesis.

Additional mechanisms by which bacterial proteinases interact with the host and result in deleterious effects have been reported. A proteolytic fraction from *P. gingivalis* was found to activate latent host MMPs as well as to induce the secretion of collagenase activity by human gingival fibroblasts (DeCarlo *et al.,* 1997; Uitto *et al.,* 1987). Moreover, a *P. gingivalis* protease with trypsin-like specificity was found to induce apoptosis of human gingival fibroblasts, a phenomenon that may contribute to cell destruction during periodontitis (Wang *et al.,* 1999). Lastly, it was shown that a proteolytic culture supernatant of *P. gingivalis* could activate plasminogen bound to *F. nucleatum* to plasmin (Darenfed *et al.,* 1999). The plasmin-coated *F. nucleatum* was found to degrade fibronectin and to inactivate tissue inhibitor of metalloproteinases, suggesting a new mechanism in tissue invasion and damage during periodontitis.

As stated above, bacterial proteinases may play a critical role in periodontitis by interacting with host cells. Some evidence for the presence of *P. gingivalis* proteinases *in vivo* is available. For instance, on the basis of analyses of serum IgG responses in periodontitis patients, it has been shown that *P. gingivalis* proteinases are important antigens (Genco *et al.,* 1999; Ismaiel *et al.,* 1988). Furthermore, a positive correlation was demonstrated between trypsin-like activity in subgingival sites and both the level of clinical disease and the levels of *P. gingivalis* (Suido *et al.,* 1988).

Other membrane components of periodontopathogens can also mediate tissue destruction observed during periodontitis. More specifically, LPS and the major surface protein (Msp) of *T. denticola* have been shown to induce release of active MMPs from polymorphonuclear leukocytes (Ding *et al.,* 1996). In a second study, incubation of polymorphonuclear leukocytes with whole cells of *F. nucleatum* or *P. gingivalis* resulted in phagocytosis of the bacteria and the release of high quantities of elastase and MMP-9, respectively (Ding *et al.,* 1997). Thus, periodontopathogens and their products may participate in extracellular matrix degradation during the course of periodontal inflammation by triggering the secretion and activation of host proteinases from PMN, macrophages and fibroblasts.

Bacterial heat-shock proteins are also components that may interact with host cells during periodontitis (Figure 3). In non-stressed cells, HSPs are present in low concentrations while in stressed cells they accumulate at higher levels. For instance, the HSP60 homolog of GroEL of *E. coli* represents 1 to 2% of the total protein content under normal conditions but its concentration is increased four to five fold under stress conditions (Shinnick, 1991). Because of the similarities between microbial and mammalian HSPs, a humoral response against these proteins can be destructive. During microbial infections, an antibody response against epitopes cross-reactive between bacterium and host may result in an autoimmune response. This phenomenon has been suggested for a number of chronic inflammatory diseases including Crohn's disease and rheumatoid arthritis (Panchapekesan *et al.*, 1992; Peetermans, 1996).

Recently, Goulhen *et al.* (1998) reported that a GroEL-like protein was localized on the cell surface of *A. actinomycetemcomitans*. The mechanism involved in translocating this HSP to the bacterial cell surface is not fully understood, since HSPs are typically cytosolic proteins that lack the specific leader sequences normally required for cell surface expression (Zügel and Kaufmann, 1999). The native GroEL protein (HSP60) from *A. actinomycetemcomitans* was found to possess a strong toxic effect on HaCaT epithelial cells (Goulhen *et al.*, 1998). It was also reported by Uitto *et al.* (1999) that this protein may play a significant role in proliferation (by activating mitogen-activated protein kinase pathways) and migration of epithelial cells during periodontitis. The importance of the GroEL-like protein of *A. actinomycetemcomitans* in invasion of mammalian cells such as epithelial cells (Meyer and Fives-Taylor, 1997) remains to be investigated. This is particularly relevant since Garduno *et al.* (1998) recently reported that a cell surface GroEL-like protein of *Legionella pneumophila* mediated adherence and invasion in a HeLa cell model.

Hinode *et al.* (1998) showed that a GroEL-like protein from *C. rectus* has a cross-reactive epitope with the equivalent human molecule and was capable of enhancing IL-6 and -8 in human gingival fibroblasts. Again, these results suggest that the stress proteins can activate the immune network of inflammation and may

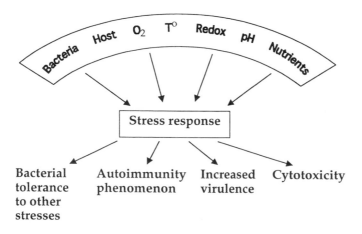

Figure 3. Role of the stress response in periodontal disease.

be considered virulence factors. During infections, bacteria can also stimulate over expression of HSPs by host cells. Saito *et al.* (1997) reported that injection of sublethal doses of *F. nucleatum* in mice resulted in increased levels of both HSP60 and HSP70 in peritoneal exudate. The exact significance of elevated host HSPs during infections needs to be clarified.

HSPs are likely to be produced during periodontitis. Ando *et al.* (1995) found that gingival homogenate samples from patients with adult periodontitis reacted with anti-human HSP60. They also reported that antibodies which reacted with bacterial HSPs (*A. actinomycetemcomitans, F. nucleatum*) were found in a serum sample from a periodontitis patient. Human serum IgG antibody response to the HSP64 (GroEL-like) from *A. actinomycetemcomitans* was twice as great in patients with localized juvenile periodontitis than in normal subjects with no periodontal lesions (Mayrand et *al.*, unpublished data). Jamarillo *et al.* (1996) found a HSP90 in *P. gingivalis* that cross-reacted with antibodies to human HSP90. They also showed the presence of antibodies cross-reactive with the HSP90 of *P. gingivalis* in serum patients. The most interesting answers are yet to be obtained concerning the importance of bacterial heat shock proteins as virulence factors in periodontitis, and the consequences of HSP expression on recognition by self-reactive antibodies for autoimmunity.

The host immune response to pathogens is based on the recognition of cell surface antigens. Recently, a number of pathogenic bacteria have been reported to produce a specific class of antigens called superantigens (Legaard *et al.,* 1991; Pincus *et al.,* 1992). These structures bind to major histocompatibility complex (MHC) class II molecules and interact with the V_β domain of the T-cell receptors. Entire subgroups of T cells expressing the appropriate V_β domain are thus stimulated. This massive activation of T cells has been suggested to contribute to the pathogenesis of bacterial infections. In addition, there is evidence that superantigens may play a role in the pathogenesis of chronic inflammatory diseases, such as rheumatoid arthritis (Paliard *et al.,* 1991). To our knowledge, no research groups have investigated superantigens in oral bacteria. Because of the potential role that these molecules could play in periodontitis, research on this particular aspect deserves consideration.

Conclusions

Microbiological and immunological studies carried out over the last years suggest the following model for the pathogenesis of periodontal diseases (Figure 4). Bacterial interactions and modification of the environment result in overgrowth of periodontopathogens including *P. gingivalis, B. forsythus, A. actinomycetemcomitans* and *T. denticola*. Adhesins expressed by these bacteria favor their establishment in subgingival sites via attachment to other bacteria or host cells. Bacteria produce proteinases that are not neutralized by plasma proteinase inhibitors found in the gingival crevicular fluid. Bacterial proteinases, either cell- or vesicle-associated, directly destroy tissue components and negatively affect the host defense system, favoring a continuous migration of inflammatory cells at the infection site. Hydrolysis of the host proteins generates nutrients (peptides, iron) to support growth of the periodontopathogens. LPS present in high concentration in diseased sites induces proinflammatory mediators such as cytokines and initiate host-mediated damage. The inability of phagocytic cells to destroy and digest all the bacteria results in a chronic influx of inflammatory cells associated with release of host proteinases including MMPs. Together, the host and bacterial proteinases severely damage the periodontal tissues.

Although the pathogenesis of periodontal diseases is now better known, there are still unresolved issues that require studies. Most particularly, do uncultivable bacteria play a significant role in the initiation and progression of periodontitis? Are there unknown virulence factors not yet identified because studies of characterization are performed *in vitro*? Our improved knowledge of the pathogenesis of periodontitis may help to develop potential new therapies. Two promising approaches in this regard would be i) to favor the establishment and growth of bacteriocin-producing nonpathogenic bacteria in the subgingival microbiota; and ii) to neutralize both bacterial and host proteinase activities with specific inhibitors.

References

Aida, Y., T. Kukita, H. Takada, K. Maeda, and M.J. Pabst. 1995. Lipopolysaccharides from periodontal pathogens prime neutrophils for enhanced respiratory burst: differential effect of a synthetic lipid A precursor IV$_A$ (LA-14-PP). J. Periodontal

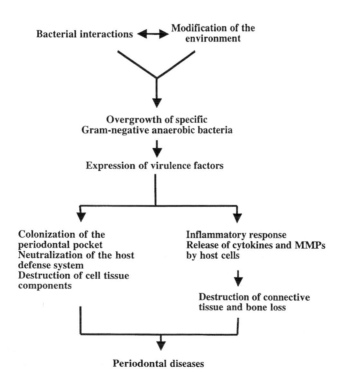

Figure 4. Critical steps leading to periodontal disease.

Res. 30: 116-123.

Amano, A., A. Sharma, H.T. Sojar, H.K. Kuramitsu, and R.J. Genco. 1994. Effects of temperature stress on expression of fimbriae and superoxide dismutase by *Porphyromonas gingivalis*. Infect. Immun. 62: 4682-4685.

Ando, T. , T. Kato, K. Ishihara, H. Ogiuchi, and K. Okuda. 1995. Heat shock proteins in the human periodontal disease process. Microbiol. Immunol. 39: 321-327.

Bartold, P.M. 1991. Modulation of gingival fibroblast function by lipopolysaccharides, p. 277-290. *In* S. Hamada, S.C. Holt and J.R. McGhee (ed.), Periodontal disease: pathogens and host immune responses. Quintessence Publishing Co., Tokyo.

Barua, P.K., D.W. Dyer, and M.E. Neiders. 1990. Effect of iron limitation on *Bacteroides gingivalis*. Oral Microbiol. Immunol. 5: 263-268.

Baumgartner, J.C., W.A. Falkler, and T. Beckerman. 1992. Experimentally induced infection by oral anaerobic microorganisms in a mouse model. Oral Microbiol. Immunol. 7: 253-256.

Beauséjour, A., N. Deslauriers, and D. Grenier. 1997. Activation of the interleukin-1β precursor by *Treponema denticola*: a potential role in chronic inflammatory periodontal diseases. Infect. Immun. 65: 3199-3202.

Birkedal-Hansen, H. 1993. Role of cytokines and inflammatory mediators in tissue destruction. J. Periodontal Res. 28: 500-510.

Black, R.A., S.R. Kronheim, M. Cantrell, M.C. Deeley, C.J. March, K.S. Prickett, J. Wignall, P.J. Conlon, D. Cosman, T.P. Hopp, and D.Y. Mochizuki. 1988. Generation of biologically active interleukin-1β by proteolytic cleavage of the inactive precursor. J. Biol. Chem. 263: 9437-9442.

Bowden, G.H.W., and I.R. Hamilton. 1998. Survival of oral bacteria. Crit. Rev. Oral Biol. 9: 54-85.

Bradshaw, D.J., P.D. Marsh, C. Allison, and K.M. Schilling. 1996. Effect of oxygen, inoculum composition and flow rate on development of mixed-culture oral biofilms. Microbiology 142: 623-629.

Bramanti, T.E., S.C. Holt, J.L. Ebersole, and T. van Dyke. 1993. Regulation of *Porphyromonas gingivalis* virulence: hemin limitation effects on the outer membrane protein (OMP) expression and biological activity. J. Periodontal Res. 28: 464-466.

Brochu, V., D. Grenier, and D. Mayrand. 1999. Involvement of proteases in utilization of human transferrin-bound iron by *Porphyromonas gingivalis*. J. Dent. Res. 78: 505.

Carlsson, J., J.F. Hofling, and G.K. Sundqvist. 1984. Degradation of albumin, haemopexin, haptoglobin and transferrin, by black-pigmented *Bacteroides* species. J. Med. Microbiol. 18: 39-46.

Childs, W.C., and R.J. Gibbons. 1988. Use of percoll density gradients for studying the attachment of bacteria to oral epithelial cells. J. Dent. Res. 67: 826-830.

Childs, W.C., and R.J. Gibbons. 1990. Selective modulation of bacterial attachment to oral epithelial cells by enzyme activities associated with poor oral hygiene J. Periodontal Res. 25: 172-178.

Choi, B.K., B.J. Paster, F.E. Dewhirst, and U.B. Göbel. 1994. Diversity of cultivable and uncultivable oral spirochetes from a patient with severe destructive periodontitis. Infect. Immun. 62: 1889-1895.

Chu, L., T.E. Bramanti, J.L. Ebersole, and S.C. Holt. 1991. Hemolytic activity in the periodontopathogen, *Porphyromonas gingivalis*: kinetics of enzyme formation and localization. Infect. Immun. 59: 1932-1940.

Cimasoni, G. 1983. Crevicular fluid updated. Monographs in Oral Science, Volume 12. Karger. Basel. 152 p.

Clewell, D.B. 1993. Bacterial sex pheromone-induced plasmid transfer. Cell 73: 9-12.

Curtis, M.A., J. Aduse-Opoku, J.M. Slaney, M. Rangarajan, V. Booth, J. Cridland, and P. Shepherd. 1996. Characterization of an adherence and antigenic determinant of the ArgI protease of *Porphyromonas gingivalis* which is present on multiple genes. Infect. Immun. 64: 2532-2539.

Daly, C.G., G.J. Seymour, and J.B. Kieser. 1990. Bacterial endotoxin: a role in chronic inflammatory periodontal disease? J. Oral Pathol. 9: 1-15.

Darenfed, H., D. Grenier, and D. Mayrand. 1999. Acquisition of plasmin activity by *Fusobacterium nucleatum ss. nucleatum* and potential contribution to tissue destruction during periodontitis. Infect. Immun. 67: 6439-6444.

DeCarlo, A.A., L.J. Windsor, M.K. Bodden, G.J. Harber, B. Birkedal-Hansen, and H. Birkedal-Hansen. 1997. Activation and novel processing of matrix metalloproteinases by a thiol-proteinase from the oral anaerobe *Porphyromonas gingivalis*. J. Dent. Res. 76: 1260-1270.

Deshpande, R.G., and M.B. Khan. 1999. Purification and characterization of hemolysin from *Porphyromonas gingivalis* A7436. FEMS Microbiol. Lett. 176: 387-394.

Dinarello, C.A. 1994. The biological properties of interleukin-1β. Eur. Cytokine Netw. 5: 517-531.

Ding, Y., M. Haapasalo, E. Kerosuo, K. Lounatmaa, A. Kotiranta, and T. Sorsa. 1997. Release and activation of human neutrophil matrix metallo- and serine proteinases during phagocytosis of *Fusobacterium nucleatum, Porphyromonas gingivalis* and *Treponema denticola*. J. Clin. Periodontol. 24: 237-248.

Ding, Y., V.-J. Uitto, M. Haapasalo, K. Lounatmaa, Y.T. Konttinen, T. Salo, D. Grenier, and T. Sorsa. 1996. Membrane components of *Treponema denticola* trigger proteinase release from polymorphonuclear leukocytes. J. Dent. Res. 75: 1986-1993.

DiRita, V.J., and J.J. Mekalanos. 1989. Genetic regulation of bacterial virulence. Annu. Rev. Genet. 23: 455-482.

Dowds, B.C.A. 1994. The oxidative stress response in *Bacillus subtilis*. FEMS Microbiol. Lett. 124: 255-264.

Dzink, J.L., S.S. Socransky, and A.D. Haffajee. 1988. The predominant cultivable microbiota of active and inactive lesions of destructive periodontal diseases. J. Clin. Periodontol. 15: 316-323.

Dzink, J.L., S.S. Socransky, and C.L. Smith. 1986. Interactions between *Bacteroides forsythus* and *Fusobacterium nucleatum*. J. Dent. Res. 65: 853.

Ebersole J.L., F. Feuille, L. Kesavalu, and S.C. Holt. 1997. Host modulation of tissue destruction caused by periodontopathogens: effects on a mixed microbial infection composed of *Porphyromonas gingivalis* and *Fusobacterium nucleatum*.

Microb. Pathog. 23: 23-32.

Ellen, R.P. 1999. Perturbation and exploitation of host cell cytoskeleton by periodontal pathogens. Microb. Infect. 1: 621-632.

Evans, D. J., D. Evans, L. Engstrand, and D.Y. Graham. 1992. Urease-associated heat-shock protein of *Helicobacter pylori*. Infect. Immun. 60: 2125-2127.

Fabricius, L., G. Dahlen, S.E. Holm, and A.J.R. Moller. 1982. Influence of combination of oral bacteria on periapical tissues of monkeys. Scand. J. Dent. Res. 90: 200-206.

Fernandez, R.C., S.M. Logan, S.H.S. Lee, and P. S. Hoffman. 1996. Elevated levels of *Legionella pneumophila* stress protein Hsp60 early in infection of human monocytes and L929 cells correlate with virulence. Infect. Immun. 64: 1968-1976.

Feuille, F., J.L. Ebersole, L. Kesavalu, M.J. Steffen, and S.C. Holt. 1996. Mixed infection with *Porphyromonas gingivalis* and *Fusobacterium nucleatum* in a murine lesion model: potential synergistic effects on virulence. Infect. Immun. 64: 2095-2100.

Fletcher, H.M., H. A. Schenkein, and F.L. Macrina. 1994. Cloning and characterization of a new protease gene (*prtH*) from *Porphyromonas gingivalis*. Infect. Immun. 62: 4279-4286.

Fletcher, J., K. Reddi, S. Poole, S. Nair, B. Henderson, P. Tabona, and M. Wilson. 1997. Interactions between periodontopathogenic bacteria and cytokines. J. Periodontal Res. 32: 200-205.

Fujimura, S. 1979. Sanguicin, a bacteriocin of oral *Streptococcus sanguis*. Antimicrob. Agents Chemother. 16: 262-265.

Garduno, R.A., E. Garduno, and P.S. Hoffman. 1998. Surface-associated Hsp60 chaperonin of *Legionella pneumophila* mediates invasion in a HeLa cell model. Infect. Immun. 66: 4602-4610.

Genco, C.A. 1995. Regulation of hemin and iron transport in *Porphyromonas gingivalis*. Adv. Dent. Res. 9: 41-47.

Genco, C.A., B.M. Odusanya, J. Potempa, J. Mikolajczyk-Pawlinska, and J. Travis. A peptide domain on gingipain R1 which confers immunity against *Porphyromonas gingivalis* infection in mice. Infect. Immun. 66: 4108-4114.

Genco, C.A., W. Simpson, R.Y. Forng, M. Egal, and B.M. Odusanya. 1995. Characterization of a Tn4351-generated hemin uptake mutant of *Porphyromonas gingivalis*: evidence for the coordinate regulation of virulence factors by hemin. Infect. Immun. 63: 2459-2466.

Genco, C.A., J. Potempa, J. Mikolajczyk-Pawlinska, and J. Travis. 1999. Role of gingipains R in the pathogenesis of *Porphyromonas gingivalis*-mediated periodontal disease. Clin. Infect. Dis. 28: 456-465.

Gharbia, S.E., H.N. Shah, and S.G. Welch. 1989. The influence of peptides on the uptake of amino acids in *Fusobacterium*: predicted interaction with *Porphyromonas gingivalis*. Curr. Microbiol. 19: 231-235

Gibbons, R.J. 1989. Bacterial adhesion to oral tissues: a model for infectious diseases. J. Dent. Res. 68: 750-760.

Gibson, M.T., D. Mangat, G. Gagliano, M. Wilson, J. Fletcher, J. Bulman, and H.N. Newman. 1994. Evaluation of the efficacy of a redox agent in the treatment of chronic periodontitis. J. Clin. Periodontol. 21: 690-700.

Goulhen, F., A. Hafezi, V.-J. Uitto, D. Hinode, R. Nakamura, D. Grenier, and D. Mayrand. 1998. Subcellular localization and cytotoxic activity of the GroEL-like protein isolated from *Actinobacillus actinomycetemcomitans*. Infect. Immun. 66: 5307-5313.

Grenier, D. 1991. Characteristics of hemolytic and hemagglutinating activities of *Treponema denticola*. Oral Microbiol. Immunol. 6: 246-249.

Grenier, D. 1991. Haemin-binding property of *Porphyromonas gingivalis* outer membranes. FEMS Microbiol. Lett. 77: 45-50.

Grenier, D. 1992. Further evidence for a possible role of trypsin-like activity in the adherence of *Porphyromonas gingivalis*. Can. J. Microbiol. 38: 1189-1192.

Grenier, D. 1992. Nutritional interactions between two suspected periodontopathogens, *Treponema denticola* and *Porphyromonas gingivalis*. Infect. Immun. 60: 5298-5301.

Grenier, D. 1994. Effect of proteolytic enzymes on the lysis and growth of oral bacteria. Oral Microbiol. Immunol. 9: 224-228.

Grenier, D. 1995. Characterization of the trypsin like activity of *Bacteroides forsythus*. Microbiology 141: 921-926.

Grenier, D. 1996. Antagonistic effect of oral bacteria towards *Treponema denticola*. J. Clin. Microbiol. 34: 1249-1252.

Grenier, D. 1996. Degradation of host protease inhibitors and activation of plasminogen by proteolytic enzymes from *Porphyromonas gingivalis* and *Treponema denticola*. Microbiology 142: 955-961.

Grenier, D., and M. Bélanger. 1991. Protective effect of *Porphyromonas gingivalis* outer membrane vesicles against bactericidal activity of human serum. Infect. Immun. 59: 3004-3008.

Grenier D., and D. Mayrand. 1986. Nutritional relationships between oral bacteria. Infect. Immun. 53: 616-620.

Grenier, D., and D. Mayrand. 1987. Selected characteristics of pathogenic and nonpathogenic strains of *Bacteroides gingivalis*. J. Clin. Microbiol. 25: 738-740.

Grenier, D., and D. Mayrand. 1993. Proteinases, p. 227-243. *In* H.N. Shah, D. Mayrand, and R.J. Genco (ed.). Biology of the species *Porphyromonas gingivalis*. CRC Press, Boca Raton.

Grenier, D., and J. Michaud. 1994. Demonstration of human immunoglobulin G Fc-binding activity in oral bacteria. Clin. Diagn. Lab. Immunol. 1: 247-249.

Grenier, D., J. Bertrand, and D. Mayrand. 1995. *Porphyromonas gingivalis* outer membrane vesicles promote bacterial resistance to chlorhexidine. Oral. Microbiol. Immunol. 10: 319-320.

Grenier, D., A. Leduc, and D. Mayrand. 1997. Interaction between *Actinobacillus actinomycetemcomitans* lipopolysaccharides and human hemoglobin. FEMS Microbiol. Lett. 151: 77-81.

Grenier, D., S. Labbé, C. Mouton, and D. Mayrand. 1994. Hydrolytic enzymes and lectin-binding activity of black-pigmented anaerobic rods. Microbiology 140: 873-878.

Guiney, D.G. 1997. Regulation of bacterial virulence gene expression by the host environment. J. Clin. Invest. 99: 565-569.

Haffajee, A.D., and S.S. Socransky. 1994. Microbial etiological agents of destructive periodontal diseases. Periodontology 2000 5: 78-111.

Hamada, S., T. Koga, T. Nishihara, T. Fujiwara, and N. Okahashi. 1988. Characterization and immunobiologic activities of lipopolysaccharides from periodontal bacteria. Adv. Dent. Res. 2: 284-291.

Hammond, B.F., S.E. Lillard, and R.H. Stevens. 1987. A bacteriocin of *Actinobacillus actinomycetemcomitans*. Infect. Immun. 55: 686-691.

Hanazawa, S., Y. Kawata, Y. Murakami, K. Naganuma, S. Amano, Y. Miyata, and S. Kitano. 1995. *Porphyromonas gingivalis* fimbria-stimulated bone resorption in vitro is inhibited by a tyrosine kinase inhibitor. Infect. Immun. 63: 2374-2377.

Hanazawa, S., Y. Murakami, K. Hirose, S. Amano, Y. Ohmori, H. Higuchi, and S. Kitano. 1991. *Bacteroides (Porphyromonas) gingivalis* fimbriae activate mouse peritoneal macrophages and induce gene expression and production of interleukin-1. Infect. Immun. 59: 1972-1977.

Henderson, B., and M. Wilson. 1998. Commensal communism and the oral cavity. J. Dent. Res. 77: 1674-1683.

Henderson, B., S. Poole, and M. Wilson. 1996. Bacterial modulins: a novel class of virulence factors which cause host tissue pathology by inducing cytokine synthesis. Microbiol. Rev. 60: 316-341.

Hesketh, L.M., J.E. Wyatt, and P.S. Handley. 1987. Effect of protease on cell surface structure, hydrophobicity and adhesion of tufted strains of *Streptococcus sanguis* biotypes I and II. Microbios 50: 131-139.

Hillman, J.D., and S.S. Socransky. 1982. Bacterial interference in the oral ecology of *Actinobacillus actinomycetemcomitans* and its relationship to human periodontitis. Archs Oral Biol. 25: 75-77.

Hillman, J.D., and S.S. Socransky. 1989. The theory and application of bacterial interference to oral diseases p. 1-17. *In* H.M. Myers (ed.), New biotechnology in oral research, Karger, Basel.

Hillman, J.D., S.S. Socransky, and M. Shivers. 1985. The relationships between streptococcal species and periodontopathic bacteria in human dental plaque. Archs Oral Biol. 30: 791-795.

Hinode, D., M. Yoshioka, S.-I. Tanabe, O. Miki, K. Masuda, and R. Nakamura. 1998. The GroEL-like protein from *Campylobacter rectus*: immunological characterization and interleukin-6 and -8 induction in human gingival fibroblast. FEMS Microbiol. Lett. 167: 1-6.

Hoover, C.I., C.Y. Ng, and J.R. Felton. 1992. Correlation of haemagglutination activity with trypsin-like protease activity of *Porphyromonas gingivalis*. Archs Oral Biol. 37: 515-520.

Hou, L.-T., C.-M. Liu, and E.F. Rossomando. 1995. Crevicular interleukin-1β in moderate and severe periodontitis patients and the effect of phase I periodontal treatment. J. Periodontol. 22: 162-167.

Hritz, M., E. Fisher, and D.R. Demuth. 1996. Differential regulation of the leukotoxin operon in highly leukotoxic and minimally leukotoxic strains of *Actinobacillus actinomycetemcomitans*. Infect. Immun. 64: 2724-2729.

Ishihara, K., and K. Okuda. 1999. Molecular analysis for pathogenicity of oral treponemes. Microbiol. Immunol. 43: 495-503.

Ishihara, K., T. Miura, H.K. Kuramitsu, and K. Okuda. 1996. Characterization of the *Treponema denticola prtP* gene encoding a prolyl-phenylalanine-specific protease (dentilisin). Infect. Immun. 64: 5178-5186.

Ismaiel, M.O., J. Greenman, and C. Scully. 1988. Serum antibodies against the trypsin-like protease of *Bacteroides gingivalis* in periodontitis. J. Periodontal Res. 23: 193-198.

Jamarillo, E., C.A. Edwards, N. Van Poperin, C.E. Shelburne, and D.E. Lopatin. 1996. Localization of a protein cross-reactive with human hsp90 in stressed *Porphyromonas gingivalis*. J. Dent. Res. 75: 203.

Jansen, J.-H., D. Grenier, and J.S. Van der Hoeven. 1995. Characterization of immunoglobulin G-degrading proteases of *Prevotella intermedia* and *Prevotella nigrescens*. Oral Microbiol. Immunol. 10: 138-145.

Johansson, A., A. Bergenholtz, and S.E. Holm. 1994. Bacterial interference *in vitro*. APMIS. 102: 810-816.

Kadri, R., D. Devine, and W. Ashraf. 1998. Purification and functional analysis of the DnaK homologue from *Prevotella intermedia* OMZ 326. FEMS Micriobiol. Lett. 167: 63-68

Kaufmann, S.H.E., and B. Schoel. 1994. Heat shock proteins as antigens in immunity against infection and self, p. 495-531. *In* R.I. Morimoto, A. Tissieres, and C. Georgopoulos (ed.), The biology of heat shock proteins and molecular chaperones, Cold Spring Harbor Laboratory Press. Cold Spring Harbor, N.Y.

Kenney, E.B., and M.M. Ash. 1969. Oxidation-reduction potential of developing plaque, periodontal pockets and gingival sulci. J. Periodontol. 40: 630-633.

Kesavalu, L., S.C. Holt, and J.L. Ebersole. 1998. Virulence of a polymicrobic complex, *Treponema denticola* and *Porphyromonas gingivalis*, in a murine model. Oral Microbiol. Immunol. 13: 373-377.

Kesavalu, L., S.C. Holt, and J.L. Ebersole. 1999. Environmental modulation of oral treponeme virulence in a murine model. Infect. Immun. 67: 2783-2789.

Kiel, R.A., K.S. Kornman, and P.B. Robertson. 1983. Clinical and microbiological effects of localized-ligature-induced periodontitis on non-ligated sites in the cynomolgus monkey. J. Periodontal Res. 18: 200-211.

Kilian, M. 1981. Degradation of immunoglobulins A1, A2, and G by suspected principal periodontal pathogens. Infect. Immun. 34: 757-765.

Kimizuka, R., T. Miura, and K. Okuda. 1996. Characterization of *Actinobacillus actinomycetemcomitans* hemolysin. Microbiol. Immunol. 40: 717-723.

Koga, T., T. Kusuzaki, H. Asakawa, H. Senpuku, T. Nishihara, and T. Noguchi. 1993. The 64-kilodalton GroEL-like protein of *Actinobacillus actinomycetemcomitans*. J. Periodontal Res. 28: 475-477.

Kolenbrander, P.E., and J. London. 1992. Ecological significance of coaggregation in oral bacteria. Adv. Microb. Ecol. 12: 183-217.

Kolodrubetz, D., T. Dailey, J.Ebersole, and E. Kraig. 1989. Cloning and expression of the leukotoxin gene from *Actinobacillus actinomycetemcomitans*. Infect. Immun. 57: 1465-1469.

Kornman, K.S., S.C. Holt, and P.B. Robertson. 1981. The microbiology of ligature-induced periodontitis in the cynomolgus monkey. J. Periodontal Res. 16: 363-371.

Kornman, K.S., R.C. Page, and M.S. Tonetti. 1997. The host response to the microbial challenge in periodontitis: assembling the players. Periodontology 2000 14: 33-

53.
Korostoff, J., J.F. Wang, I. Kieba, M. Miller, B.J. Shenker, and E.T. Lally. 1998. *Actinobacillus actinomycetemcomitans* leukotoxin induces apoptosis in HL-60 cells. Infect. Immun. 66: 4474-4483.

Kuramitsu, H.K. 1998. Proteases of *Porphyromonas gingivalis:* what don't they do? Oral Microbiol. Immunol. 13: 263-270.

Labbé, S., and D. Grenier. 1995. Characterization of the human immunoglobulin G Fc-binding activity of *Prevotella intermedia.* Infect. Immun. 63: 2785-2789.

Lathigra, R.B., P.D. Butcher, T.R. Garbe, and D.B. Young. 1991. Heat-shock proteins as virulence factors of pathogens. Curr. Top. Microbiol. Immunol. 167: 125-143.

Latifi, A., M.K. Winson, M. Foglino, B.W. Bycroft, G.S. Stewart, A. Lazdunski, and P. Williams. 1995. Multiple homologues of LuxR and LuxI control expression of virulence determinants and secondary metabolites through quorum sensing in *Pseudomonas aeruginosa* PAO1. Mol. Microbiol. 17: 333-343.

Legaard, P.K., R.D. Legrand, and M.L. Misfeldt. 1991. The superantigen *Pseudomonas* exotoxin A requires additional functions from accessory cells for T lymphocyte proliferation. Cell. Immunol. 135: 372-382.

Leke, N., D. Grenier, M. Goldner, and D. Mayrand. 1999. Effects of hydrogen peroxide on growth and selected properties of *Porphyromonas gingivalis.* FEMS Microbiol. Lett. 174: 347-353.

Li, J., R.P. Ellen, C.I. Hoover, and J.R. Felton. 1991. Association of proteases of *Porphyromonas (Bacteroides) gingivalis* with its adhesion to *Actinomyces viscosus.* J. Dent. Res. 70: 82-86.

Lindquist, S. 1986. The heat-shock response. Annu. Rev. Biochem. 55: 1151-1191.

Litwin, C.M., and S.B. Calderwood. 1993. Role of iron in regulation of virulence genes. Clin. Microbiol. Rev. 6: 137-149.

Lokensgard, I., V. Bakken, and K. Schenck. 1994. Heat-shock response in *Actinobacillus actinomycetemcomitans.* FEMS Immunol. Med. Microbiol. 8: 321-328.

London, J. 1999. Oral biofilm: a molecular odyssey, p. 53-65. *In* E. Rosenberg (ed.), Microbial ecology and infectious diseases. American Society for Microbiology, Washington, DC.

London, J., and P.E. Kolenbrander. 1996. Coaggregation: enhancing colonization in a fluctuating environment, p. 249-279. *In* M. Fletcher (ed.). Molecular and ecological diversity of bacterial adhesion. John Wiley & Sons, New York.

Loomer, P.M., R.P. Ellen, and H.C. Tenenbaum. 1998. Effects of *Porphyromonas gingivalis* 2561 extracts on osteogenic and osteoclastic cell function in co-culture. J. Periodontol. 69: 1263-1270.

Lu, B., and B. C. McBride. 1994. Stress response of *Porphyromonas gingivalis.* Oral Microbiol. Immunol. 9: 166-173.

Lu, B., and B.C. McBride. 1998. Expression of the *tpr* protease gene of *Porphyromonas gingivalis* is regulated by peptide nutrients. Infect. Immun. 66: 5147-5156.

Lunsford, R.D. 1998. Streptococcal transformation: essential features and applications of a natural gene exchange system. Plasmid 39: 10-20.

Mäkinen, K.K., and P.L. Mäkinen. 1996. The peptidolytic capacity of the spirochete system. Med. Microbiol. Immunol. 185: 1-10.

Marsh, P.D. 1989. Host defenses and microbial homeostasis: role of microbial interactions. J. Dent. Res. 68: 1567-1575.

Marsh, P.D., A.S. McDermid, A.S. McKee, and A. Baskerville. 1994. The effect of growth rate on the virulence and proteolytic activity of *Porphyromonas gingivalis* W50. Microbiology 140: 861-865.

Mayrand, D., and D. Grenier. 1989. Biological activities of outer membrane vesicles. Can. J. Microbiol. 35: 607- 613.

McDermid, A.S., A.S. McKee, and P.D. Marsh. 1988. Effect of environmental pH on enzyme activity and growth of *Bacteroides gingivalis* W50. Infect. Immun. 56: 1096-1100.

McKee, A.S., A.S. McDermid, A. Baskerville, B. Dowsett, D.C. Ellwood, and P.D. Marsh. 1986. Effect of hemin on the physiology and virulence of *Bacteroides gingivalis* W50. Infect. Immun. 52: 349-355.

Meyer, D.H., and P.M. Fives-Taylor. 1997. The role of *Actinobacillus actinomycetemcomitans* in the pathogenesis of periodontal disease. Trends Microbiol. 5: 224-228.

Mikx, F.H.M., J.C. Maltha, and G.J. Campen. 1990. Spirochetes in early lesions of necrotizing ulcerative gingivitis experimentally induced in beagles. Oral Microbiol. Immunol. 5: 86-89.

Miller, J.F., J.J. Mekalanos, and S.F. Falkow. 1989. Coordinate regulation and sensory transduction in the control of bacterial virulence. Science 243: 916-922.

Moore, W.E.C., R.R. Ranney, and L.V. Holdeman. 1982. Subgingival microflora in periodontal disease: cultural studies, p. 13-26. *In* R.J. Genco and S.E. Mergenhagen (ed.), Host parasite interactions in periodontal disease. American Society for Microbiology, Washington DC.

Mosley, B., S.K. Dower, S. Gillis, and D. Cosman. 1987. Determination of the minimum polypeptide lengths of the functionally active sites of human interleukins 1α and 1β. Proc. Natl. Acad. Sci. 84: 4572-4576.

Mukherjee, S. 1985. The role of crevicular fluid iron in periodontal disease. J. Periodontol. 56: 22-27.

Naito, Y., and R.J. Gibbons. 1988. Attachment of *Bacteroides gingivalis* to collagenous substrata. J. Dent. Res. 67: 1075-1080.

Nakamura, T., S. Fujimura, and N. Kanagawa. 1980. Antibacterial activity of the black pigment (haematin) of *Bacteroides melaninogenicus*. Matsumoto Shigaku 6: 100-108.

Nakamura, T., S. Fujimura, N. Obata, and N. Yamazaki. 1981. Bacteriocin-like substance (melaninocin) from oral *Bacteroides melaninogenicus*. Infect. Immun. 31: 28-32.

Nakamura, T., Y. Suginaka, N. Obata, N. Yamazaki, and I. Takazoe. 1978. Growth inhibition of *Streptococcus mutans* by the black pigment (haematin) of *Bacteroides melaninogenicus*. Archs Oral Biol. 23: 593-595.

Nakayama, K., T. Kadowaki, K. Okamoto, and K. Yamamoto. 1995. Construction and characterization of arginine-specific cysteine proteinase (Arg-gingipain)-deficient mutants of *Porphyromonas gingivalis*. J. Biol. Chem. 270: 23619-23626.

Neiders, M.E., P.B. Chen, H. Suido, H.S. Reynolds, J.J. Zambon, M. Shlossman, and R.J. Genco. 1989. Heterogeneity of virulence among strains of *Bacteroides gingivalis*. J. Periodontal Res. 24: 192-198.

Nilius, A.M., S.C. Spencer, and L.G. Simonson. 1993. Stimulation of *in vitro* growth of *Treponema denticola* by extracellular growth factors produced by *Porphyromonas gingivalis*. J. Dent. Res. 72: 1027-1031.

Nowotny, A., U.H. Behling, B. Hammond, C.-H. Lai, M. Listgarten, P.H. Pham, and F. Sanavi. 1982. Release of toxic microvesicles by *Actinobacillus actinomycetemcomitans*. Infect. Immun. 37: 151-154.

Ogawa, T., H. Uchida, and S. Hamada. 1994. *Porphyromonas gingivalis* fimbriae and their synthetic peptides induce proinflammatory cytokines in human peripheral blood monocyte culture. FEMS Microbiol. Lett. 116: 237-242.

Ohshima, M., K. Otsuka, and K. Suzuki. 1994. Interleukin-1β stimulates collagenase production by cultured human periodontal ligament fibroblasts. J. Periodontal Res. 29: 421-429.

Okuda, K., T. Kato, K. Ishihara, and Y. Naito. 1991. Adherence to experimental pellicle of rough-type lipopolysaccharides from subgingival plaque bacteria. Oral Microbiol. Immunol. 6: 241-245.

Ower, P.C., M. Ciantar, H.N. Newman, M. Wilson, and J.S. Bulman. 1995. The effects on chronic periodontitis of a subgingivally-placed redox agent in a slow release device. J. Clin. Periodontol. 22: 494-500.

Paliard, X., S.G. West, J.A. Lafferty, J.R. Clement, J.W. Kappler, P. Marrack, and B.L. Kotzin. 1991. Evidence for the effects of a superantigen in rheumatoid arthritis. Science 253: 325-329.

Panchapekesan, J., M. Daglis, and P. Gatenby. 1992. Antibodies to 65 kDa and 70 kDa heat shock proteins in rheumatoid arthritis and systemic lupus erythematosus. Immunol. Cell Biol. 70: 295-300.

Park, Y.S., B. Lu, C. Mazur, and B.C. McBride. 1997. Inducible expression of a *Porphyromonas gingivalis* W83 membrane-associated protease. Infect. Immun. 65: 1101-1104.

Peetermans, W.E. 1996. Expression of and immune response to heat shock protein 65 in Crohn's disease, p. 197-211. *In* W. vanEden and D.B. Young (ed.), Stress proteins in medicine. Marcel Dekker Inc., New York, N.Y.

Percival, R.S., P.D. Marsh, D.A. Devine, M. Rangarajan, J. Aduse-Opoku, P. Shepherd, and M.A. Curtis. 1999. Effect of temperature on growth, hemagglutination, and protease activity of *Porphyromonas gingivalis*. Infect. Immun. 67: 1917-1921.

Pike, R., W. McGraw, J. Potempa, and J. Travis. 1994. Lysine- and arginine-specific proteinases from *Porphyromonas gingivalis*. Isolation, characterization and evidence for the existence of complexes with hemagglutinins. J. Biol. Chem. 269: 406-411.

Pincus, S.H., P.A. Rosa, G.J. Spangrude, and J.A. Heinemann. 1992. The interplay of microbes and their hosts. Immunol. Today 13: 471-473.

Potempa, J., N. Pavloff, and J. Travis. 1995. *Porphyromonas gingivalis*: a proteinase/gene accounting audit. Trends Microbiol. 3: 430-434.

Preiss, D.S., and J. Meyle. 1994. Interleukin-1β concentration of gingival crevicular fluid. J. Periodontol. 65: 423-428.

Privalle, C.T., and I. Fridovich. 1987. Induction of superoxide dismutase in *Escherichia coli* by heat shock. Proc. Natl. Acad. Sci. USA. 84: 2723-2726.

Rietschel, E.T., T. Kirikae, F.U. Schade, U. Mamat, G. Schmidt, H. Loppanow, A.J. Ulmer, U. Zahringer, U. Seydel, F. Di Padova, M. Schreiner, and H. Brade. 1994. Bacterial endotoxin: molecular relationships of structure to activity and function. FASEB 8: 217-225.

Riviere, G.R., K.S. Smith, N. Carranza, E. Tzagaroulaki, S.L. Kay, M. Dock, X. Zhu, and T.A. De Rouen. 1996. Associations between *Porphyromonas gingivalis* and oral treponemes in subgingival plaque. Oral Microbiol. Immunol. 11: 150-155.

Rosen, G., R. Naor, E. Rahamim, R. Yishai, and M.N. Sela. 1995. Proteases of *Treponema denticola* outer sheath and extracellular vesicles. Infect. Immun. 63: 3973-3979.

Saglie, F.R., J.C. Ferreira, C.T. Smith, P.L. Valentin, F.A. Carranza, and M.G. Newman. 1985. Identification of bacteria by studying one section under light microscopy, scanning, and transmission microscopy. J. Electr. Microscop. Tech. 2: 581-588.

Saito, A, H.T. Sojar, and R.J. Genco. 1996. *Porphyromonas gingivalis* surface components induce interleukin-1 release and tyrosine phosphorylation in macrophages. FEMS Immunol. Med. Microbiol. 15: 51-57.

Saito, K., H. Katsuragi, M. Mikami, C. Kato, M. Miyamaru, and K. Nagaso. 1997. Increase of heat shock protein and induction of γ/δ T cells in peritoneal exudate of mice after injection of live *Fusobacterium nucleatum*. Immunology 90: 229-235.

Scannapieco, F.A., S.J. Miller, H.S. Reynolds, J.J. Zambon, and M.J. Levine. 1987. Effect of anaerobiosis on the surface ultrastructure and surface proteins of *Actinobacillus actinomycetemcomitans* (*Haemophilus actinomycetemcomitans*). Infect. Immun. 55: 2320-2323.

Shah, H.N., and S.E. Gharbia. 1993. Batch culture and physiological properties, p. 85-103. *In* H.N. Shah, D. Mayrand, and R.J. Genco (ed.), Biology of the species *Porphyromonas gingivalis*. CRC Press, Boca Raton.

Shapira, L., C. Champagne, T.E. van Dyke, and S. Amar. 1998. Strain-dependent activation of monocytes and inflammatory macrophages by lipopolysaccharides of *Porphyromonas gingivalis*. Infect. Immun. 66: 2736-2742.

Shinnick, T.M. 1991. Heat shock proteins as antigens of bacterial and parasitic pathogens. Curr. Top. Microbiol. Immunol. 167: 145-160.

Shivers, M., J.D. Hillman, and S.S. Socransky. 1987. *In vivo* interactions between *Streptococcus sanguis* and *Actinobacillus actinomycetemcomitans*. J. Dent. Res. 66: 195.

Simonson, L.G., K.T. McMahon, D.W. Childers, and H.E. Morton. 1992. Bacterial synergy of *Treponema denticola* and *Porphyromonas gingivalis* in a multinational population. Oral Microbiol. Immunol. 7: 111-112.

Simpson, W., B.M. Odusanya, R.Y. Forng, M.O. Lassiter, and C.A. Genco. 1994. Characterization of the hemin uptake defect in a Tn4351 mutant of *Porphyromonas gingivalis*. J. Dent. Res. 73: 349.

Slots, J., H.S. Reynolds, and R.J. Genco. 1980. *Actinobacillus actinomycetemcomitans* in human periodontal disease: a cross-sectional microbiological investigation. Infect. Immun. 29: 1013-1020.

Smalley, J.W., D. Mayrand, and D. Grenier. 1993. Vesicles, p. 259-292. *In* H.N. Shah, D. Mayrand, and R.J. Genco (ed.). Biology of the species *Porphyromonas gingivalis*. CRC Press, Boca Raton.

Smalley, J.W., A.J. Birss, A.S. McKee, and P.D. Marsh. 1991. Haemin-restriction influences haemin-binding, haemagglutination and protease activity of cells and extracellular membrane vesicles of *Porphyromonas gingivalis* W50. FEMS Microbiol. Lett. 90: 63-68.

Smejkal, R.M., R. Wolff, and J.G. Olenick. 1988. *Leishmania braziliensis panamensis*: increased activity resulting from heat-shock. Exp. Parasitol. 63: 322-331.

Socransky, S.S. 1979. Criteria for infectious agents in dental caries and periodontal disease. J. Clin. Periodontol. 61: 16-21.

Socransky, S.S., A.D. Haffajee, J.L. Dzink, and J.D. Hillman. 1988. Associations between microbial species in subgingival plaque samples. Oral Microbiol. Immunol. 3: 1-7.

Socransky, S.S., W.J. Loesche, C. Hubersak, and J.B. MacDonald. 1964. Dependency of *Treponema microdentium* on other oral organisms for isobutyrate, polyamines and a controlled oxidation-reduction potential. J. Bacteriol. 88: 200-209.

Socransky, S.S., A.D. Haffajee, M.A. Cugini, C. Smith, and R.L. Kent. 1997. Microbial complexes in subgingival plaque. J. Dent. Res. 76: 51.

Sorsa, T., T. Ingman, K. Suomalainen, M. Haapasalo, Y.T. Konttinen, O. Lindy, H. Saari, and V.-J. Uitto. 1992. Identification of proteases from periodontopathogenic bacteria as activators of latent human neutrophil and fibroblast-type interstitial collagenases. Infect. Immun. 60: 4491-4495.

Spitznagel, J., E. Kraig, and D. Kolodrubetz. 1995. The regulation of leukotoxin production in *Actinobacillus actinomycetemcomitans* strain JP2. Adv. Dent. Res. 9: 48-54.

Spitznagel, J., E. Kraig, and D. Kolodrubetz. 1991. Regulation of leukotoxin in leukotoxic and nonleukotoxic strains of *Actinobacillus actinomycetemcomitans*. Infect. Immun. 59: 1394-1401.

Sreenivasan, P.K., D.H. Meyer, and P.M. Fives-Taylor. 1993. Factors influencing the growth and viability of *Actinobacillus actinomycetemcomitans*. Oral Microbiol. Immunol. 8: 361-369.

Stevens, R.H., and B.F. Hammond. 1988. The comparative cytotoxicity of periodontal bacteria. J. Periodontol. 59: 741-749.

Stevens, R.H., S.E. Lillard, and B.F. Hammond. 1987. Purification and biochemical properties of a bacteriocin from *Actinobacillus actinomycetemcomitans*. Infect. Immun. 55: 692-697.

Straley, S.C., and R.D. Perry. 1995. Environmental modulation of gene expression and pathogenesis in *Yersinia*. Trends Microbiol. 3: 310-317.

Suido, H., J.J. Zambon, P.A. Mashimo, R. Dunford, and R.J. Genco. 1988. Correlations between gingival crevicular fluid enzymes and the subgingival microflora. J. Dent. Res. 67: 1070-1074.

Sundqvist, G., J. Carlsson, B. Hermann, and A. Tärnvik. 1985. Degradation of human immunoglobulins G and M and complement factors C3 and C5 by black-pigmented

Bacteroides. J. Med. Microbiol. 19: 85-94.

Taichman, N.S., J.E. Klass, B.J. Shenker, E.J. Macarak, H. Boehringer, and C.C. Tsai. 1984. Suspected periodontopathic organisms alter in vitro proliferation of endothelial cells. J. Periodontal Res. 19: 583-586.

Takada, H., J. Mihara, I. Morisaki, and S. Hamada. 1991. Production of cytokines by human gingival fibroblasts, p. 265-276. *In* S. Hamada, S.C. Holt and J.R. McGhee (ed.), Periodontal disease: pathogens and host immune response. Quintessence Publishing Co., Tokyo.

Takahashi, N., and C.F. Schachtele. 1990. Effect of pH on the growth and proteolytic activity of *Porphyromonas gingivalis* and *Bacteroides intermedius.* J. Dent. Res. 69: 1266-1269.

Takazoe, I., T. Nakamura, and K. Okuda. 1984. Colonization of the subgingival area by *Bacteroides gingivalis.* J. Dent. Res. 63: 422-426.

Takemoto, T, H. Kurihara, and G. Dahlen. 1997. Characterization of *Bacteroides forsythus* isolates. J. Clin. Microbiol. 35: 1378-1381.

Tanner, A. 1991. Microbial succession in the development of periodontal disease, p. 13-25. *In* S. Hamada, S.C. Holt and J.R. McGhee (ed.), Periodontal disease: pathogens and host immune response. Quintessence Publishing Co., Tokyo.

Tatakis, D.N. 1993. Interleukin-1 and bone metabolism: a review. J. Periodontol. 64: 416-431.

Ter Steeg, P.F., and J.S. van der Hoeven. 1990. Growth stimulation of *Treponema denticola* by periodontal microorganisms. Antonie van Leeuwenhoek. 57: 63-70.

Ter Steeg, P.F., J.S. van der Hoeven, M.H. de Jong, P.J.J. van Munster, and M.J.H. Jansen. 1988. Modelling the gingival pocket by enrichment of subgingival microflora in human serum in chemostats. Microb. Ecol. Health Dis. 1: 73-84.

Tokuda, M., W. Chen, T. Karunakaran, and H.K. Kuramitsu. 1998. Regulation of protease expression in *Porphyromonas gingivalis.* Infect. Immun. 66: 5232-5237.

Tokuda, M., M. Duncan, M.-I. Cho, and H.K. Kuramitsu. 1996. Role of *Porphyromonas gingivalis* protease activity in colonization of oral surfaces. Infect. Immun. 64: 4067-4073.

Tolo, K., and K. Helgeland. 1991. Fc-binding components: a virulence factor in *Actinobacillus actinomycetemcomitans.* Oral Microbiol. Immunol. 6: 373-377.

Uitto, V.-J., D. Grenier, E.C. Chan, and B.C. McBride. 1988. Isolation of a chymotrypsin enzyme from *Treponema denticola.* Infect. Immun. 56: 2717-2722.

Uitto, V.-J., A. Hafezi, L. Zhang, S. Paju, F. Goulhen, D. Grenier, D. Mayrand, and J. Heino. 1999. Effects of GroEL-like heat-stress protein of *Actinobacillus actinomycetemcomitans* on epithelial migration and integrin expression. J. Dent. Res. 78: 136.

Uitto, V.-J., H. Larjava, J. Heino, and T. Sorsa. 1987. A protease of *Bacteroides gingivalis* degrades cell surface and matrix glycoproteins of cultured gingival fibroblasts and induces secretion of collagenase and plasminogen activator. Infect. Immun. 57: 213-218.

Uitto, V.-J., Y.-M. Pan, W.K. Leung, H. Larjava, R.P. Ellen, B.B. Finlay, and B.C. McBride. 1995. Cytopathic effects of *Treponema denticola* chymotrypsin-like proteinase on migrating and stratified epithelial cells. Infect. Immun. 63: 3401-3410.

van Steenbergen, T.J.M., L.M. van der Mispel, and J. de Graaff. 1986. Effects of ammonia and volatile fatty acids produced by oral bacteria on tissue culture cells. J. Dent. Res. 65: 909-912.

van Winkelhoff, A.J., N. Kippuw, and J. de Graaff. 1987. Cross-inhibition between black-pigmented *Bacteroides* species. J. Dent. Res. 66: 1663-1667.

Vayssier, C., D. Mayrand, and D. Grenier. 1994. Detection of stress proteins in *Porphyromonas gingivalis* and other oral bacteria by Western immunoblotting analysis. FEMS Microbiol. Lett. 121: 303-308.

Visick, J.E., and S. Clarke. 1995. Repair, refold, recycle: how bacteria can deal with spontaneous and environmental damage to proteins. Mol. Microbiol. 16: 835-845.

Wang, P.-L., S. Shirasu, M. Shinohara, M. Daito, M. Oido, Y. Kowashi, and K. Ohura. 1999. Induction of apoptosis in human gingival fibroblasts by a *Porphyromonas gingivalis* protease preparation. Archs Oral Biol. 44: 337-342.

Watson, K. 1990. Microbial stress proteins. Adv. Microbiol. Physiol. 31: 183-223.

Weinberg, A., and S.C. Holt. 1990. Interaction of *Treponema denticola* TD-4, GM-1, and MS25 with human gingival fibroblasts. Infect. Immun. 58: 1720-1729.

Wennstrom, J.L., G. Dahlen, K. Grondahl, and L. Heijl. 1987. Periodic subgingival antimicrobial irrigation of periodontal pockets. I. Microbiological and radiographical observations. J. Clin. Periodontol. 14: 573-580.

Whittaker, C.J., C.M. Klier, and P.E. Kolenbrander. 1996. Mechanisms of adhesion by oral bacteria. Annu. Rev. Microbiol. 50: 513-552.

Wilson, M., and B. Henderson. 1995. Virulence factors of *Actinobacillus actinomycetemcomitans* relevant to the pathogenesis of inflammatory periodontal diseases. FEMS Microbiol. Rev. 17: 365-379.

Wilson, M., K. Reddi, and B. Henderson. 1996. Cytokine-inducing components of periodontopathogenic bacteria. J. Periodontal Res. 31: 393-407.

Wirth, R., A. Muscholl, and G. Wanner. 1996. The role of pheromones in bacterial interactions. Trends Microbiol. 45: 96-103.

Wolff, L.F., W.F. Liljemark, C.G. Bloomquist, and B.L. Philstrom. 1985. The distribution of *Actinobacillus actinomycetemcomitans* in human plaque. J. Periodontal Res. 20: 237-250.

Wolff, L.F., B.L. Pihlstrom, M.B. Bakdash, E.M., Schaffer, D.M. Aeppli, and C.L. Bandt. 1989. Four-year investigation of salt and peroxide regimen compared with conventional oral hygiene. J. Am. Dent. Ass. 118: 67-72.

Yoshimura, A., Y. Hara, T. Kaneko, and I. Kato. 1997. Secretion of IL-1β, TNF-α, IL-8 and IL-1ra by human polymorphonuclear leukocytes in response to lipopolysaccharides from periodontopathic bacteria. J. Periodontal Res. 32: 279-286.

Zambon, J.J., C. DeLuca, J. Slots, and R.J. Genco. 1983. Studies of leukotoxin from *Actinobacillus actinomycetemcomitans* using the promyelocytic HL-60 cell line. Infect. Immun. 40: 205-212.

Zhou, L., R. Srisatjaluk, D.E. Justus, and R.J. Doyle. 1998. On the origin of membrane vesicles in Gram-negative bacteria. FEMS Microbiol. Lett. 163: 223-228.

Zügel, U., and S.H.E. Kaufmann. 1999. Role of heat shock proteins in protection from and pathogenesis of infectious diseases. Clin. Microbiol. Rev. 12: 19-39.

Index